Nanotechnology and Nanomedicine in Diabetes

Nanotechnology and Nanomedicine in Diabetes

Editors

Lan-Anh Le MBBS
Principal GP
Rosemead Surgery
Maidenhead
Berkshire
UK

Ross J. Hunter MBBS MRCP PhD
Cardiology Research Fellow
St Bartholomew's Hospital
London
UK

Victor R. Preedy PhD DSc
Professor of Nutritional Biochemistry
School of Medicine
King's College London
and
Professor of Clinical Biochemistry
King's College Hospital
UK

Published by Science Publishers, an imprint of Edenbridge Ltd.
- St. Helier, Jersey, British Channel Islands
- P.O. Box 699, Enfield, NH 03748, USA

E-mail: *info@scipub.net*   Website: *www.scipub.net*

Marketed and distributed by:

CRC Press
Taylor & Francis Group
an Informa business
www.taylorandfrancisgroup.com

6000 Broken Sound Parkway, NW
Suite 300, Boca Raton, FL 33487
711 Third Avenue
New York, NY 10017
2 Park Square, Milton Park
Abingdon, Oxon OX14 4RN, UK

Copyright reserved © 2012
ISBN 978-1-57808-729-7

Cover Illustrations: Reproduced by kind courtesy of the undermentioned authors:
- Figure No. 3 from Chapter 4 by Novella Rapini, Nunzio Bottini and Massimo Bottini
- Scheme 2 from Chapter 6 by Kwang Pill Lee, Anantha Iyengar Gopalan and Shanmugasundaram Komathi
- Figure No. 5 from Chapter 7 by Muhammad H. Asif, Magnus Willander, Peter Strålfors and Bengt Danielsson
- Figure No. 3 from Chapter 13 by Silke Krol, Ana Maria Waaga-Gasser and Piero Marchetti
- Figure No. 3 from Chapter 14 by Rhishikesh Mandke, Ashwin Basarkar and Jagdish Singh

```
         Library of Congress Cataloging-in-Publication Data

Nanotechnology and nanomedicine in diabetes / editors, Lan-Anh Le,
Ross J. Hunter, Victor R. Preedy.
       p. cm.
  Includes bibliographical references and index.
  ISBN 978-1-57808-729-7 (hardback)
  1. Diabetes. 2. Nanomedicine. I. Le, Lan Anh, 1976- II. Hunter,
Ross J. (Ross Jacob), 1977- III. Preedy, Victor R.
  RC660.N357 2012
  616.4'62--dc23
                                                         2011038911
```

The views expressed in this book are those of the author(s) and the publisher does not assume responsibility for the authenticity of the findings/conclusions drawn by the author(s). No responsibility is assumed by the publisher for any injury and/or damage to persons or property as a matter of products liability, negligence or otherwise, or from any use or operation of any methods, products, instructions or ideas contained in the material herein. Because of rapid advances in the medical sciences, in particular, independent verification of diagnoses and drug dosages should be made.

All rights reserved. No part of this publication may be reproduced, stored in a retrieval system, or transmitted in any form or by any means, electronic, mechanical, photocopying or otherwise, without the prior permission of the publisher, in writing. The exception to this is when a reasonable part of the text is quoted for purpose of book review, abstracting etc.

This book is sold subject to the condition that it shall not, by way of trade or otherwise be lent, re-sold, hired out, or otherwise circulated without the publisher's prior consent in any form of binding or cover other than that in which it is published and without a similar condition including this condition being imposed on the subsequent purchaser.

Printed in the United States of America

Preface

Diabetes mellitus is a spectrum of diseases characterised by the imbalance of serum glucose and insulin, resulting in hyperglycaemia and affecting all bodily systems. It's contributory role in the aetiology of coronary artery disease is arguably it's most serious manifestation, since this is the leading cause of death in the Western world and accounts for a third of all premature deaths. Diabetic renal disease is the most common cause of renal failure requiring dialysis, and diabetic retinopathy remains the leading cause of blindness globally. Diabetic foot also leads to more amputations worldwide than any other cause.

With a prevalence of 4–5% of the population, diabetes accounts for a devastating 5% of all deaths globally each year and is a massive cause of morbidity and suffering. Diabetes therefore poses a huge challenge for patients, healthcare professionals and governments alike. The burden of the disease is clearly a global phenomenon as 80% of those with diabetes live in low to middle income countries. An ageing population and an explosive increase in obesity are behind a forthcoming diabetes epidemic, since the prevalence is expected to double between 2005 and 2030. Sadly, despite various national and global initiatives, the death rate from the condition is expected to double in the next 10 years and warrants an immediate comprehensive global strategy.

Although the importance of glucose control for the prevention of adverse sequelae is well established, there remain several barriers to overcome. Our understanding of diabetes and it's manifestations in the different organ systems remains arguably rather crude. The aetiologies of the various forms of diabetes are incompletely understood and there are major limitations to current oral hypoglycaemic drug therapy and to insulin therapy which still requires injection. Furthermore, our management of the various disease processes that result from diabetes is limited mostly to blood glucose control and management of cardiovascular risk factors, and does little to address these disease processes directly.

Advances in medical sciences, particularly in the realm of nano-medicine, offer great promise for complex diseases such as diabetes. Improved understanding of the aetiologies of the different forms of diabetes and the pathophysiology underlying the resultant multi-organ disease process that results from hyperglycaemia may define new therapeutic targets for the prevention or perhaps cure of this condition, and better treatment of the multi-organ pathology that it causes. New methods of drug delivery using nano technology may help optimise diabetic control, and alternatives to injecting insulin are now on the horizon. Likewise, the possibility of directly addressing the pathological processes that complicate diabetes may allow real advances in reducing the morbidity and mortality associated with this condition.

The nanosciences encompass a variety of technologies ranging from particles to networks and nanostructures. For example, nanoparticles have been proposed to be suitable carriers of therapeutic agents whilst nanostructures provide suitable platforms for sub-micro bioengineering. However, understanding the importance of nanoscience and technology is somewhat problematical as a great deal of text can be rather technical in nature with little consideration to the novice. In this book **Nanotechnology and Nanomedicine in Diabetes** we aim to disseminate the information in a readable way by having unique sections for the novice and expert alike. This enables the reader to transfer their knowledge base from one discipline to another or from one academic level to another. Each chapter has an abstract, "key facts", and a "mini-dictionary" of key terms and phrases within each chapter. Finally, each chapter has a series of summary points.

We cover for example, introductions to the field, an overview on nanosciences, lipid matrix nanoparticles, atomic force microscopy, nanocarriers of antisense oligonucleotides, carbon nanotubes coupled to siRNA, glucose sensors, nanocomposites, zinc oxide nanorods, nanoprobes to monitor cell processes, mucoadhesive nanoparticles, insulin-nanoparticles, carbon nanomaterials for MALDI-TOF MS analysis, loaded biodegradable nanoparticles, nano-encapsulation strategies, nano-immuno science, nanolabeled diabetogenic T cells, nanoparticle-delivery in retinopathy, nanofiber matrices as diabetic wound dressings, poly-n-acetyl glucosamine nanofibers and nanotechnology footsocks.

This book is intended to impart some of the key advances in nanomedicine in diabetes. These reviews will bring the reader to the forefront of this exciting field. It is hoped that by condensing these key developments in the field into this short book, that we may clarify the future direction of this rapidly evolving specialty and highlight the promise of the nanosciences in the field of diabetes. Contributors to **Nanotechnology and Nanomedicine in Diabetes** are all either international or national experts, leading authorities or are carrying out ground breaking and innovative work on their subject. The book is essential reading for research scientists, medical doctors, health care professionals, pathologists, biologists, biochemists, chemists and physicists, general practitioners as well as those interested in disease and nano sciences in general. **Nanotechnology and Nanomedicine in Diabetes** is part of a collection of books on *Nanoscience Applied to Health and Medicine*.

Dr Lan-Anh Le MBBS
Dr Ross Hunter MBBS PhD
Professor Victor R. Preedy PhD DSc FRCPath

Contents

Preface   v

Section 1: General Aspects

1. Nanotechnology and Nanomedicine in Diabetes: An Overview   3
   Denison J. Kuruvilla and *Aliasger K. Salem*

2. Lipid Matrix Nanoparticles in Diabetes   14
   Eliana B. Souto, Joana F. Fangueiro and Selma S. Souto

3. Atomic Force Microscopy: An Enabling Nanotechnology for Diabetes Research   34
   Brian J. Rodriguez and Suzanne P. Jarvis

4. Nanocarriers of Antisense Oligonucleotides in Diabetes   59
   Novella Rapini, Nunzio Bottini and Massimo Bottini

Section 2: Glucose

5. Carbon Nanotubes Coupled to siRNA Generates Efficient Transfection and is a Tool for Examining Glucose Uptake in Skeletal Muscle   81
   Johanna T. Lanner and *Håkan Westerblad*

6. Glucose Sensors Based on Functional Nanocomposites   98
   Kwang Pill Lee, Anantha Iyengar Gopalan and *Shanmugasundaram Komathi*

7. Zinc Oxide Nanorods and Their Application to Intracellular Glucose Measurements   120
   Muhammad H. Asif, Magnus Willander, Peter Strålfors and *Bengt Danielsson*

Section 3: Insulin

8. **Nanoprobes to Monitor Cell Processes in the Pancreas** — 143
   Claire Billotey, Caroline Aspord, Florence Gazeau, Pascal Perriat, Olivier Tillement, Charles Thivolet and Marc Janier

9. **Mucoadhesive Nanoparticles for Oral Delivery of Insulin** — 165
   Sajeesh S., Chandra P. Sharma and Christine Vauthier

10. **Insulin-nanoparticles for Transdermal Absorption** — 186
    Jiangling Wan, Huibi Xu and Xiangliang Yang

11. **Applications of Carbon Nanomaterials for MALDI-TOF-MS and Electrochemical Analysis of Insulin** — 202
    Stefan A. Schönbichler, Lukas K.H. Bittner, Johannes D. Pallua, Verena A. Huck-Pezzei, Christine Pezzei, Günther K. Bonn and Christian W. Huck

Section 4: Drugs and Treatments

12. **Second Generation Sulfonylurea Glipizide Loaded Biodegradable Nanoparticles in Diabetes** — 227
    Swarnlata Saraf, Shailendra Saraf and Lan-Anh Le

13. **Immune Protection for Transplanted Pancreatic Islets by Nano-Encapsulation Strategies: A Chemist's Insight** — 248
    Silke Krol, Ana Maria Waaga-Gasser and Piero Marchetti

14. **Nanoparticles, Interleukin-10 and Autoimmune Diabetes** — 270
    Rhishikesh Mandke, Ashwin Basarkar and Jagdish Singh

15. ***In vivo* MR Imaging of Nanolabeled Diabetogenic T cells** — 287
    Amol Kavishwar, Zdravka Medarova and Anna Moore

16. **Nanoparticle-Mediated Delivery of Angiogenic Inhibitors in Diabetic Retinopathy** — 304
    Krysten M. Farjo, Rafal Farjo, Ronald Wassel and Jian-xing Ma

17. **Drug Loaded Nanofiber Matrices as Diabetic Wound Dressings** — 325
    William Gionfriddo and Lakshmi S. Nair

18. **Poly-N-acetyl Glucosamine Nanofibers Derived from a Marine Diatom: Applications in Diabetic Wound Healing and Tissue Regeneration** — 345
    John N. Vournakis, Thomas Fischer, Haley Buff Lindner, Marina Demcheva, Arun Seth and Robin C. Muise-Helmericks

19.	**Nanotechnology Footsocks for Diabetic Foot** *Alberto Piaggesi, Elisabetta Iacopi, Elisa Banchellini* and *Laura Ambrosini Nobili*	365
20.	**Nanosciences, Diabetes and the Patient** *Martin C.R., Le L., Hunter R., Patel V.B.* and *Preedy V.R.*	380

Index   385
About The Editors   389
Color Plate Section   391

Section 1: General Aspects

1

Nanotechnology and Nanomedicine in Diabetes: An Overview

Denison J. Kuruvilla[1] and Aliasger K. Salem[1,a,]*

INTRODUCTION

Diabetes is a rapidly growing chronic disease that affects more than 220 million people around the world (WHO, factsheet 2011). The serious complications of the disease include lower limb amputations, blindness and cardiovascular disease (WHO, factsheet 2011). Frequent monitoring of blood glucose levels is effective in reducing disease related complications. The current technique for monitoring blood glucose levels involves using the finger prick blood glucose test. The limitations of this test are that it is painful and is unable to detect the large fluctuations in blood glucose levels between sampling time points.

Nanomedicine refers to the application of materials at the nano scale level (1–100 nm) to specific medical problems (Pickup et al. 2008). Nanomedicine in diabetes therapy involves the application of nanotechnology in glucose monitoring, insulin delivery, drug delivery and wound healing.

---

[1]Department of Pharmaceutical Sciences and Experimental Therapeutics, College of Pharmacy, University of Iowa, Iowa City, Iowa, 52242, USA.
[a]E-mail: aliasger-salem@uiowa.edu
*Corresponding author

GLUCOSE MONITORING

Monitoring glucose concentrations is crucial for preventing diabetic complications in patients. The most commonly used finger prick capillary blood glucose method is limited by potential non-compliance among patients. In addition, the intermittent results obtained from multiple blood samples are not suitable for detecting dangerous fluctuations in blood glucose concentrations between tests.

Continuous glucose monitoring (CGM) systems are favored, since they respond instantaneously to changes in patient's blood glucose levels. These can be used even when the patients are sleeping, which could not be done with the conventional finger prick method (Pickup et al. 2008). Microdialysis and enzyme tipped catheters have been developed for providing continuous information on glucose concentrations (Shibata et al. 2010). However, these were found to cause discomfort and increased risk of infection (Shibata et al. 2010).

Nanosensors have been used for glucose detection. The earlier sensors developed for glucose detection were based on enzymes like glucose oxidase (Xu et al. 2002). The nanosensors incorporated glucose oxidase, an oxygen sensitive fluorescent indicator and a fluorescent dye insensitive to oxygen as a reference (Xu et al. 2002). Glucose gets oxidized to gluconic acid resulting in depletion of oxygen in a nearby region, which can then be measured by the oxygen sensitive fluorescent dye (Xu et al. 2002). Another version of the glucose sensor was based on the reversible competitive binding between glucose and flourescein labeled dextran to sugar binding sites of concavalin-A (Meadows and Schultz 1993). However, these sensors were found to have shortened lifetimes in the nano scale range (Billingsley et al. 2010). Since only a low volume of glucose oxidase is contained in the sensor, any regional depletion of resources or degradation significantly reduced the function of the sensor. The nanosensor components encapsulated within a lipophilic core were found to reduce the non-specific binding of proteins to the sensor (Dubach et al. 2007). In addition, it also improved the lifetime and stability of the nanosensors (Dubach et al. 2007).

Several other types of nanosensors have also been developed. The binding of glucose to hexokinase induces a conformational change in the protein, which causes a 25% reduction in its intrinsic fluorescence (Hussain et al. 2005). Yeast hexokinases entrapped in silica sol gel have been investigated for glucose monitoring (Hussain et al. 2005). Lipophilic boronic acid (BA) derivatives have affinity for glucose molecules and this property has been used for developing nanosensors (Billingsley et al. 2010). Lipophilic BA reacts with alizarin to form highly fluorescent boronate esters (Springsteen et al. 2002). When glucose is present, it forms glucose derived boronate ester and non-florescent alizarin (Billingsley et al. 2010). The reduction in fluorescent intensity is related to changes in glucose concentration (Billingsley et al. 2010).

Bacterial glucose-binding protein (GBP) is a single polypeptide chain that has two domains connected by a hinge. These proteins are engineered by site-directed mutagenesis to attach flourophores like acrylodan (ex 392 nm, em 520 nm) or (((2-(iodoacetoxy)ethyl)methyl)amino)-7-nitrobenz-2-oxa-1,3-diazole (IANBD, ex 469 nm, em 540 nm) (Marvin and Hellinga 1998). The binding of glucose to GBP results in a decrease in fluorescence intensity which is then correlated to glucose concentration. For example, GBP with acrylodan attached at position 255 has a two-fold decrease in fluorescence intensity with addition of glucose (2 mM) (Marvin and Hellinga 1998).

INSULIN DELIVERY

Insulin is secreted by the beta cells of pancreatic islet of langerhans (Trehan and Ali 1998). It is a peptide hormone commonly used for the treatment of diabetes mellitus. Insulin has been made medically available since 1921. Since then, insulin delivery has been used as a primary therapy for all forms of diabetes (Sabetsky and Ekblom 2010). Even with the emergence of oral antidiabetic drugs, it has been found that these drugs cannot be used as a standalone therapy (Calvert et al. 2007). These drugs are usually administered along with endogenous or exogenous insulin. In addition, insulin has been shown to promote insulin secretion and also preserve islets from apoptosis.

Most insulin delivery in the early years was carried out using parenteral routes of administration. The administration of insulin by multiple subcutaneous injections daily has been the most common therapy for patients (Carino et al. 2000; Lowman et al. 1999). Lack of compliance due to several factors like weight gain, pain and life style restrictions has affected the efficacy of parenteral insulin therapy (Cefalu 2004). Researchers focused on oral administration of insulin for improving the compliance to insulin therapy.

One of the major difficulties in insulin delivery is its low oral bioavailability. Insulin is a peptide hormone that is susceptible to strong acidic environments in the stomach and proteolytic digestion in the intestine. In addition, insulin has a short half-life and low permeation rates. In order to improve the oral bioavailability of insulin, the use of permeation enhancers, enteric coatings, polymeric formulations and protease inhibitors have been investigated with limited success (Carino et al. 2000).

Nanoparticles based on biodegradable and biocompatible polymeric systems have been developed to improve the bioavailability of insulin (Damgé et al. 2008). Hydrophobic, negatively charged nanoparticles that are smaller than 1 µm show the best oral absorption rates (Jung et al. 2000). Insulin loaded nanoparticles delivered orally demonstrate a sustained effect of decreasing the blood glucose level over a longer period of time when compared to subcutaneous injections (Lin et al. 2007). Polyacrylic acid, chitosan based delivery systems improved the uptake of insulin by opening the epithelial tight junctions, facilitating paracellular transport (Artursson et al. 1994; Aungst 2000). These polymeric systems also protect the insulin from gastric degradation due to their pH dependent release mechanism (Artursson et al. 1994). Insulin molecules tend to aggregate together and can affect its therapeutic efficacy (Lovatt et al. 1996). Complexation with β-cyclodextrin prevents the self aggregating nature of insulin at neutral pH, thereby improving its stability (Lovatt et al. 1996).

Pulmonary delivery of insulin by inhalation is a promising alternative for insulin delivery (Uchenna et al. 2001). Inhaled insulin provides rapid acting insulin that can be used for the treatment of both type 1 and type 2 diabetes (Skyler et al. 2001; Cefalu et al. 2001). Lungs have large surface areas and large vascularisation that facilitate systemic delivery of insulin (Uchenna et al. 2001). Insulin absorption across the alveolar capillary and epithelial cells is assumed to be by transcytosis (Klonoff 1999). The maximum delivery efficiency achieved using current technology is about 30% (Klonoff 1999). The aerodynamic diameter of the particles must be less than 2 µm for the particles to be targeted to the alveoli (Lucas et al. 1999).

The transdermal delivery of insulin has also been investigated (Zhao et al. 2010). The main advantages of this route of delivery are the low proteolytic activity of skin and the prolonged effects achieved by continuous absorption (Higaki et al. 2006). However, it is difficult for the drug to penetrate through the skin (Higaki et al. 2006). The outermost layer of skin, the stratum corneum, acts as the main barrier for tranderdermal delivery (Higaki et al. 2006). Several attempts have been made to improve the transdermal penetration (Higaki et al. 2006). These include physical methods like skin poration and use of chemical absorption enhancers (Higaki et al. 2006; Coulman et al. 2009). Insulin nanoparticles showed high permeation rates and have significant potential for diabetic therapy (Zhao et al. 2010).

Gold nanoparticles have been investigated as carriers for transmucosal insulin delivery (Joshi et al. 2005). Gold nanoparticles are non-toxic and biocompatible (Joshi et al. 2005). The presence of an amino acid layer between insulin and gold nanoparticles was found to improve the uptake of insulin (Joshi et al. 2005). Insulin bound to gold nanoparticles was found to achieve blood glucose control comparable to those of subcutaneous insulin injections (Joshi et al. 2005).

Another application of nanotechnology is in the design of new insulin loaded nanocomposite membranes which are able to self regulate the rate of insulin permeation based on blood glucose levels (Gordijo et al. 2011). The membrane contains glucose sensors that trigger insulin release (Gordijo et al. 2011). In the presence of high glucose concentrations (hyperglycemia), the oxidation of glucose to gluconic acid reduces the pH in the microenvironment which causes the hydrogel nanoparticles to shrink, allowing faster insulin permeation across the membrane (Gordijo et al. 2011).

DRUGS & TREATMENTS

In addition to insulin delivery, several types of drugs have been investigated for the control of hyperglycemia in diabetic patients.

Insulin secretogogues increase insulin release from pancreatic islets by closing the ATP-dependent potassium channel (K_{ATP}) in the plasma membrane of pancreatic β-cells (Kahn et al. 2005). The closure of the channel causes depolarization of the plasma membrane, resulting in the opening of the calcium channels (Kahn et al. 2005). Calcium ions enter the cell and the rise in cytosolic calcium ion concentration stimulates insulin secretion (Kahn et al. 2005). Sulfonylurea drugs are a specific class of insulin secretogogues that include chlorpropamide, glyburide, glipizide and glimepiride (Kahn et al. 2005). Repaglinide and nateglinide are non-sulfonylurea insulin secretogogues that have similar mechanisms of action (Kahn et al. 2005).

Biguanides and thiazolidinediones are used to treat insulin resistance and are important for treatment of type 2 diabetes (Kahn et al. 2005). Metformin, a biguanide, acts to reduce hepatic glucose production and has been used since the 1960s (Kahn et al. 2005; Moller 2001). Thiazolidinediones discovered in the late 1970s were noted for their efficacy in reducing hyperglycemia and reducing insulin resistance (Kahn et al. 2005). They bind to peroxisome proliferator-activated receptor- γ (PPAR γ) which enhances insulin action (Moller 2001).

α-glucosidase inhibitors are classic competitive inhibitors that interfere with gut glucose absorption (Moller 2001). They compete with the oligosaccharides for the binding site of α-glucosidase enzyme in the gut, thereby reducing glucose absorption from the gut (Kahn et al. 2005). Abdominal discomfort and elevated plasma hepatic transaminase levels are the major side effects of this class of inhibitors (Akkati et al. 2010).

The use of nanoparticles for prolonged or controlled drug delivery, as well as to improve the bioavailability and stability of the drug makes it desirable for diabetic therapy.

Repaglinide has been shown to have very good anti diabetic properties, however, the short half life (< 1 hr) and poor absorption characteristics limit its therapeutic efficacy (Jain and Saraf 2009). In addition, it produces hypoglycemia after oral administration (Dhanalekshmi et al. 2010). Nanodrug delivery systems release the drug in a controlled manner for a prolonged period of time, thereby reducing the adverse effects of the conventional dose (Dhanalekshmi et al. 2010). Repaglinide loaded poly (methyl methacrylate) nanoparticles were shown to prolong the release of the drug and also showed minimal *in vivo* toxicity (Dhanalekshmi et al. 2010). Surface modified biodegradable nanoparticles like pegylated polylactic-co-glycolic acid (mPEG-PLGA) nanoparticles loaded with repaglinide were found to have a long duration of action and are an effective delivery system for decreasing dosing frequency and for improving patient compliance (Jain and Saraf 2009).

Gliclazide, a second generation of sulfonylureas, is widely used in the treatment of non-insulin-dependent diabetes mellitus (Wang et al. 2010). However, gliclazide has poor dissolution properties and also poor permeability across the GI membrane, leading to poor absorption of the drug across the gastrointestinal tract (Hong et al. 1998). Gliclazide loaded eudragit (L100 and RS) nanoparticles were shown to enhance bioavailability and provide sustained release of the drug (Devarajan and Sonavane 2007). In addition, the nanoparticle formulation was found to be stable at the end of 6 months and showed better activity of the drug *in vivo* as compared to conventional gliclazide formulations (Devarajan and Sonavane 2007). Chitosan microparticles have also shown promise in providing sustained delivery of gliclazide for diabetes treatment (Barakat and Almurshedi 2010).

TISSUE DAMAGE & WOUND HEALING

Diabetes mellitus is associated with a broad range of effects on almost every organ system (Kahn et al. 2005). A classic wound associated with diabetes is the non-healing foot ulcer (Kahn et al. 2005), which continues to be the most common contributor for lower limb amputations in Europe and the USA. This has been primarily due to the development of peripheral neuropathy which leads to the loss of protective sensations of the feet (Fisher et al. 2010). In addition, ischemia in diabetic patients leads to a higher risk of developing atherosclerosis of the lower extremities (Laing 1998). It is found to start at a younger age and at an accelerated rate when compared to non-diabetic patients (Laing 1998).

The use of nanofibres for diabetic wound healing has been reported (Choi 2008). Epidermal growth factor (rhEGF) conjugated nanofibres were investigated for *in vivo* wound healing of diabetic ulcers in mice (Choi 2008). RhEGF nanofibres demonstrated much higher wound closure rates in comparison to controls (Choi 2008). Curcumin, an active ingredient of turmeric has been shown to assist in wound healing in diabetic mice (Merrell 2009). Curcumin treatment enhanced the biosynthesis of extracellular matrix proteins and also increased formation of granulation tissue (Merrell 2009). However, the low *in vivo* stability and low bioavailability of curcumin makes it difficult for oral administration (Merrell 2009). Nanofiber matrices are able to mimic the diameter of collagen fibrils in the extracellular matrix. Curcumin loaded poly(ε-caprolactone) nanofibres were shown to reduce inflammation and enhanced wound closure *in vivo* in a diabetic mouse model (Merrell 2009). Poly-N-acetyl glucosamine nanofibres (sNAG) were shown to be effective in wound healing and are biodegradable (Scherer 2009). sNAG treatment enhanced wound closure and upregulated angiogenesis and cell proliferation in new tissues (Scherer 2009).

A nanotechnology based footsock for efficient control of diabetic foot preulcer has also been investigated (Elisa et al. 2008). The sock was made from polyammide fibres loaded with an active moisturizer, which contained a nanoemulsion of liposomes (Elisa et al. 2008). The positively charged nanoemulsions are attracted to the negative charge of the skin and are therefore more effective (Elisa et al. 2008). When the patients wore the socks, there was a constant release of the moisturizer for about 36 hours, thereby eliminating the need to moisturize the skin (Elisa et al. 2008). The skin was found to be more hydrated and was less hard when compared to controls (Elisa et al. 2008).

CONCLUSION

The application of nanotechnology in diabetes treatment is a promising approach that needs to be extensively explored. Advances in the field of biomaterials and engineering, improved glucose monitoring and insulin delivery systems will prove to be important next steps in improving treatment of diabetes.

Summary Points

- Diabetes affects more than 220 million people around the world.
- The application of nanotechnology in diabetes includes glucose monitoring, insulin delivery, drug delivery and wound healing.
- Fluorescent glucose nanosensors provide continuous glucose monitoring in contrast to conventional finger prick tests.
- The low oral bioavailability and short half-life of insulin can be overcome by encapsulating it in nanoparticles.
- Insulin secretogogues increase insulin release from pancreatic β cells.
- Many drugs have short half-lives and poor absorption characteristics. Nanodrug delivery systems can release the drug in a controlled manner for extended periods of time.
- Nanofibers exhibit higher wound healing rates in comparison to controls.
- Poly-N-acetyl glucosamine biodegradable nanofibres (sNAG) are highly effective in wound healing.
- Use of nanotechnology for improved glucose monitoring and insulin delivery is crucial for improved diabetic therapy.

References

Akkati, S., K.G. Sam and G. Tungha. 2011. Emergence of Promising Therapies in Diabetes Mellitus. The Journal of Clinical Pharmacology 51: 796–804.

Artursson, P., T. Lindmark, S.S. Davis and L. Illum. 1994. Effect of Chitosan on the Permeability of Monolayers of Intestinal Epithelial Cells (Caco-2). Pharmaceutical Research 11: 1358–61.

Aungst, B.J. 2000. Intestinal permeation enhancers. Journal of Pharmaceutical Sciences 89: 429–42.

Barakat, N.S., and A.S. Almurshedi. 2010. Design and development of gliclazide-loaded chitosan for oral sustained drug delivery: *In vitro/in vivo* evaluation. Journal of Microencapsulation. 0: 1–12.

Billingsley, K., M.K. Balaconis, J.M. Dubach, N. Zhang, E. Lim, K.P. Francis and H.A. Clark. 2010. Fluorescent Nano-Optodes for Glucose Detection. Analytical Chemistry 82: 3707–13.

Calvert M.J., R.J. McManus and N. Freemantle. 2007. The management of people with type 2 diabetes with hypoglycaemic agents in primary care: retrospective cohort study. Fam. Pract. 24: 224–9.

Carino, G.P., J.S. Jacob and E. Mathiowitz. 2000. Nanosphere based oral insulin delivery. Journal of Controlled Release 65: 261–9.

Cefalu, W.T. 2004. Concept, Strategies, and Feasibility of Noninvasive Insulin Delivery. Diabetes Care 27: 239–46.

Cefalu, W.T., J.S. Skyler, I.A. Kourides, W.H. Landschulz, C.C. Balagtas, S-L. Cheng and R.A. Gelfand. 2001. Inhaled Human Insulin Treatment in Patients with Type 2 Diabetes Mellitus. Annals of Internal Medicine 134: 203–7.

Choi, J.S., K.W. Leong and H.S. Yoo. 2008. *In vivo* wound healing of diabetic ulcers using electrospun nanofibers immobilized with human epidermal growth factor (EGF). Biomaterials 29: 587–96.

Coulman, S.A., A. Anstey, C. Gateley, A. Morrissey, P. McLoughlin, C. Allender and J.C. Birchall. 2009. Microneedle mediated delivery of nanoparticles into human skin. International Journal of Pharmaceutics 366: 190–200.

Damgé, C., C.P. Reis and P. Maincent. 2008. Nanoparticle strategies for the oral delivery of insulin. Expert Opinion on Drug Delivery 5: 45–68.

Devarajan, P.V., and G.S. Sonavane. 2007. Preparation and *In Vitro/In Vivo* Evaluation of Gliclazide Loaded Eudragit Nanoparticles as a Sustained Release Carriers. Drug Development and Industrial Pharmacy 33: 101–11.

Dhanalekshmi, U.M., G. Poovi, N. Kishore and P.N. Reddy. 2010. *In vitro* characterization and *in vivo* toxicity study of repaglinide loaded poly (methyl methacrylate) nanoparticles. International Journal of Pharmaceutics 396: 194–203.

Diabetes. 2011. Fact sheet N°312. http://www.who.int/mediacentre/factsheets/fs312/en/.

Dubach, J.M., D.I. Harjes and H.A. Clark. 2007. Fluorescent Ion-Selective Nanosensors for Intracellular Analysis with Improved Lifetime and Size. Nano Letters 7: 1827–31.

Elisa, B., M. Silvia, D. Valentina, R. Loredana, T. Anna, S. Alessia, G. Chiara, C. Fabrizio, R. Marco and A. Piaggesi. 2008. Use of Nanotechnology-Designed Footsock in the Management of Preulcerative Conditions in the Diabetic Foot: Results of a Single, Blind Randomized Study. The International Journal of Lower Extremity Wounds 7: 82–7.

Fisher, T.K., C.L. Scimeca, M. Bharara, J.L. Mills Sr and D.G. Armstrong. 2010. A step-wise approach for surgical management of diabetic foot infections. Journal of Vascular Surgery 52: 72S–5S.

Gordijo, C.R., K. Koulajian, A.J. Shuhendler, L.D. Bonifacio, H.Y. Huang, S. Chiang, G.A. Ozin, A. Giacca and X.Y. Wu. 2011. Nanotechnology-Enabled Closed Loop Insulin Delivery Device: *In Vitro* and *In Vivo* Evaluation of Glucose-Regulated Insulin Release for Diabetes Control. Advanced Functional Materials 21: 73–82.

Higaki, M., M. Kameyama, M. Udagawa, Y. Ueno, Y. Yamaguchi, R. Igarashi, T. Ishihara and Y. Mizushima. 2006. Transdermal Delivery of CaCO3-Nanoparticles Containing Insulin. Diabetes Technology & Therapeutics 8: 369–74.

Hong, S.S., S.H. Lee, Y.J. Lee, S.J. Chung, M.H. Lee and C.K. Shim. 1998. Accelerated oral absorption of gliclazide in human subjects from a soft gelatin capsule containing a PEG 400 suspension of gliclazide. Journal of Controlled Release 51: 185–92.

Hussain, F., D.J.S. Birch and J.C. Pickup. 2005. Glucose sensing based on the intrinsic fluorescence of sol-gel immobilized yeast hexokinase. Analytical Biochemistry 339: 137–43.

Jain, S., and S. Saraf. 2009. Repaglinide-loaded long-circulating biodegradable nanoparticles: Rational approach for the management of type 2 diabetes mellitus. Journal of Diabetes 1: 29–35.

Joshi, H.M., D.R. Bhumkar, K. Joshi, V. Pokharkar and M. Sastry. 2005. Gold Nanoparticles as Carriers for Efficient Transmucosal Insulin Delivery. Langmuir 22: 300–5.

Jung, T., W. Kamm, A. Breitenbach, E. Kaiserling, J.X. Xiao and T. Kissel. 2000. Biodegradable nanoparticles for oral delivery of peptides: is there a role for polymers to affect mucosal uptake? European Journal of Pharmaceutics and Biopharmaceutics 50: 147–60.

Kahn, C.R., C.W. Gordon, L.K. George, M.J. Alan, C.M. Alan and J.S. Robert. 2005. Joslin's Diabetes Mellitus: Lippincott Williams & Wilkins 1224 pp.

Klonoff, D.C. 1999. Inhaled insulin. Diabetes Technol. Ther. 1: 307–13.

Laing, P. 1998. The development and complications of diabetic foot ulcers. The American Journal of Surgery 176: 11S–9S.

Lin Y-H., C.T. Chen, H.F Liang, A.R. Kulkarni, P.W. Lee, C.H. Chen and H.W. Sung. 2007. Novel nanoparticles for oral insulin delivery via the paracellular pathway. Nanotechnology 18: 105102.

Lovatt, M., A. Cooper and P. Camilleri. 1996. Energetics of cyclodextrin-induced dissociation of insulin. European Biophysics Journal 24: 354–7.

Lowman, A.M., M. Morishita, M. Kajita, T. Nagai and N.A. Peppas. 1999. Oral delivery of insulin using pH-responsive complexation gels. Journal of Pharmaceutical Sciences 88: 933–7.

Lucas, P., K. Anderson, U.J. Potter and J.N. Staniforth. 1999. Enhancement of Small Particle Size Dry Powder Aerosol Formulations using an Ultra Low Density Additive. Pharmaceutical Research 16: 1643–7.

Marvin, J.S., and H.W. Hellinga. 1998. Engineering Biosensors by Introducing Fluorescent Allosteric Signal Transducers: Construction of a Novel Glucose Sensor. Journal of the American Chemical Society 120: 7–11.

Meadows, D.L., and J.S. Schultz. 1993. Design, manufacture and characterization of an optical fiber glucose affinity sensor based on an homogeneous fluorescence energy transfer assay system. Analytica Chimica Acta. 280: 21–30.

Merrell, J.G., S.W. McLaughlin, L. Tie, C.T. Laurencin, A.F. Chen and L.S. Nair. 2009. Curcumin-loaded poly(ε-caprolactone) nanofibres: Diabetic wound dressing with anti-oxidant and anti-inflammatory properties. Clinical and Experimental Pharmacology and Physiology 36: 1149–56.

Moller, D.E. 2001. New drug targets for type 2 diabetes and the metabolic syndrome. Nature 414: 821–7.

Pickup, J.C., Z.L. Zhi, F. Khan, T. Saxl and D.J. Birch. 2008. Nanomedicine and its potential in diabetes research and practice. Diabetes Metab. Res. Rev. 24: 604–10.

Sabetsky, V., and J. Ekblom. 2010. Insulin: a new era for an old hormone. Pharmacol. Res. 61: 1–4.

Scherer, S.S., G. Pietramaggiori, J. Matthews, S. Perry, A. Assmann, A. Carothers, M. Demcheva, R.C. Muise-Helmericks, A. Seth, J.N. Vournakis, R.C. Valeri, T.H. Fischer, H.B. Hechtman and D.P. Orgill. 2009. Poly-N-Acetyl Glucosamine Nanofibers: A New Bioactive Material to Enhance Diabetic Wound Healing by Cell Migration and Angiogenesis. Annals of Surgery 250: 322–30.

Shibata, H., Y.J. Heo, T. Okitsu, Y. Matsunaga, T. Kawanishi and S. Takeuchi. 2010. Injectable hydrogel microbeads for fluorescence-based *in vivo* continuous glucose monitoring. Proceedings of the National Academy of Sciences 107: 17894–8.

Skyler, J.S., W.T. Cefalu, I.A. Kourides, W.H. Landschulz, C.C. Balagtas, S.L. Cheng and R.A. Gelfand. 2001. Efficacy of inhaled human insulin in type 1 diabetes mellitus: a randomised proof-of-concept study. The Lancet 357: 331–5.

Springsteen, G., and B. Wang. 2002. A detailed examination of boronic acid-diol complexation. Tetrahedron 58: 5291–300.

Trehan, A., and A. Ali. 1998. Recent approaches in insulin delivery. Drug Dev. Ind. Pharm. 24: 589–97.

Uchenna, A.R., U.M. Ikechukwu, M. Armand, R. Kinget and N. Verbeke. 2001. The lung as a route for systemic delivery of therapeutic proteins and peptides. Respiratory Research 2: 198 – 209.

Wang, L., J. Wang, X. Lin and X. Tang. 2010. Preparation and *in vitro* evaluation of gliclazide sustained-release matrix pellets: formulation and storage stability. Drug Development and Industrial Pharmacy 36: 814–22.

Xu, H., J.W. Aylott and R. Kopelman. 2002. Fluorescent nano-PEBBLE sensors designed for intracellular glucose imaging. Analyst 127: 1471–7.

Zhao, X., Y. Zu, S. Zu, D. Wang, Y. Zhang and B. Zu. 2010. Insulin nanoparticles for transdermal delivery: preparation and physicochemical characterization and *in vitro* evaluation. Drug Development and Industrial Pharmacy 36: 1177–85.

2

Lipid Matrix Nanoparticles in Diabetes

Eliana B. Souto,[1,a,]* Joana F. Fangueiro[1,b] and Selma S. Souto[2]

ABSTRACT

The use of drug delivery systems (DDS) based on lipid matrices is useful for the delivery of anti-diabetic drugs. Since the peptide suffers denaturation under gastric pH and enzymatic proteolysis, it cannot be administered oral. In addition, the gut has low permeability for large biomolecules. Nowadays, the only treatment for diabetes is the subcutaneous injection of insulin. Subcutaneous administration is reported as a problematic route because of low patient compliance and lipoatrophy in local injection. DDS composed of lipids provide alternative routes, such as oral and pulmonary, to overcome the traditional subcutaneous injection of insulin. This strategy allows the delivery enhancement and the stabilization of the incorporated drug, which leads

---

[1]Faculty of Health Sciences, University Fernando Pessoa, Rua Carlos da Maia, 296, 4200-150,Porto, Portugal.
[a]E-mail: eliana@ufp.edu.pt
[b]E-mail: 14224@ufp.edu.pt
[2]Department of Endrocrinology, Hospital São João, Alameda Professor Hernâni Monteiro, 4200-319. Porto, Portugal; E-mail: sbsouto.md@gmail.com
*Corresponding author

List of abbreviations after the text.

to a higher bioavailability. The lipids used are physiological preventing toxicological reactions. DDS based on a lipid matrix include solid lipid nanoparticles, composed of solid lipids only; nanostructures lipid carriers, similar to solid lipid nanoparticles, but composed of a liquid lipid and a solid lipid; lipid drug conjugates that are composed of lipids, usually fatty acids with esters; phospholipid micelles composed of phospholipids but represent sterically stabilized micelles; liposomes are composed by one or more phospholipid bilayers and self-emulsifying drug delivery systems are composed also of lipids (liquids or solids) that creat microemulsions. These DDS have already been tested for the treatment of diabetes and appear to be good alternatives to the traditional injection.

1. INTRODUCTION

The design and development of novel DDS such as lipid matrix nanoparticles consist of delivering the drug precisely and safely to its target site at the right period of time to have a controlled release and to achieve the maximum therapeutic effect (Mishra et al. 2010). Majority of drugs show high aqueous solubility and low permeability due to their biopharmaceutical properties. Thus, to overcome some failures related to absorption, distribution and toxicity the use of carrier systems to increase the stability of a variety of drugs is imperative. Furthermore, their nanometer size (Table 1) is also suitable to target organs and cells efficiently (Martins et al. 2009).

Table 1. Different types of DDS and their mean particle size.

Types of DDS	Particle size (nm)
SLN	50–1000
NLC	50–1000
LDC	< 1000
Phospholipid micelles	4–100
Liposomes	25–3500
SEDDS	50–300

Description of the mean particle size of the several DDS described.

Also, physiological lipids or lipids composed of physiological compounds (acylglycerols, phospholipids) are used as matrix material (Table 2) and are well tolerated by the organism (Souto et al. 2004). Ideally, a DDS should be able of providing extended blood circulation and deliver the drug moiety at the targeted site. DDS consist of macromolecular materials (typically of lipids or polymers) with the drug either dissolved within a polymeric matrix, entrapped inside the lipid, encapsulated, or adsorbed onto surface of particles (Mishra et al. 2010).

Table 2. Examples of lipids used in lipid matrix systems.

Lipids	Type of system	References
Solid lipids		
Stearic acid	SLN	Liu et al. 2008; Liu et al. 2007
Cetyl palmitate	SLN	Sarmento et al. 2007a
Dynasan®116 (glyceryl tripalmitate)	SLN, NLC	Souto et al. 2004
Glyceril dioleate	SMEDDS	Gursoy and Benita 2004
Liquid lipids		
Plurol® Oleique (Polyglyceryl Oleate)	SEDDS	Atef and Belmonte 2008
Palmitic acid	SLN	Liu et al. 2008; Liu et al. 2007
Labrafil® M 1944 CS (unsaturated polyglycolysed glycerides)	Phospholipid micelles	Cilek et al. 2005
Labrafac® (Propylene Glycol Dicaprylocaprate)	SEDDS	Atef and Belmonte 2008
Lauroglycol®FCC (Propylene Glycol Laurate)	SEDDS	Atef and Belmonte 2008
Phospholipids		
DSPE-PG8G (distearoylphosphatidylethanolamine-polyglyceline),	Liposomes	Hanato et al. 2009
Phosphatidylcholine	Liposomes	Kisel et al. 2001; Chono et al. 2009
Phosphatidylinositol	Liposomes	Kisel et al. 2001
Soybean phosphatidylcholine	Phospholipid micelles	Wang et al. 2010

In this table are described several lipids that were used to perform the lipid matrix systems. The lipids were separated by solid lipids (triglycerides, complex glyceride mixtures, waxes and fatty acids) that are used to produce all kind of systems, but specially SLN. The liquid lipids used in SEDDS, NNC and SLN. The phospholipids usually used for produce liposomes and phospholipid micelles.

The use of solid lipid matrices for the controlled release of drugs is known in pharmacy for many years. The interest in lipid matrices for the delivery of peptides and proteins by oral intake has also been the focus of research on novel DDS. In parallel with increasing incidence of diabetes worldwide, new methods of insulin therapy are been exploited. Nowadays, the diabetes treatment is a conventional insulin injection. However, this classical procedure is in conflict with the natural system for the maintenance of glucose homeostasis since subcutaneous injection does not provide immediately delivery. Insulin is not able to reach the target due to biological shortcomings (such as enzymatic and proteolytic degradation) and because the peptide has a relatively high molecular weight (Kisel et al. 2001). In this section the various alternative systems for the diabetes treatment are described to improve the delivery of anti-diabetic drugs.

2. SOLID LIPID NANOPARTICLES (SLN) AND NANOSTRUCTURE LIPID CARRIERS (NLC)

2.1 Definitions and Properties

SLN were developed in the beginning of the 1990s and first introduced by Müller et al. in 1993. They consist of colloidal particles of a lipid matrix that is solid at body temperature and are made from a solid lipid only (Fig. 1). By definition, the lipids can be triglycerides, complex glyceride mixtures or waxes. In addition to lipid, surfactants are used as stabilizers in aqueous dispersion (Joshi and Müller 2009; Sarmento et al. 2007a; Wissing et al. 2004). SLN show several advantages for drug delivery, such as good tolerability, low cytotoxicity (since physiological and biodegradable lipids are used), fast and effective production process including the possibility of large industrial scale, protection of incorporated labile drugs from enzymatic and chemical degradation, possibility to produce highly concentrated lipid suspensions and site-specific targeting. However, the drug loading capacity of conventional SLN is limited due to the formation of a perfect lipid crystal matrix by using purified lipid, which leaves little space for drug loading. Usually, drugs are incorporated between the fatty acid chains, alternatively in between lipid layers or in amorphous clusters in crystal imperfections. Thus, the more perfect densely packed the crystal is, the fewer drug molecules can be incorporated (Fig. 2). The complexity of the physical state of the lipid derived from transformation between different modifications, possibility of supercooled melts which

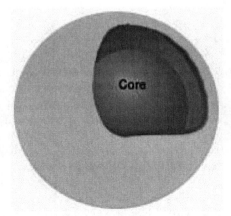

Fig. 1. Illustration of a conventional SLN (Unpublished). SLN has a circular shape and a solid lipid core. The drug is encapsulated in the core with the lipid and these particles have usually nanometer size. The encapsulation of the drugs into the core, protect their physicochemical properties.

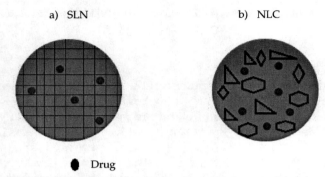

Fig. 2. Inner structure of the lipid matrix systems: a) SLN and b) NLC (Unpublished). In SLN, lipid matrix is a perfect crystal where as in NLC the crystal structure has many imperfections. In SLN drug expulsion can occur easier due the lipid matrix transforms from high energy modification, characterized by the presence of many imperfections, to the β-modification forming a perfect crystal with no room for guest molecules. In NLC the lipid matrix is composed not only of solid lipids, but also of liquid lipids which can be mixed in such combination that the particle solidifies upon cooling but does not recrystallize and remains in the amorphous state. This results in a structure with many imperfections to accommodate drug and thus improve loading capacity.

cause stability problems during storage or administration (e.g., gelation, particle size increase, drug expulsion) and the biodegradation for lipolytic enzymes with parenteral administration (Martins et al. 2007; Muchow et al. 2008; Sarmento et al. 2007a; Souto et al. 2004; Wissing et al. 2004).

NLC have been introduced in the end of the 1990s to overcome disadvantages of SLN described above. Named also as solid lipid nanoparticles of second generation, NLC are generally used for the delivery of lipophilic drugs. Identical to SLN, the particle matrix is still solid at body temperature, but is produced from a blend of a solid lipid with a liquid lipid (generally oils) leading to certain advantages compared to SLN. The advantages over other systems are the protection of chemically labile compounds against degradation, increased bioavailability and drug loading capacity, and minimization or avoidance of drug expulsion during storage (Table 6). The use of a lipid mixture with very differently structured (sized) molecules distorts the formation of a perfect crystal. The particle matrix contains many imperfections providing space to accommodate the drug in molecular form or as amorphous clusters (Fig. 2). One could state that the perfectness of NLC is the imperfectness in its crystalline structure (Muchow et al. 2008; Muller et al. 2007).

Production methods for these novel DDS are very well described in literature. Methods include the high pressure homogenization (HPH) technique and can be performed at high temperature (hot HPH) or at below room temperature (cold HPH) (Fig. 3). The hot HPH technique may also be suitable for thermolabile drugs because the exposure at high temperatures is relatively short. The cold HPH is convenient for labile drugs or hydrophilic drugs (Almeida and Souto 2007; Joshi and Muller 2009). Other productions methods include via microemulsions, solvent emulsification-evaporation and also the multiple emulsion (w/o/w) method (Gallarate et al. 2009; Wissing et al. 2004).

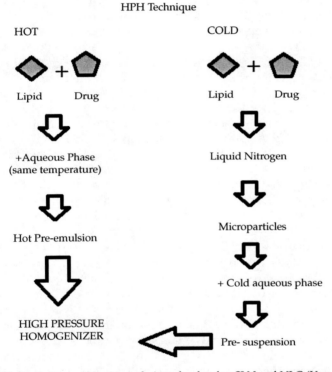

Fig. 3. High pressure homogenization technique for develop SLN and NLC (Unpublished). Preparation of SLN and NLC by high pressure homogenization (HPH) technique and can be performed at high temperature (hot HPH technique) or at below room temperature (cold HPH technique). For both techniques the drug is dissolved or solubilized in the lipid being melted at approximately 5–10°C above its melting point. For the hot HPH, the lipid and the drug are melted and combined with an aqueous surfactant solution of identical temperature. A hot pre-emulsion is formed by high speed stirring. The hot pre-emulsion is then processed in a temperature controlled high pressure homogenizer, generally a maximum of three cycles at 500 bar are sufficient. Then the lipid recrystallizes and leads to SLN or NLC. Of course, care needs to be taken that recrystallization of the lipid occurs. In the cold HPH the lipid and drug are melted together and then rapidly ground by moltar milling under liquid nitrogen forming solid lipid microparticles (approximately 50–100 µm). A pre-suspension is formed by high speed stirring of the particles in a cold aqueous surfactant solution. This pre-suspension is then homogenized at or below room temperature forming SLN or NLC, the homogenizing conditions are generally five cycles at 500 bar.

2.2 Applications in Diabetes

Insulin is used for the treatment of diabetes type 1 or insulin-dependent diabetes. The oral administration is considered the most convenient and comfortable way for administration of insulin for less invasive and painless diabetes management, leading to higher patient compliance (Table 4). The principal barriers include enzymatic activity and the physical suroundings. The enzymatic barrier provided by the gastrointestinal tract (GIT) involves protein digestion in the stomach by proteases, namely pepsins and by other enzymes from the pancreas such as trypsin, chymotrypsin and carboxypeptidases. Additionally, there also exists a specific cytosolic enzyme called insulin-degrading enzyme which accomplishes the delivery of insulin. Encapsulation of labile proteins improves their protection against enzymatic attack and harsh environment of GIT, providing also a controlled release profile. Physical suroundings involve intestinal epithelium. Drugs such insulin, are not able to diffuse across epithelial cells through lipid-bilayer cell membranes to the bloodstream. The microvilli present in the epithelium increase the absorptive area of the GIT, however these structures also contain digestive enzymes (Martins et al. 2007). Chitosan (CS) has been used as intestinal permeation enhancer. Agnihotri et al. 2004 have reported that CS nanoparticles may prolong the protein residence time in the small intestine, infiltrate into the mucus layer and subsequently mediate transiently opening the tight junctions between epithelial cells while becoming unstable and broken apart due to their pH sensitivity and/or degradability. Another study was approached by Sonaje et al. 2009 where positively and the negatively charged SLN were obtained by surfacing the particles with CS and poly-γ-glutamic acid (γ-PGA) respectively for oral insulin administration. Prepared nanoparticles demonstrated safety after oral administration and pharmacodynamic and pharmacokinetic evaluation in diabetic rats indicated that the intestinal absorption of insulin was significantly enhanced. Liu et al. 2007 investigated the use of sodium cholate and soybean phosphatidylcholine as solubilizers and stabilizing agents to produce SLN loaded with insulin by a novel reverse micelle-double emulsions. The study showed a successfully encapsulation of insulin in SLN of small particle size, excellent physical stability and good loading capacity. Sarmento et al. 2007b produced insulin-loaded SLN for oral administration. The authors obtained SLN with acceptable physiochemical parameters (shape, zeta potential values and association efficiency) for long term stability. The SLN obtained were able to reduce glucose levels on rats, protect insulin against chemical degradation in the GIT and promote its absorption. Gallarate et al. 2009 produced insulin SLN by w/o/w emulsion technique. The SLN obtained revealed good diameters of SLN and proved that encapsulation of high amounts of insulin is limited because of the acidic medium.

Other routes are also been investigated for insulin delivery. Pulmonary route is a novel non-invasive approach and is emerging as a viable alternative to injectable insulin. Advantages of this route include large absorptive area offered by the lungs, extensive vasculature, and high permeability of the membrane due the tightness of alveolar epithelium allowing rapid drug absorption. Also the low extracellular and intracellular enzymatic activity leads to the avoidance of the first-pass metabolism provided by the GIT (Liu et al. 2008). Liu et al. 2008 studied a novel nebulizer-compatible for insulin loaded-SLN pulmonary delivery produced by micelle-double emulsion method. The stability tests results (e.g., size, morphology) as well as insulin concentration before and after nebulization demonstrated good tolerance for this route. Plasma glucose levels in rats could be effectively reduced after pulmonary insulin delivery using SLN as carrier.

3. LIPID DRUG CONJUGATES (LDC)

3.1 Definitions and Properties

LDC are used to improve the oral drug delivery of hydrophilic molecules. Many hydrophilic drugs show poor oral bioavailability due to their explicit hydrophilic character. The major advantages of these systems include the protection against degradation, increase the lipophilicity (the drug shows an improved permeation through the GIT wall). Promotion of oral absorption occurs by the lipids present in the LDC nanoparticles. The LDC are poorly water soluble, they typically have a melting range of approximately 50–100°C and can be transformed in nanoparticles using a HPH method. LDC can be obtained by salt formation (e.g., with fatty acid) or by covalent linking (e.g., to esters or ethers) or by the covalent linking (Fig. 4) (Martins et al. 2009; Muchow et al. 2008; Wissing et al. 2004).

3.2 Applications in Diabetes

Ikumi et al. 2008 investigated a poly (□-glutamic acid) (□-PGA) modified with plhoridzin (PRZ), which is an inhibitor of the Na^+/glucose cotransporter (SGLT1) that was synthesized via a □-amino triethylene glycol linker. *In vitro* and *in vivo* experiments suggest that PGA-PRZ could strongly inhibit the glucose transport and remarkably suppress the increase of blood glucose level after oral administration of glucose in rats, while free phloridzin scarcely affected glucose induced hyperglycemic effect. The study showed that PGA-PRZ is a good candidate as oral anti-diabetic drug.

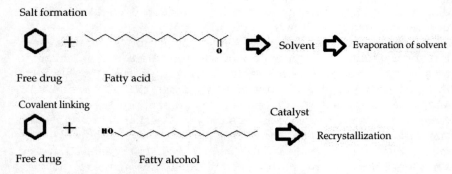

Fig. 4. Preparation methods for the formation of LDC (Unpublished). In the salt formation process, the free drug base and fatty acid are dissolved in a suitable solvent. The solvent is then consequently evaporated under reduced pressure. For the covalent linking, the drug (salt) and a fatty alcohol react together in presence of a catalyst and the LDC bulk is then purified by recrystallization. The obtained LDC bulk is then processed with an aqueous surfactant solution to a nanoparticle formulation using HPH described before for SLN and NLC.

4. PHOSPHOLIPID MICELLES

4.1 Definitions and Properties

Phospholipid micelles has been introduced as safe, biocompatible for the delivery of poorly water-soluble drugs and are sterically stabilized with polyethylene glycol (PEGylated) phospholipids. They typically have a spherical shape and the size in the nanometer range (5–100 nm). Micelles are association of colloids formed by amphiphilic compounds (Fig. 5). This is a thermodynamically stable state that many amphiphilic compounds adopt in aqueous media and are spontaneously formed under certain concentration and temperature which consist of two clearly distinct regions with opposite affinities toward a given solvent. At low concentration, these amphiphilic molecules exist separately. However, as their concentration increase, aggregation takes place within a rather narrow concentration interval (Torchilin 2007).

Phospholipid micelles should be stable *in vivo* for a sufficiently long time, not induce any biological reactions, release a free drug upon contact with target tissues or cells and, finally, the carriers components should be easily removed from the body after finishing the therapeutic function. Phospholipid micelles show high solubilization potential, long term stability and low *in vitro* cytotoxicity (Mishra et al. 2010). Also, the solubilization of drugs using micelle-forming surfactants results in an increased water solubility of sparingly soluble drugs improving their

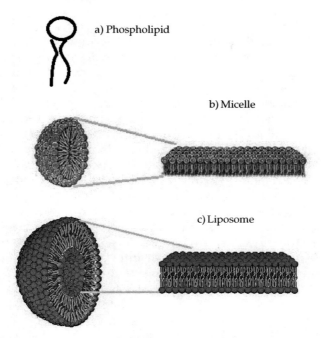

Fig. 5. Illustration of the structure of: a) Phospholipid; b) micelles. and c)liposome (Unpublished). Phospholipid, usually represented by the head with lipophilic character and tails with hydrophilic character. In micelles (b) there is only one phospholipid layer and in lipossomes (c) there is at least one phospholipid bilayer that resembles the cell membranes Both systems are association colloids formed by amphiphilic compounds. They consist of two clearly distinct regions with opposite affinities toward a given solvent. Hydrophobic fragments of amphiphilic molecules form the core of a micelle, which can solubilize poorly soluble drugs, while hydrophilic fragments form micelle's corona. In aqueous systems, non-polar molecules are solubilized within the micelle core, polar molecules will be adsorbed on the micelle surface and substances with intermediate polarity will be distributed along surfactant molecules in intermediate positions.

Color image of this figure appears in the color plate section at the end of the book.

bioavailability, reduction of toxicity and other adverse effects, enhanced permeability across the physiological barriers, and substantial and favorable changes in drug biodistribution (Torchilin 2007).

Phospholipid micelles can be prepared by co-precipitation and reconstitution of drug and lipids (Mishra et al. 2010). The formation of micelles is driven by the decrease of free energy in the system because of the removal of hydrophobic fragments from the aqueous environment and reestablishing the hydrogen bond network in water. Additional energy gain results from formation of van der Waals bonds between hydrophobic blocks in the core of the formed micelles. Conventional surfactants usually form micelles spontaneously (Torchilin 2007).

4.2 Applications in Diabetes

Cilek et al. 2005 tested the oral absorption of recombinant human insulin dissolved in the aqueous phase of w/o microemulsions in streptozotocin-induced diabetic male Wistar rats. The authors demonstrated significant improvement in oral pharmacological availability compared with insulin solution, although it was ≈ 0.1% compared with subcutaneous administration. Cui et al. 2006 formulated porcine insulin into polymeric nanoparticles by a novel strategy. Polylactic acid (PLA) and polylactic-co-glycolic acid (PLGA) were used as biodegradable polymers. Insulin was successfully formulated into biodegradable nanoparticles by formation of a phospholipid complex. Results of *in vivo* evaluation allow concluding that these systems are able to markedly improve the intestinal absorption of insulin. Wang et al. 2010 described a novel procedure for the preparation of hydrophilic peptide-containing oily formulations involving the freeze-drying of w/o emulsions. The results have clearly demonstrated that hydrophilic insulin can be solubilized in oily formulations forming an anhydrous reverse micelle system. This system showed drug release which may have practical applications for oral insulin delivery with sustained release profile.

5. LIPOSOMES

5.1. Definitions and Properties

Liposomes were introduced as DDS in the 1970s. Also named as (phospho) lipid vesicles, these are self-assembled colloidal particles that occur naturally and can be prepared artificially. At first, they were used to study biological membranes. Nowadays, they are a very useful model, extensively used as drug carriers (Martins et al. 2007).

Liposomes resemble cell membranes in their structure and composition (Fig. 5), composed by spherical vesicles of one or more phospholipid bilayers (in most cases phosphatidylcholine). They are typically made from natural, biodegradable, nontoxic and nonimmunogenic lipid molecules and can load or bind a variety of drug molecules into or onto their membranes. Lipophilic drugs can be incorporated within the lipid bilayers, while hydrophilic drugs are solubilized in the inner aqueous core. Liposomes are formed when thin lipid films or lipid cakes are hydrated and stacks of liquid crystalline bilayers become fluid and swell. During stirring, hydrated lipid sheets detach and self associate to form vesicles, which prevent interaction of water with the hydrocarbon core of the bilayers at the edges. They are widely used because of their size (Table 3), amphiphilic character and tissue biocompatibility. Some properties such as drug release, *in vivo* stability and biodistribution are determined by the size of the vesicles, their surface charge, surface hydrophobicity and membrane fluidity (Muller et al. 2004). Production methods are described in Fig. 6.

Table 3. Variety of liposomes, mean diameters and microencapsulation volume.

Type	Diameter (nm)	microencapsulation volume (µl/mg lipid)
SUV	25–50	0.5
LUV	100–2000	13.7
GUV	1000–3000	n.a.*
MLV (10 layers)	1000–3500	4.1

*n.a.–Not apllied

Lipid vesicles in liposomes can be multi-, oligo- or unilamelllar, containing many, a few or one bilayer shells, respectively. The diameter of the lipid vesicles may vary between about 20 nm and a few 3500 nm. SUV, LUV and GUV are surrounded by single lipid layer and MLV by multilayers (10 layers).

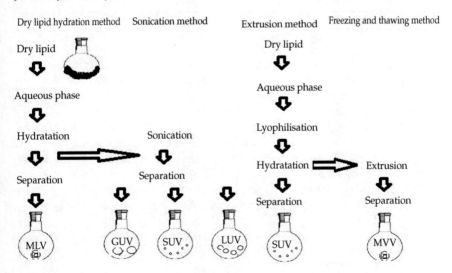

Fig. 6. Representative illustration of the several production methods for liposomes (Unpublished). Preparation methods of liposomes are well documented in literature and include the dry lipid hydration method; the extrusion method; the freezing and thawing method which transforms MLV into multivesicular vesicles (MVV) by physicochemical modifications; the sonication method transforms MLV into small SUV and others methods such as the dehydration-rehydration method; the reverse-phase evaporation method and the double emulsification. All these methods occur above the phase transition temperature (PTT) (Table 7).

Table 4. Key facts of patient compliance.

1. Chronic diseases involve pharmacoterapeutic monitoring for the maintenance of the correct therapeutic level of drug.
2. Diabetes is a disease that needs a rigorous control by patients and doctors.
3. Administration route, respect the schedule and indications of the doctor influence the treatment.
4. The patient should understand, accept and agree with the treatment.
5. Parenteral routes, such as oral, facilitate the patient compliance for being painless, easiest and accommodative.

Table 5. Key facts of lipoatrophy.

1. The treatment of diabetes type 1 is an injection of insulin several times for day, according to the treatment.
2. The subcutaneous injection means under the skin, so it penetrate the membrane and is usually applied in the arm, abdomen or legs.
3. Repeated applications in the same location can lead to lipid modification and damage cells.
4. Furthermore, insulin absorption became compromised.
5. Lipoatrophy can be avoiding by alternate local injection.

5.2 Applications in Diabetes

Kisel et al. 2001 developed liposomal insulin preparations using both high-melting and negatively charged dipalmitoyl phosphatidylethanol as the main lipid component. Liposomes prepared significantly decreased the plasma glucose level in rats by oral administration. Zhang et al. 2005 incorporated insulin in liposomes. Lectin-modified insulin liposomes decreased blood glucose level in diabetic mice or rats after oral administration. There was a linear relationship between blood glucose level and serum insulin concentration. The results suggested that lectin-modified liposomes could enhance intestinal absorption of insulin. Hanato et al. 2009 developed liposomal formulations of glucagon-like peptide-1(GLP-1) for improving pharmacological effect. This drug is a potent candidate for the treatment of diabetes. However, its clinical applications are highly limited because of rapid enzymatic degradation by dipeptidyl-peptidase IV (DPP IV). The GLP-1 liposomal formulations were successfully obtained and the pharmacological effects in rats showed the improvement of insulin secretion. These liposomal formulations might therefore provide clinical benefit in patients with type 2 diabetes. Chono et al. 2009 studied the influence of phospholipids on the pulmonary delivery of insulin. They showed that aerosolized liposomes with dipalmytoylphosphatidylcholine (DPPC) enhance pulmonary insulin delivery in rats by opening the epithelial cells space in pulmonary mucosa and not mucosal cell damages and, also, a smaller liposomal particle size is advantageous for enhanced pulmonary delivery. These findings show that aerosolized liposomes with

Table 6. Key facts of bioavailability.

1. In pharmacologic terms, the drug administered is not entirely absorbed into the systemic circulation.
2. There is a lack of drug that can penetrate the membranes and is not able to perform is action.
3. Only by intravenous route is possible to achieve 100% bioavailability.
4. In other routes, such as oral, drugs suffer enzymatic degradation which decreases the amount of drug absorbed.
5. So, bioavailability is the amount of unchanged drug that reaches the systemic circulation or target site action.

Table 7. Key facts of PTT.

1. Phase transition means the transformation of a thermodynamic system from one phase or state of matter to another.
2. There are three possible states: gas, liquid and solid.
3. These transitions between phases are controlled by temperature and pressure.
4. The PTT is the necessary temperature for the lipids change their actual phase.
5. So, liquid lipids can became solid and solid lipids can became liquids according to the increase/decrease of the temperature.

DPPC could be useful as a pulmonary delivery system for insulin. Mohanraj et al. 2010 prepared novel hybrid insulin-loaded silica nanoparticles coated liposomes under low temperature conditions, by templating 1,2-dipalmitoyl-sn-glycero-3-phosphocholine/cholesterol liposomes with silica nanoparticles via electrostatic interaction to facilitate encapsulation, protection and controlled release of insulin. Liposomes obtained showed excellent physicochemical characteristics and exhibited increased insulin encapsulation efficiency (70%) compared to insulin-loaded liposomes reported in literature. This system might be a promising candidate for the storage and delivery of proteins and peptides.

6. SELF-EMULSIFYING DRUG DELIVERY SYSTEM (SEDDS)

6.1 Definitions and Properties

SEDDS are isotropic mixtures of natural or synthetic oils and solid or liquid surfactants that have the ability to form o/w emulsions or microemulsions (SMEDDS) upon gentle stirring following dilution with the aqueous phase. SEDDS typically produce emulsions with a droplet size between 100 and 300 nm while SMEDDS form transparent microemulsions with a droplet size lower than 50 nm. Their properties render SEDDS good candidates for oral delivery of hydrophobic drugs with adequate solubility in oils or oil/surfactant blends. As SEDDS self-emulsifies in the stomach and presents the drug in small droplets of oil (<5 μm), it improves drug dissolution through providing a large interfacial area for partitioning of the drug between the oil and gastrointestinal fluid. Other advantages include increased stability of drug molecules, and the possibility of administering the final product as gelatin capsules. For drugs subjected to dissolution rate limiting absorption, SEDDS presents a possibility for enhancement in both the rate and the extent of drug absorption and the reproducibility of the plasma concentration profile (Gursoy and Benita 2004).

The oily/lipid component is generally a fatty acid ester or a medium/long chain saturated, partially unsaturated or unsaturated hydrocarbon, in liquid, semisolid or solid form at room temperature. Examples include mineral oil, vegetable oil, silicon oil, lanolin, refined animal oil, fatty acids, fatty alcohols, and mono-/di-/triglycerides. The most widely recommended surfactants are non-ionic surfactants with a relatively high hydrophilic-lipophilic balance (HLB) value. The surfactant concentration ranges between 30% and 60% (w/w) in order to form stable SEDDS (Tang et al. 2008).

The mechanism by which self-emulsification occurs is not yet well understood. It is thought that self-emulsification takes place when the entropy change favoring dispersion is greater than the energy required to increase the surface area of the dispersion. The free energy of a conventional emulsion formulation is a direct function of the energy required to create a new surface between the oil and water phases. The two phases of the emulsion tend to separate with time to reduce the interfacial area and thus the free energy of the systems. The conventional emulsifying agents stabilize emulsions resulting from aqueous dilution by forming a monolayer around the emulsion droplets, reducing the interfacial energy and forming a barrier to coalescence. On the other hand, emulsification occurs spontaneously with SEDDS because the free energy required to form the emulsion is either low, and positive or negative. This is required for the interfacial structure to show no resistance against surface shearing so that emulsification can take place. The efficiency of drug incorporation into SEDDS is generally specific to each case depending on the physicochemical compatibility of the drug/system. The addition of solvents including ethanol and PEG, contributes to the improvement of drug solubility in the lipid vehicle (Gursoy and Benita 2004).

6.2 Applications in Diabetes

Singnurkar and Gidwani 2008 reports the use of egg yolk for stabilization of insulin in self microemulsifying dispersions. These dispersions were prepared by lyophilization followed by dispersion into self microemulsifying vehicle. Insulin self microemulsifying dispersions has protected insulin from enzymatic degradation *in vitro* in presence of chymotripsin and high encapsulation efficiency (98.2 ± 0.9 %). Egg yolk encapsulated insulin was bioactive, demonstrated through both *in vivo* and *in vitro*.

7. Conclusions

Nanomedicine is still in an early stage of development, nevertheless DDS with lipid matrix have a bright, expansive and multi-purpose future in the delivery of anti-diabetic drugs. Novel applications of the various DDS described above for oral and pulmonary delivery can reach clinical trials by a prudent use the nanotechnology. Although diabetes has many remaining problems, the use of DDS would result in a concomitant improvement in the quality, efficacy and safety profile of antidiabetic drugs and the use of this technology is likely to be the key to solve these problems.

Definitions

Emulsion: A mixture of two or more immiscible (unblendable) liquids consisting in two-phase systems of matter called colloids. In an emulsion, one liquid (the dispersed phase) is dispersed in the other (the continuous phase). Emulsions are stabilized by surfactants that reduce the interfacial tension.

Hydrophilic-lipophilic Balance (HLB): The HLB of an emulsifier for water and oil is the determination by the chemical composition and ionization characteristics of a given emulsifier. The HLB of an emulsifier is not directly related to solubility, but it determines the type of emulsion that tends to be formed. It is an indication of the behavioral characteristics and not an indication of emulsifier efficiency.

Lipid drug conjugate (LDC): A LDC is an insoluble drug-lipid conjugate. This kind of DDS is obtained or by salt formation or by covalent linking. After their formation, they can be converted in SLN by adding an aqueous surfactant solution.

Liposome: Liposome is a microscopic vesicle consisting of an aqueous core enclosed in one or more phospholipid layers, used to convey pharmaceutical drugs or other substances to target cells or organs.

Micelle: A round cluster of hydrocarbon chains formed when the amount of surfactant in an aqueous solution that reaches a critical point. Those aggregates include several dozens of amphiphilic molecules and usually have a shape close to spherical. Micelles are able to surround and dissolve droplets of water or oil, forming an emulsion.

Nanostructured lipid carriers (NLC): NLC are submicron-sized spherical lipid carriers produced using blends of solid lipids and liquid lipids (oils). To obtain the blends for the particles matrix, solid lipids are mixed with liquid lipids. Because of the oil present in these mixtures, a melting point depression compared to the pure solid lipid is observed, but the blends obtained are also solid at room and body temperatures.

Pharmacodynamic: The study of the biochemical and physiological effects of drugs on the body that includes the mechanisms of drug action and the relationship between drug concentration and effect.

Pharmacokinetic: The study of the mechanisms of absorption and distribution of a drug, that includes time which a drug begins action and the duration of the effect, the chemical changes of the substance in the body and the effects and routes of excretion of the metabolites of the drug.

Phospholipid: Long molecules derived from glycerol falling carbonated fatty acid chains, each containing 10 to 24 carbon atoms with 0–6 double bonds. They have also a nitrogen-containing base, e.g., lecithin. A phospholipid is a complex fatty material found in all living cells that acts as an emollient, antioxidant, natural emulsifier and spreading agent.

Solid lipid nanoparticles (SLN): SLN are submicron-sized spherical lipid carriers of lipid-soluble drug molecules. They are typically formed by heating an aqueous lipid mixture above the melting point of the lipid, adding drug, homogenizing, then cooling to freeze the drug within the solid lipid spheres.

Summary Points

- DDS with solid lipid matrices allow encapsulation of drugs.
- Diabetes is an autoimmune disease predominately treated by insulin injections.
- DDS are been nowadays investigated to overcome the conventional insulin injection.
- SLN have a solid matrix where drugs can be incorporated.
- NLC have more space to encapsulate drugs due the use of liquid lipids.
- LDC can be transformed into SLN by the HPH technique.
- Liposomes have at least a phospholipid bilayer instead micelles that have only one.
- Liposomes are amphiphiles because they have a hydrophobic and a hydrophilic region.
- SEDDS are initially micromulsions usually for the encapsulation of hydrophobic drugs.
- Diabetes could be treated in a early future by DDS with a solid lipid matrix.

Abbreviations

CS	:	Chitosan
DDS	:	Drug delivery systems
DPP IV	:	Dipeptidyl peptidase IV
DPPC	:	Dipalmytoylphosphatidylcholine
GIT	:	Gastrointestinal tract
GPL-1	:	Glucagon-like-peptide-1
GUV	:	Giant unilamellar vesicle
HPH	:	High pressure homogenization
HLB	:	Hydrophile-lipophile balance
LDC	:	Lipid drug conjugate
LUV	:	Large unilamellar vesicle
MLV	:	Multilamellar vesicle
MVV	:	Multivesicular vesicle
N-glu-PE	:	N-glutamine-phosphatidylethanolamine
NLC	:	Nanostructured lipid carrier
PEG	:	Polyethylene glycol
PEG–PEs	:	Polyethylene glycol–phosphatidylethanolamine conjugates
PGA-PRZ-	:	Poly-□-glutamic acid phloridzin conjugates
PLA	:	polylactic acid
PLGA	:	polylactic-co-glycolic acid
PTT	:	Phase transition temperature
PRZ	:	Plhoridzin
□- PGA	:	Poly-□-glutamic acid
SEDDS	:	Self-emulsifying drug delivery systems
SLN	:	Solid lipid nanoparticle
SMEDDS	:	Self-microemulsifying drug delivery systems
SUV	:	Small unilamellar vesicle

References

Agnihotri, S.A., N.N. Mallikarjuna and T.M. Aminabhavi. 2004. Recent advances on chitosan-based micro- and nanoparticles in drug delivery. J. Control Release 100: 5–28.

Almeida, A.J., and E. Souto. 2007. Solid lipid nanoparticles as a drug delivery system for peptides and proteins. Adv. Drug Deliv. Rev. 59: 478–90.

Atef, E., and A.A. Belmonte. 2008. Formulation and *in vitro* and *in vivo* characterization of a phenytoin self-emulsifying drug delivery system (SEDDS). Eur. J. Pharm. Sci. 35: 257–63.

Chono, S., R. Fukuchi, T. Seki and K. Morimoto. 2009. Aerosolized liposomes with dipalmitoyl phosphatidylcholine enhance pulmonary insulin delivery. J. Control Release 137: 104–9.

Cilek, A., N. Celebi, F. Tirnaksiz and A. Tay. 2005. A lecithin-based microemulsion of rh-insulin with aprotinin for oral administration: Investigation of hypoglycemic effects in non-diabetic and STZ-induced diabetic rats. Int. J. Pharm. 298: 176–85.

Cui, F., F. Qian and C. Yin. 2006. Preparation and characterization of mucoadhesive polymer-coated nanoparticles. Int. J. Pharm. 316: 154–61.

Gallarate, M., M. Trotta, L. Battaglia and D. Chirio. 2009. Preparation of solid lipid nanoparticles from W/O/W emulsions: preliminary studies on insulin encapsulation. J. Microencapsul. 26: 394–402.

Gursoy, R.N., and S. Benita. 2004. Self-emulsifying drug delivery systems (SEDDS) for improved oral delivery of lipophilic drugs. Biomed. Pharmacother. 58: 173–82.

Hanato, J., K. Kuriyama, T. Mizumoto, K. Debari, J. Hatanaka, S. Onoue and S. Yamada. 2009. Liposomal formulations of glucagon-like peptide-1: improved bioavailability and anti-diabetic effect. Int. J. Pharm. 382: 111–6.

Ikumi, Y., T. Kida, S. Sakuma, S. Yamashita and M. Akashi. 2008. Polymer-phloridzin conjugates as an anti-diabetic drug that inhibits glucose absorption through the Na+/glucose cotransporter (SGLT1) in the small intestine. J. Control Release 125: 42–9.

Joshi, M.D., and R.H. Muller. 2009. Lipid nanoparticles for parenteral delivery of actives. Eur. J. Pharm. Biopharm. 71: 161–72.

Kisel, M.A., L.N. Kulik, I.S. Tsybovsky, A.P. Vlasov, M.S. Vorob'yov, E.A. Kholodova and Z.V. Zabarovskaya. 2001. Liposomes with phosphatidylethanol as a carrier for oral delivery of insulin: studies in the rat. Int. J. Pharm. 216: 105–14.

Liu, J., T. Gong, H. Fu, C. Wang, X. Wang, Q. Chen, Q. Zhang, Q. He and Z. Zhang. 2008. Solid lipid nanoparticles for pulmonary delivery of insulin. Int. J. Pharm. 356: 333–44.

Liu, J., T. Gong, C. Wang, Z. Zhong and Z. Zhang. 2007. Solid lipid nanoparticles loaded with insulin by sodium cholate-phosphatidylcholine-based mixed micelles: preparation and characterization. Int. J. Pharm. 340: 153–62.

Martins, S., D. Ferreira and E.B. Souto. 2009. Lipid Nanoparticle-Based Systems for Delivery of Biomacromolecule Therapeutics. In Delivery Technologies for Biopharmaceuticals: Peptides, Proteins, Nucleic Acids and Vaccines. L.J.a.H.M. Nielsen, editor. John Wiley & Sons, Ltd.

Martins, S., B. Sarmento, D.C. Ferreira and E.B. Souto. 2007. Lipid-based colloidal carriers for peptide and protein delivery—liposomes versus lipid nanoparticles. Int. J. Nanomedicine 2: 595–607.

Mishra, B., B.B. Patel and S. Tiwari. 2010. Colloidal nanocarriers: a review on formulation technology, types and applications toward targeted drug delivery. Nanomedicine 6: 9–24.

Mohanraj, V.J., T.J. Barnes and C.A. Prestidge. 2010. Silica nanoparticle coated liposomes: a new type of hybrid nanocapsule for proteins. Int. J. Pharm. 392: 285–93.

Muchow, M., P. Maincent and R.H. Muller. 2008. Lipid nanoparticles with a solid matrix (SLN, NLC, LDC) for oral drug delivery. Drug Dev. Ind. Pharm. 34: 1394–405.

Muller, M., S. Mackeben and C.C. Muller-Goymann. 2004. Physicochemical characterisation of liposomes with encapsulated local anaesthetics. Int. J. Pharm. 274: 139–48.

Muller, R.H., R.D. Petersen, A. Hommoss and J. Pardeike. 2007. Nanostructured lipid carriers (NLC) in cosmetic dermal products. Adv. Drug Deliv. Rev. 59: 522–30.

Sarmento, B., S. Martins, D. Ferreira and E.B. Souto. 2007a. Oral insulin delivery by means of solid lipid nanoparticles. Int. J. Nanomedicine 2: 743–9.

Sarmento, B., A. Ribeiro, F. Veiga, P. Sampaio, R. Neufeld and D. Ferreira. 2007b. Alginate/chitosan nanoparticles are effective for oral insulin delivery. Pharm. Res. 24: 2198–206.

Singnurkar, P.S., and S.K. Gidwani. 2008. Insulin-egg yolk dispersions in self microemulsifying system. Indian J. Biochem. Biophys. 70: 727–732.

Sonaje, K., Y.H. Lin, J.H. Juang, S.P. Wey, C.T. Chen and H.W. Sung. 2009. *In vivo* evaluation of safety and efficacy of self-assembled nanoparticles for oral insulin delivery. Biomaterials 30: 2329–39.

Souto, E.B., S.A. Wissing, C.M. Barbosa and R.H. Muller. 2004. Development of a controlled release formulation based on SLN and NLC for topical clotrimazole delivery. Int. J. Pharm. 278: 71–7.

Tang, B., G. Cheng, J.C. Gu and C.H. Xu. 2008. Development of solid self-emulsifying drug delivery systems: preparation techniques and dosage forms. Drug Discov. Today. 13: 606–12.

Torchilin, V.P. 2007. Micellar nanocarriers: pharmaceutical perspectives. Pharm. Res. 24: 1–16.

Wang, T., N. Wang, A. Hao, X. He, T. Li and Y. Deng. 2010. Lyophilization of water-in-oil emulsions to prepare phospholipid-based anhydrous reverse micelles for oral peptide delivery. Eur. J. Pharm. Sci. 39: 373–9.

Wissing, S.A., O. Kayser and R.H. Muller. 2004. Solid lipid nanoparticles for parenteral drug delivery. Adv. Drug Deliv. Rev. 56: 1257–72.

Zhang, N., Q.N. Ping, G.H. Huang and W.F. Xu. 2005. Investigation of lectin-modified insulin liposomes as carriers for oral administration. Int. J. Pharm. 294: 247–59.

3

Atomic Force Microscopy: An Enabling Nanotechnology for Diabetes Research

Brian J. Rodriguez[1,]* *and Suzanne P. Jarvis*[2]

ABSTRACT

Atomic force microscopy (AFM), a scanning probe microscopy-based technique with atomic resolution and piconewton force sensitivity, has been an enabling tool for nanotechnology and nanobiotechnology since its invention in 1986. In AFM, a nanometer-sized probe tip attached to the end of a microfabricated cantilever is scanned near or in contact with a surface in order to construct an image and to determine various properties of the surface by measuring the interaction force between the tip and the surface. AFM can be operated in air and liquid and can be applied to soft biological samples in physiological environments. Thus, it has vast applications in the developing field of nanomedicine. Diabetes is a group of diseases characterized by abnormally high

---

[1]Lecturer in Nanoscience, Conway Institute of Biomolecular and Biomedical Research, University College Dublin, Belfield, Dublin 4, Ireland; E-mail: brian.rodriguez@ucd.ie
[2]Professor of Biophysics, Conway Institute of Biomolecular and Biomedical Research, University College Dublin, Belfield, Dublin 4, Ireland; E-mail: suzi.jarvis@ucd.ie
*Corresponding author

List of abbreviations after the text.

blood glucose resulting from insulin resistance and/or impaired or absent insulin secretion, and the field of nanomedicine can contribute in terms of understanding the development and progression of the disease and also in designing nanotherapies to treat the disease. In this chapter, we discuss the role of AFM in understanding and designing therapies for diabetes and the impact of AFM on diabetes research, from the nano- to the microscale, and from single molecules to macromolecular assemblies and individual cells. In particular, the role of AFM in understanding pathophysiological changes and the pathology of the disease, and the role of AFM in the treatment of diabetes are presented.

1. INTRODUCTION

Recent advances in nanotechnology have been enabled by tools capable of fabricating and imaging nanoscale features, measuring materials properties at the nanoscale, and manipulating nanosized objects. One family of techniques in particular, namely, scanning probe microscopy (SPM), has made a significant impact on a wide variety of fields due primarily to its versatility, resolution, and sensitivity. The field of SPM has evolved from the inventions of the scanning tunneling microscope (STM) in 1982 (Binnig et al. 1982) and the atomic force microscope (AFM) in 1986 (Binnig et al. 1986) to an impressive measurement and manipulation platform capable of measuring chemical recognition events and, e.g., structural, mechanical, electrical, and magnetic, properties of materials, and of manipulating at the single-molecule and atomic levels. Whereas the STM can be used to construct atomic scale images of conducting materials based on the tunneling current between tip and sample, the AFM can be applied to any material, and images are constructed from the interaction force between the tip and sample. These tools have been identified as ushering in the nanometer age (Rohrer 1994) and have been essential to the development of nanotechnology, i.e., the measurement, manipulation, and exploitation of materials and materials properties at the nanoscale. Nanotechnologies applied to biology and medicine are thus nanobiotechnology and nanomedicine, and can be described as the characterization and manipulation of biological systems and the diagnosis and therapy of disease at the nanoscale using nanomaterials and nanotools. The use of nanomedicine for, e.g., biosensing, drug-design, and the design and characterization of nanomaterials for drug delivery and power-generators holds the promise for future breakthroughs in therapies and diagnostics.

Atomic force microscopy is well-suited for studying biology at the nanoscale as it has the ability to (i) image nanostructures, proteins, biological macromolecules, and cells in (ii) air, liquid, and physiological environments, (iii) measure local interaction properties and materials properties at specified positions, i.e., AFM force spectroscopy, and (iv) be integrated with complementary imaging modalities. The basic operating principles of AFM are described in Table 1 and illustrated in Fig. 1. AFM, and to a lesser extent, STM, have been applied to biological systems for

Table 1. Key Features of AFM.

1. AFM is a nanotechnology tool based on monitoring the interaction force between a surface and a nanometer-sized probe tip at the end of a microfabricated cantilever beam.
2. The interaction force is measured by monitoring the deflection of the AFM cantilever.
3. The deflection of the cantilever is detected using an optical beam deflection method in which a laser or laser diode light source is reflected off the backside of the cantilever and into a photodetector.
4. The photodetector thus relates the position of the reflected light to the distance the cantilever is deflected.
5. The tip and sample are brought together using a motor positioning system with course position control and a piezoelectric element with precise position control.
6. A feedback loop is employed which keeps, e.g., a constant force between the tip and sample by adjusting the tip-sample separation using a piezoelectric element.
7. The tip moves in a raster-scan pattern with respect to the sample allowing an image of the surface topography to be recorded.

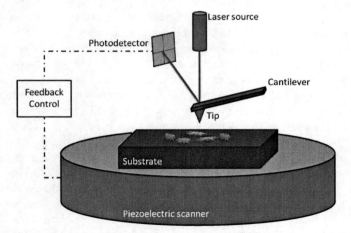

Fig. 1. Schematic depicting the basic components of an AFM. In AFM, the cantilever deflection is measured using a photodetector to quantify changes in the position of a laser reflected off the backside of the cantilever. A feedback loop is used to maintain a constant force between the tip and the sample by adjusting the tip-sample separation using a piezoelectric scanner, which can also be used to move the sample with respect to the tip, thus allowing an image of tip-sample interaction force to be recorded.

some time (Hansma et al. 1988; Drake et al. 1989), but the breakthrough for imaging soft biomaterials came with the invention of dynamic AFM modes (Zhong et al. 1993) and the first application of these modes in a liquid environment (Hansma et al. 1994) and on cells in solution (Putman et al. 1994). For over two decades now, AFM has been used to study biology, as demonstrated by early review papers on spectroscopy and imaging (Rademacher et al. 1992; Hansma and Hoh 1994; Heinz and Hoh 1999) through to more recent papers specifically addressing the position of AFM in nanobiotechnology (Müller and Dufrêne 2008). Advances in AFM for biology continue to be made, particularly in the areas of frequency modulation AFM (FM-AFM) in liquid (Jarvis et al. 2008), multifrequency AFM (Proksch 2011), high-speed AFM (Ando et al. 2001; Kodera et al. 2010), and nanomedicine (Riehemann et al. 2009; Lal and Arnsdorf 2010; Sitterberg et al. 2010). One disease of particular importance for nanomedicine, with applications in diagnosis and therapy, and in understanding the mechanisms of the disease, is diabetes (Pickup et al. 2008).

Diabetes is a disease affecting millions of individuals throughout the world and enormous socioeconomic burdens are associated with diabetes and the treatment of diabetic complications. Most patients with diabetes have non-insulin dependent diabetes mellitus or type 2 diabetes. Insulin dependent diabetes mellitus is known as type 1 or juvenile onset diabetes. Both diseases are characterized by high blood glucose either because the body does not produce (enough) insulin, the hormone responsible for glycemic control, or because the body is unable to use the insulin that is produced by the body effectively. In general, normal blood glucose is maintained through the balance of insulin production and action. In type 2 diabetes, insulin insensitivity or resistance, triggered partly by environmental (e.g., obesity, overeating, physical inactivity) and genetic (i.e., heredity) factors, and pancreatic β-cell dysfunction, are key mechanisms in the development of the disease. Insulin resistance leads to increased β-cell dysfunction over time and without diagnosis and intervention, the disease will progress; early stages of insulin resistance or pre-diabetes often go undiagnosed (Harris et al. 1992). Impaired insulin secretion, loss of β-cell mass, increased β-cell apoptosis, and the presence of islet cell amyloid are also associated with type 2 diabetes (Haataja et al. 2008). Type 1 diabetes generally results from an autoimmune attack on β-cells thought to be triggered by an environmental factor in susceptible individuals, leading to β-cell destruction and complete insulin deficiency. Thus, early screening, improved insulin administration and or action, blood glucose monitoring, and effective therapies for glycemic control are important factors for treating and preventing the progression and complications of diabetes. Poor glycemic control can lead to or is associated with a variety of serious short- and long-term complications, often exacerbated by risk factors such as hypertension and high cholesterol, including retinopathy, neuropathy, nephropathy, and cardiovascular disease, which can lead to blindness, amputation, renal failure, and death. Type 1 diabetes is further associated with other autoimmune conditions such as thyroid disease. Currently, there is no cure for diabetes.

In this chapter, the applications of AFM to areas of health and disease in diabetes research are presented. In particular, the role of AFM in understanding pathophysiological changes in diabetic tissues, proteins, and cells through measurements of structural and mechanical properties are discussed in Section 2; The role of AFM in understanding the pathology of the disease through the characterization of fibrils, proteins, and cells relevant to diabetes is discussed in Section 3; In Section 4, the role AFM plays in the treatment of diabetes is presented, including in the design and characterization of nanomaterials such as nanoparticles and nanocoatings, and nanodevices; Finally, in Section 5, future prospects for AFM as a nanotechnology for diabetes research are outlined.

Applications to Areas of Health and Disease

Diabetes is a disease associated with elevated blood glucose levels, which, over time can lead to a number of complications and to increased health risks. AFM is a nanotechnology tool with a demonstrated ability to aid in understanding pathophysiological changes and the pathology of diabetes, and to contribute to the development and characterization of nanomaterials for the treatment of the disease.

2. ROLE OF AFM IN THE ASSESSMENT OF THE PATHOPHYSIOLOGY OF DIABETES

Microvascular disease and neuropathy often work in tandem to prolong wound healing in patients with diabetes, which in extreme cases can lead to, e.g., foot ulcers, and result in amputation. Poor glycemic control resulting in hyperglycemia leads to normal enzyme-mediated glycosylation of the blood hemoglobin and to (non-enzymatic) glycation of the hemoglobin and of other proteins, the former of which leads to elevated glycated hemoglobin levels, while the latter can lead to the formation of advanced glycation end-products (AGEs) and impaired biomolecular function (Brownlee et al. 1984). Delayed wound healing in diabetics is often associated with the presence of AGEs, which have been further shown to be associated with changes in the mechanical properties of tissues.

Furthermore, diabetes is known to accelerate the aging process, and many changes in tissue properties attributed to aging can also manifest in patients with diabetes, including reduced elasticity in collagenous tissues. As collagen is present throughout the body, e.g., from skin and bones, to teeth, eyes, and arterial walls, this change in the mechanical properties of collagen can have a significant impact on health. While a number of tools have been used to study structural and ultrastructural changes in tissues and cells, AFM has several key advantages in this regard as (i) AFM does not require any special sample preparation, i.e., no metal coating of the sample is required, as is the case for electron microscopy; (ii) AFM can be operated in liquid and physiological environments; (iii) AFM can be used to measure mechanical properties, such as elasticity; and (iv) AFM can be used to measure properties locally, on the nanoscale, an important consideration when dealing with complex, often hierarchically-structured biomaterials. In this section, we address the use of AFM to study the structural and mechanical properties of tissues, fibrils, membrane proteins, and cells affected by diabetes.

2.1 Tissues and Fibrils

Collagen, specifically collagen from diabetic rats, has been studied extensively by AFM, typically revealing structural alterations in collagen fibrils. In a study by Odetti et al., tendons from diabetic and non-diabetic rats and tendons from non-diabetic rats which had been subsequently incubated in a phosphate buffered saline (PBS) solution containing glucose for 15 days were investigated by AFM (Odetti et al. 2000). Diabetic and glucose-incubated collagen fibrils were found to have similar radii and gap depths, and larger radii and gap depths than non-diabetic collagen fibrils, as illustrated in Fig. 2. The diabetic and glucose-incubated collagens were further analyzed to determine fructosomine and pentosidine levels by a commercial assay and using high pressure liquid chromatography, respectively, allowing the authors to conclude that extended exposure to glucose both *in vivo* and *in vitro* leads to structural changes in collagen fibrils. This work provides evidence that glycation through cross-links plays a significant role in structural changes. It also further corroborates earlier work by some of the same authors addressing the role of AGEs on structural changes in collagen fibrils from aged rat-tail tendon (Odetti et al. 1998). A 2003 article by Wang et al., comparing collagen fibril structure in nerve and tendon collagens from diabetic and non-diabetic rats, revealed that collagen diameters were consistently larger for the collagens harvested from diabetic rats, while the typical 67 nm periodicity of the non-diabetic collagen fibrils remained unchanged in the diabetic collagen fibrils (Wang et al. 2003).

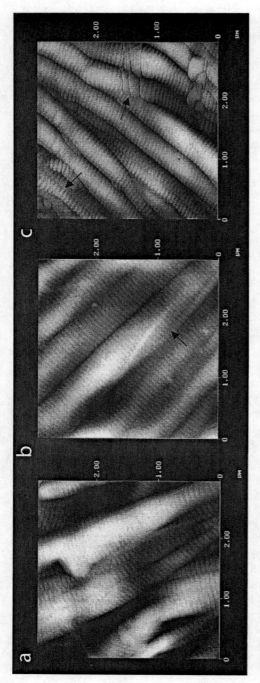

Fig. 2. AFM of non-diabetic, diabetic, and glucose-incubated collagen. AFM topography images of (a) non-diabetic, (b) diabetic, and (c) glucose-incubated rat tail tendon. The fibrils in (b) and (c) have larger diameters than those in (a). Arrows indicate the merging of fibrils. Reprinted from Odetti et al., copyright 2000, with permission from John Wiley & Sons, Ltd. (Odetti et al. 2000).

Collagen is not the only protein that may be affected by glycation, nor is it the only structural protein with an important role in microvascular disease, a secondary effect of diabetes. In particular, fibrillin, which is known to play an important role in the elasticity of healthy tissues, has also been shown to become modified in type 1 diabetic tissues (Akhtar et al. 2010). Akhtar et al. used AFM to measure structural changes in normal and diabetic fibrillin microfibrils isolated from a rat aorta, and reported that while the length of the microfibrils remained unaltered, the periodicity of the fibrils was reduced significantly in diabetic rats, an ultrastructural change thought to be related to observed changes in macroscopic mechanical properties in diabetic rat aortas.

2.2 Membrane Proteins

While high-resolution AFM is capable of obtaining stunning images of, e.g., membrane proteins and lipid bilayers, it has found few applications in diabetes research. In 2009, Mangenot et al. used high-resolution AFM to address the formation of junctional microdomains from normal bovine and human cataract lens membranes (Mangenot et al. 2009). The cataract lens was extracted from the surgery debris of a type 2 diabetes patient, while the normal lens membranes were harvested from calves. Junctional microdomains form junctions between cells in the lens facilitating the flow of ions, waste, etc. between them, allowing cells deep within the lens to be nourished, evacuate waste, and remain healthy (Donaldson et al. 2001), thus preventing the lens opacification associated with cataracts. In the normal lens membranes, it was found that aquaporin-0 formed well-registered, square-lattice array structures often bordered by connexons, as shown in Fig. 3a. In the cataract lens membranes, however, the aquaporin-0 microdomains appear less continuous, forming in rows, and there is a notable absence of connexons, as can be seen in Fig. 3b. The authors conclude that the lack of connexons in the cataract lens from the type 2 diabetes patient contributed to both the malformation of the aquaporin-0 microdomains and to the ability of the junctions to effectively circulate nutrients and waste, and note a similar observation in aged cataract lens membranes (Buzhynskyy et al. 2007).

2.3 Cells

A cellular-level effect of the toll of diabetes on the body has also been highlighted by AFM measurements of mechanical properties of red blood cells. Just as variations in elasticity in red blood cells infected with the *Plasmodium falciparum* parasite responsible for malaria have been observed by micropipette aspiration (Suresh et al. 2005), highlighting a possible route for nanodiagnostics of disease, several groups have reported

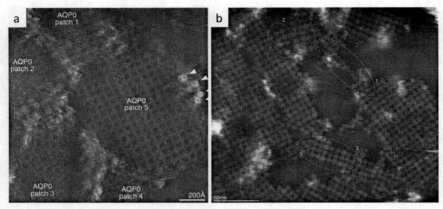

Fig. 3. AFM images of junctional microdomains in calf and human cataract lenses. (a) AFM topography image of healthy calf lens membranes. Arrows indicate the location of connexons and aquaporin-0 patches are labeled AQP0 patch 1–5. (b) AFM topography image of membranes from a cataract lens of a patient with type 2 diabetes. Single, double, and triple rows of aquaporin-0 tetramers are highlighted with dashed-ellipses 1–3. Reproduced with permission and copyright 2008, Springer (Mangenot et al. 2009).

increased stiffness of diabetic red blood cells (Fornal et al. 2006; Hekele et al. 2008; Jin et al. 2010). The observation that erythrocytes from type 2 diabetic patients have greater Young's moduli, and thus lower elasticity, is particularly troubling when taken with the knowledge that the same patients likely suffer from arterial stiffening as a result of ultrastructural changes in collagen and/or fibrillin; the combination of the two further impairs circulation, thus increasing wound healing time.

3. ROLE OF AFM IN UNDERSTANDING THE PATHOGENESIS AND PATHOLOGY OF DIABETES

As discussed previously, AFM has been used extensively to address problems of biological interest, and in limited cases, to directly assess the structural and mechanical changes associated with diabetes. Type 1 and 2 diabetes are both associated with β-cell dysfunction or destruction, although the exact mechanisms remain unknown. The long list of secondary diseases and complications associated with diabetes suggests the importance of understanding the effect of high blood glucose in cells and tissues throughout the body, in particular in the pancreas and kidneys. As AFM has been applied in a wide variety of biologically-relevant studies, including cell mechanotransduction, lipid bilayer and raft formation, protein membrane and channel formation, protein misfolding and fibrillization, binding forces between molecules, and protein nanomechanics, AFM has the potential to play a significant role in understanding the development and effect of diseases such as diabetes. In this section, we discuss the use of AFM to study tissues relevant to diabetes.

3.1 Fibrils

Both insulin and amylin, or islet amyloid polypeptide, are peptide hormones secreted by islet β-cells in the pancreas, and both peptides are known to fibrillate under certain conditions to form structurally similar amyloid fibrils. The formation of amyloid fibrils is associated with many neurodegenerative diseases and the fundamentals of protein folding and misfolding have been the topics of extensive research (Dobson 2003; Chiti and Dobson 2006). In relation to diabetes, amyloid deposits have been found in pancreatic islet β-cells of type 2 diabetes patients (and elsewhere including the cataract lens exterior) and are considered a pathological hallmark of the disease, perhaps contributing to the dysfunction of the β-cells or merely a byproduct of the disease (Haataja et al. 2008). Additionally, amylin amyloid fibrils have been shown to be toxic to cultured β-cells (Lorenzo et al. 1994). Understanding the conditions leading to the formation of such fibrils and unraveling the structure of such deposits are important steps in determining the role amyloid fibrils play in the progression of diabetes, a course of study aptly suited for AFM.

Goldsbury et al. studied the assembly of individual amylin fibrils formed from synthetic human amylin *in vitro* on an atomically-flat substrate using time-lapse AFM (i.e., by recording sequential images) (Goldsbury et al. 1999). Unlike electron microscopy techniques which require dehydrated fibrils, AFM measurements in liquid allow the fibrils to be visualized in physiologically relevant conditions, and furthermore, time-lapse AFM allows the dynamics of the formation process (i.e., the growth rate) to be measured. The authors report that both fibrils formed in a test tube and subsequently deposited onto a substrate and those assembled directly onto the substrate are polymorphic, likely consisting of multiple protofibrils. Time-lapse AFM studies further reveal that while the fibrils grow bidirectionally on a surface and with similar growth rates, as shown in Fig. 4, the fibrils are less coiled than those prepared in a test tube and the density of protofibrils is higher, presumably due to the interaction with the surface. Furthermore, controlled tweaking of the growth environment could reveal which conditions accelerate or retard formation, leading to targeted therapies. Sedman et al. also looked at the formation of amylin amyloid fibrils formed *in vitro* for 6 and 18 hours, looking specifically at a stretch of residues in amylin (20–29) thought to be crucial to fibril formation (Sedman et al. 2005). The kinetics of human

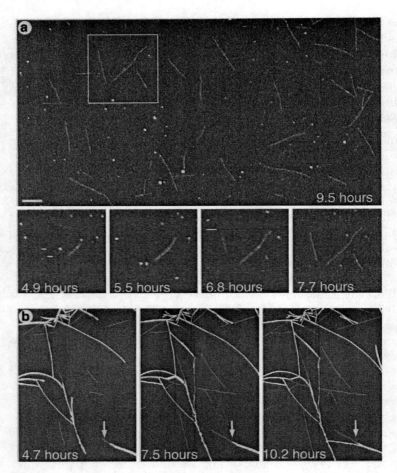

Fig. 4. Time-lapse AFM images of amylin fibrils growing on mica. (a) AFM image of protofibrils formed after 9.5 hours. The bidirectional growth of protofibrils in the selected area is shown in the panel of four images at the bottom of (a). (b) Protofibrils can be seen forming higher order fibrils over time in a different experiment. Arrows indicates where one protofibril appears to grow from a higher order fibril. Reprinted from Goldsbury et al., copyright 1999, with permission from Elsevier (Goldsbury et al. 1999).

insulin amyloid fibril formation has also been addressed by AFM (Jansen et al. 2005) and using time-lapse AFM (Podestà et al. 2006) and also as a function of temperature and insulin concentration at low pH (Mauro et al. 2007). More recently, Fukuma et al. used liquid FM-AFM to determine the surface structure of isolated amylin amyloid fibrils with Ångström resolution (Fukuma et al. 2008). As shown in Fig. 5, the results reveal the cross-β strand structure of amylin fibrils.

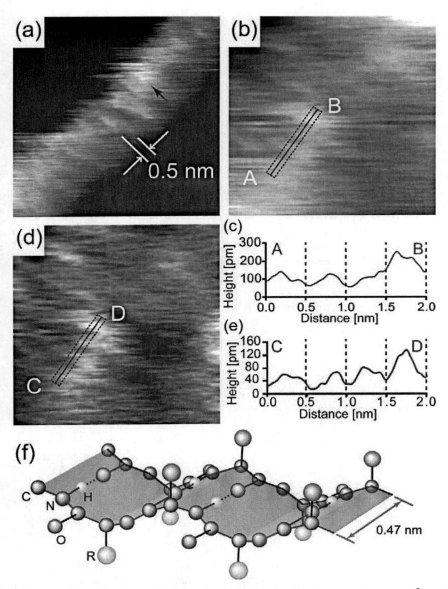

Fig. 5. High resolution FM-AFM images of an amylin amyloid fibril. (a) 10 x 10 nm^2 and (b) 5 x 5 nm^2 scans of an amylin fibril on a mica substrate in PBS. The arrow in (a) indicates the location of the scan in (b). (c) Cross-section line profile plot along the line A-B in (b). (d) A quadratic line-by-line flattened image of (b). (e) Cross-section line profile plot along the line C-D in (d), which corresponds to A-B in (b). (f) Model showing the observed beta strand structure with 0.47 nm repeats. Reprinted from Fukuma et al., copyright 2008, with permission from IOP Publishing Ltd. (Fukuma et al. 2008).

Glucagon is also an amyloid fibril forming hormone secreted by the pancreas that is involved in the regulation of blood glucose. The fibrillation process has been studied by AFM (Dong et al. 2006; De Jong et al. 2006), and force spectroscopy measurements have revealed information about the amyloid structure (Dong et al. 2008) and glucagon/anti-glucagon binding (Lin et al. 2007). Glucagon preparations are prescribed as a treatment of hypoglycemia. In general, amyloid fibril formation and functional properties of amyloid have been intensely studied by AFM adding to the understanding of the role of amyloid in diabetes.

3.2 Protein-membrane Interactions

While amyloid deposits are typically found in islet cells of patients with type 2 diabetes, and amyloids have been identified with toxicity, it is not yet clear to which aggregation state the toxicity can be attributed (Dobson 2003). Green et al. used time-lapse AFM to demonstrate that the amylin peptide penetrates and disrupts mica-supported lipid bilayers, as shown in Fig. 6, and thus membrane-permeant oligomers may be responsible for cytotoxicity (Green et al. 2004). Cho et al. reported the formation of amyloid deposits after exposing a lipid bilayer to amylin (Cho et al. 2008).

3.3 Cells

While diabetes is associated mainly with β-cell dysfunction, very few attempts at β-cell imaging by AFM have been made (Girasole et al. 2007), however, high blood glucose can affect the functioning of other cells and the means by which cells interact with their environment. It has recently been shown that pancreatic acinar cells can be reprogrammed *in vivo* into insulin producing β-cells (Zhou et al. 2008; German 2008). Living acinar cells have also been imaged by AFM, enabling the membrane dynamics of exocytosis to be visualized (Schneider et al. 1997; Cho et al. 2002). Exocytosis is the process by which a cell moves the contents of secretory vesicles across the cell membrane, including insulin and amylin in β-cells, and furthermore, the expression of exocytotic proteins can be impaired by diabetes (Abdel-Halim et al. 1996). To investigate impaired membrane fusion in the liver of rats with chronic pancreatitis, a condition which is related to or may lead to diabetes, the binding forces of hepatic microsomal and plasma membrane proteins in normal and pancreatitic rats has also been studied using AFM force spectroscopy (Slezak et al. 1999). The kidney is responsible for filtering waste from blood, and kidney disease is common in patients with diabetes. Recently, decreased insulin action in kidney podocytes, a type of epithelial cell, has been linked to diabetic complications (Welsh et al. 2010; Rask-Madsen and King 2010).

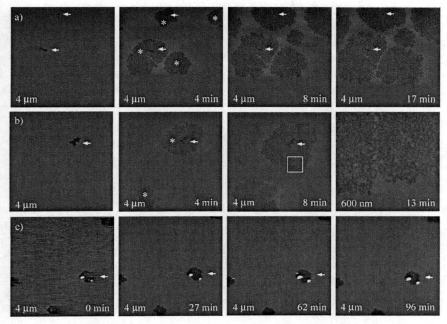

Fig. 6. Time-lapse AFM images of a protein-membrane interaction. AFM images taken after the introduction of (a) 10 mM human amylin, (b) 10 mM rat amylin, and (c) 10 mM rat amylin with 300 mM Congo red onto a palmitoyl oleoyl phosphatidylcholine/palmitoyl oleoyl phosphatidylglycerol (3:1) lipid bilayer. Defects present in the original layer are indicated by arrows. Lipid bilayer disruption is indicated by stars. The rightmost panel in (b) shows a close-up of the membrane being disrupted. Reprinted from Green et al., copyright 2004, with permission from Elsevier (Green et al. 2004).

Podocytes have been imaged in tissue sections by AFM (Ushiki et al. 1994), and the elasticity of epithelial cells has been shown to decrease with aging (Berdyyva et al. 2005).

4. NANOTHERAPIES: ROLE OF AFM IN THE TREATMENT OF DIABETES

Diabetes treatments generally involve lifestyle and drug therapies. For type 2 patients, the drugs are designed to improve the effectiveness of the insulin secreted by the islet cells, and in some cases, additional insulin is prescribed, while in type 1 patients, the primary therapy is the administration of insulin. A key component to delivering correct dosages is implementing an effective closed feedback loop of glucose measurement and insulin action (Imanishi and Ito 1995; Hovorka 2005), in effect replacing the role of the dysfunctional and or destroyed islet cells. In this section, the role of AFM in the development of therapies related to glucose monitoring and sensing, and to insulin delivery are presented. Clearly, AFM has been employed to study the size, structure, and properties of many nanomaterials related to diabetes research discussed in this book, but this discussion is largely beyond the scope of this chapter.

4.1 Glucose Sensing

Ideally glucose monitoring should be continuous and non-invasive (Pickup et al. 2005). The development of glucose biosensors is an active field (Wang 2008) and AFM has or can play a role in many aspects, from understanding binding forces involved in glucose biosensing to characterizing the nanomaterials and coatings used in glucose biosensing. Notably, several attempts have been made to fabricate glucose sensors directly onto AFM tips (de Souze Pereira 2000; Kueng et al. 2005). In addition, glucose/galactose receptor binding has been studied using AFM force spectroscopy revealing a change in stiffness in the activated form, which may be relevant for glucose biosensor design and implementation (Sokolov et al. 2006).

Layer-by-layer assembly (Johnston et al. 2006), the assembly of oppositely charged polyelectrolyte or polymer layers, can be investigated using AFM, which has been employed extensively for the study of polymers and is sensitive to electrostatic forces. Glucose sensors have been prepared via a layer-by-layer approach (Ferreira et al. 2004; Zhao et al. 2005). Other approaches for the development of implantable glucose biosensors incorporate nitric-oxide xerogel microarrays into polymer membranes and layers (Oh et al. 2005). Nitric oxide is an effective inhibitor of platelet and bacterial adhesion and such devices may exhibit improved biocompatibility and lifetimes (Oh et al. 2005).

4.2 Insulin

Insulin is generally delivered, primarily in type 1 diabetic patients, as a solution or a suspension subcutaneously via syringe or an extracorporeal insulin pump, or through an insulin pump implanted in the peritoneal cavity. Transdermal, intranasal, pulmonary, and oral routes are at various stages of development or are commercially available. One limitation of insulin delivery is that it does not effectively mimic the patterns of insulin secretion in the body, making it difficult to achieve glycemic control. As discussed in this subsection, significant efforts have been made in designing insulin for tailored properties and delivery routes, while also designing encapsulation approaches for coating insulin for delivery and islet cells for transplantation.

AFM has been used to study the self-association and crystallization of insulin formulations which have been modified by specific alterations of wild-type insulin to reduce monomeric association and influence self-assembly to develop, e.g., extended time action microcrystalline insulin suspension therapies. The rate of dissociation to the bioactive insulin monomer will directly affect the pharmacokinetics of the insulin and will depend on the structure of the crystals, which can be determined using AFM (Yip et al. 1998a). Yip et al. have released a series of papers using time-lapse AFM to study the influence of sequence variation and insulin analogs on insulin crystallization, as shown in Fig. 7 (Yip et al. 1996, 1998a, 1998b, 2000). Time-lapse AFM has also been used to study the morphology and growth of bovine insulin crystals (Waizumi et al. 2003). Interestingly, Yip et al. have additionally studied the mechanisms of insulin dissociation using AFM force spectroscopy to probe insulin monomer-monomer binding, which is crucial to understanding the efficacy of insulin therapies, observing both dimer dissociation and disentanglement events in the force curves (Yip et al. 1998c).

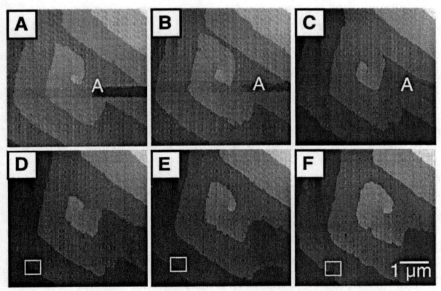

Fig. 7. Time-lapse images of an insulin crystal. Images were recorded at (a) 0, (b) 3605, (c) 7210, (d) 10,815, (e) 18,025, and (f) 40,590 seconds. A region damaged by the AFM tip (and subsequently repaired) is marked 'A.' In (d) a protein aggregate is indicated by a white box, which persists as a defect in the growing crystal. Reprinted from Yip et al., copyright 1998, with permission from Elsevier (Yip et al. 1998a).

Many formulations of insulin, including encapsulated or coated nanoparticles, have been considered for alternative, i.e., oral, delivery routes (Fan et al. 2006; Ye et al. 2006). AFM can be used to image nano- and nanoencapsulated particles, providing information on the process of nanoparticle formation and encapsulation processes (Olivia et al. 2002, 2003), nanoparticle size, nanoparticle degradation, etc. Sitterberg et al. have written a recent review of the uses of AFM for characterizing nanoscale drug delivery systems (Sitterberg et al. 2010).

While these efforts address the delivery of insulin, the development of a bioartificial pancreas (Kizilel et al. 2005; Opara et al. 2010) via, e.g., islet nanoencapsulation, holds promise for long-term treatment of type 1 diabetes (Beck et al. 2007). To address the problem of rejection of transplanted islet cells, nanoencapsulation has developed as a method to achieve immunoisolation. The material should be biocompatible, and isolate the cells from the immune system while still permitting normal exo- and endocytosis. AFM has been used to characterize such micro- to nanoencapsulation devices (Xu et al. 1998) and multilayer nanoencapsulation has been used to coat islet cells (Krol et al. 2006).

5. FUTURE PROSPECTS

While AFM has already made a significant impact on diabetes research, there are several areas where further developments are expected, particularly in live cell imaging, in imaging dynamic processes with high-speed AFM, in correlating structure and function, and in nanocharacterization support of the vast efforts in the design of biosensors, drugs, drug delivery systems, nanoencapsulation and other nanodiagnostic and nanotherapeutic approaches. Combined AFM and confocal imaging approaches may help understand signaling processes associated with diabetes, and allow mechano- and electrotransduction in relevant cells to be studied. AFM recognition imaging for visualizing binding sites and high-speed AFM for imaging dynamic processes, such as fusion pore formation, have potential applications as well. Ultra high-resolution AFM has potential applications in understanding interactions between sugars and cell membranes and for visualizing the interfacial structure at the surface of insulin microcrystals. Additionally, by combining complementary AFM techniques for measuring and comparing elasticity, viscoelasticity, electromechanical coupling, and electrostatic interactions of diabetic and healthy tissues, differences between the tissues can be better understood. Notably, electromechanical coupling has already been studied at the nanoscale in amyloid fibrils in a liquid environment (Kalinin et al. 2007; Nikiforov et al. 2010).

6. SUMMARY

Atomic force microscopy has found many applications in nanotechnology and nanomedicine, including the understanding of diabetes, a disease with increasing incidence rates, which affects millions of people. In particular, AFM has been used to observe ultrastructural differences in collagenous tissues and differences in mechanical properties of red blood cells between diabetic and non-diabetic animals and patients. These observations provide insight into the effect diabetes has on the entire body and stress the importance of glycemic control in preventing long-term complications. While the results suggest AFM could be used for diagnosis and monitoring the progression of the disease, there are more suitable techniques. AFM has also been used to investigate the role the amylin peptide and amylin fibrils has on diabetes and has the potential for revealing the relationship between amylin and β-cell dysfunction. AFM has also played a key role in the development of nanoparticles and nanoencapsulation for biosensing and drug delivery, and in understanding and designing different insulin formulations for less-invasive insulin delivery and rapid or slow acting insulin. In the 25 years since the birth of AFM there have been many contributions to the fields of nanotechnology and nanomedicine, and with the continuing evolution of novel imaging modes and improved instrument and probe design, AFM will continue to make a significant impact.

Summary Points

- AFM can be operated in physiological environments and can be combined with optical microscopy techniques, making it applicable to a wide variety of topics in biology and nanomedicine.
- Prolonged exposure to high glucose levels as a result of diabetes can lead to ultrastructural changes in collagen fibrils, which have been visualized using AFM.
- AFM has revealed an increase in the stiffness of red blood cells of diabetic vs. non-diabetic patients.
- Amylin, insulin, and glucagon are all peptides which fibrillate to form amyloid fibrils. AFM has been used to visualize the formation and ultrastructure of such macromolecules, and has been used to observe the interactions of such peptide monomers with lipid bilayers.
- AFM has provided significant nanocharacterization support for the design of biosensors, drugs, drug delivery systems, nanoencapsulation and other nanodiagnostic and nanotherapeutic approaches for treating diabetes.

- AFM as a nanotechnology continues to be developed and has the potential to be an indispensable tool for understanding and treating diabetes.

Definitions

Contact mode: In contact mode AFM, the cantilever is scanned with the tip in contact with the sample surface. A feedback loop is employed which keeps a constant force between the tip and sample by adjusting the tip-sample separation, thus allowing an image of the surface topography to be recorded.

Force Spectroscopy: AFM force spectroscopy refers to the local measurement of the interaction force between an AFM tip and a sample as a function of the distance between the two. In the case when a molecule is attached between the tip and a substrate, the force required to break molecular bonds can be measured. When the tip is used to plastically deform a material, local elasticity can be measured. Spectroscopy can also refer to the local bias-dependent response (e.g., current, electromechanical coupling) of a sample in contact with a tip.

Glycation: Glycation is the non-enzymatic binding of a sugar molecule to a protein or lipid.

Multifrequency: Whereas in tapping mode, the cantilever is oscillated at a single resonant frequency, multifrequency AFM techniques excite the cantilever at several frequencies simultaneously or across a band of frequencies.

Tapping mode: Tapping mode or intermittent contact AFM is a dynamic AFM mode of operation whereby the cantilever is oscillated near the resonance frequency, and the interaction forces between the tip and sample are monitored using a feedback loop, which keeps the amplitude of the cantilever oscillation constant as it moves laterally relative to the sample surface. Tapping mode AFM reduces the lateral friction forces inherent in contact mode AFM and also overcomes the capillary forces of surfaces in air.

Acknowledgement

This work was supported by Science Foundation Ireland (grant nos. 10/RFP/MTR2855 and 07/IN1/B931).

Abbreviations

| AGEs | : | Advanced Glycation End-Products |
| AFM | : | Atomic Force Microscopy |

FM	:	Frequency Modulation
PBS	:	phosphate buffer saline
SPM	:	Scanning Probe Microscopy
STM	:	Scanning Tunneling Microscopy

References

Abdel-Halim, S.M., A. Guenifi, A. Khan, O. Larsson, P.O. Berggren, C.G. Ostenson and S. Efendić. 1996. Impaired coupling of glucose signal to the exocytotic machinery in diabetic GK rats: a defect ameliorated by cAMP. Diabetes 45: 934–940.

Akhtar, R., J.K. Cruickshank, N.J. Gardiner, B. Derby and M.J. Sherratt. 2010. The effect of type 1 diabetes on the structure and function of fibrillin microfibrils. Mater. Res. Soc. Symp. Proc. 1274: 1274-QQ05-17.

Ando, T., N. Kodera, E. Takai, D. Maruyama, K. Saito and A. Toda. 2001. A high-speed atomic force microscope for studying biological macromolecules. Proc. Natl. Acad. Sci. USA 98: 12468–12472.

Beck, J., R. Angus, B. Madsen, D. Britt, B. Vernon and K.T. Nguyen. 2007. Islet encapsulation: strategies to enhance islet cell function. Tissure Engineering 13: 589–599.

Berdyyva, T.K., C.D. Woodworth and I. Sokolov. 2005. Human epithelial cells increase their rigidity with ageing *in vitro*: direct measurements. Phys. Med. Biol. 50: 81–92.

Binnig, G., C.F. Quate and Ch. Gerber. 1986. Atomic force microscope. Phys. Rev. Lett. 56: 930–933.

Binnig, G., H. Rohrer, Ch. Gerber and E. Weibel. 1982. Surface studies by scanning tunneling microscopy. Phys. Rev. Lett. 49: 57–61.

Brownlee, M., H. Vlassara and A. Cerami. 1984. Nonenzymatic glycosylation and the pathogenesis of diabetic complications. Ann. Int. Med. 101:527–537.

Buzhynskyy, N., J.-F. Girmens, W. Faigle and S. Scheuring. 2007. Human cataract lens membrane at subnanometer resolution. J. Mol. Biol. 374: 162–169.

Chiti, F., and C.M. Dobson. 2006. Protein misfolding, functional amyloid, and human disease. Annu. Rev. Biochem. 75: 333–366.

Cho, S.-J., A.S. Quinn, M.H. Stromer, S. Dash, J. Cho, D.J. Taatjes and B.P. Jena. 2002. Structure and dynamics of the fusion pore in live cells. Cell Bio. Intl. 26: 35–42.

Cho, W.-J., B.P. Jena and A.M. Jeremic. 2008. Nano-scale imaging and dynamics of amylin-membrane interactions and its implication in type II diabetes mellitus. Methods Cell Biol. 90:267–286.

De Jong, K.L., B. Incledon, C.M. Yip and M.R. DeFelippis. 2006. Amyloid fibrils of glucagon characterized by high-resolution atomic force microscopy. Biophys. J. 91: 1905–1914.

de Souza Pereira. 2000. Detection of the absorption of glucose molecules by living cells using atomic force microscopy, Fed. Eur. Biochem. Soc. Lett. 475: 43–46.

Dobson, C.M. 2003. Protein folding and misfolding. Nature 426: 884–890.

Donaldson, P., J. Kistler and R.T. Mathias. 2001. Molecular solutions to mammalian lens transparency. News Physiol. Sci. 16: 118–123.

Dong, M., M.B. Hovgaard, S. Xu, D.E. Otzen and F. Besenbacher. 2006. AFM study of glucagon fibrillation via oligomeric structures resulting in interwoven fibrils. Nanotechnology 17: 4003–4009.

Dong, M., M.B. Hovgaard, W. Mamdouh, S. Xu, D.E. Otzen and F. Besenbacher. 2008. AFM-based force spectroscopy measurements of mature amyloid fibrils of the peptide glucagon. Nanotechnology 19: 384013.

Drake, B., C.B. Prater, A.L. Weisenhorn, S.A.C. Gould, T.R. Albrecht, C.F. Quate, D.S. Cannell, H.G. Hansma and P.K. Hansma. 1989. Imaging crystals, polymers, and processes in water with the atomic force microscope. Science 243: 1586–1589.

Fan, Y.F., Y.N. Wang and J.B. Ma. 2006. Preparation of insulin nanoparticles and their encapsulation with biodegradable polyelectrolytes via the layer-by-layer adsorption. Int. j. Pharm. 324: 158–167.

Ferreira, M., P.A. Fiorito, O.N. Oliveira, Jr. and S.I. Córdoba de Torresi. 2004. Enzyme-mediated amperometric biosensors prepared with the layer-by-layer adsorption technique. Biosensors and Bioelectronics 19: 1611–1615.

Fornal, M., M. Lekka, G. Pyka-Fościak, K. Lebed, T. Grodzicki, B. Wizner and J. Styczeń. 2006. Erythrocyte stiffness in diabetes mellitus studied with atomic force microscope. Clin. Hemorheo. Microcirc. 35: 273–276.

Fukuma, T., A.S. Mostaert, L.C. Serpell and S.P. Jarvis. 2008. Revealing molecular-level surface structure of amyloid fibrils in liquid by means of frequency modulation atomic force microscopy. Nanotechnology 19: 384010.

German M.S. 2008. New β-cells from old acini. Nature Biotech. 26: 1092–1093.

Girasole, M., A. Cricenti, R. Generosi, G. Longo, G. Pompeo, S. Contest and A. Congui-Castellano. 2007. Different membrane modifications revealed by atomic force/lateral force microscopy after doping of human pancreatic cells with Cd, Zn, or Pb. Microsc. Res. Tech. 70: 912–917.

Goldsbury, C., J. Kistler, U. Aebi, T. Arvinte and G.J.S. Cooper. 1999. Watching amyloid fibrils grow by time-lapse atomic force microscopy. J. Mol. Biol. 285: 33–39.

Green, J.D., L. Kreplak, C. Goldsbury, X. Li Blatter, M. Stolz, G.S. Cooper, A. Seelig and U. Aebi. 2004. Atomic force microscopy reveals defects within mica supported lipid bilayers induced by the amyloidogenic human amylin peptide. J. Mol. Biol. 342: 877–887.

Haataja, L., T. Gurlo, C.J. Huang and P.C. Butler. 2008. Islet amyloid in type 2 diabetes, and the toxic oligomer hypothesis. Endocrine Rev. 29:303–316.

Hansma, H.G., and J.H. Hoh. 1994. Biomolecular imaging with the atomic force microscope. Annu. Rev. Biophys. Biomol. Struct. 23: 115–139 (1994).

Hansma, P.K., J.P. Cleveland, M. Rademacher, D.A. Walters, P.E. Hillner, M. Bezanilla, M. Fritz, D. Vie, H.G. Hansma, C.B. Prater, J. Massie, L. Fukunaga, J. Gurley and V. Elings. 1994. Tapping mode atomic force microscopy in liquids. Appl. Phys. Lett. 64: 1738–1740.

Hansma, P.K., V.B. Elings, O. Marti and C.E. Bracker. 1988. Scanning tunneling microscopy and atomic force microscopy: application to biology and technology. Science 242: 209–216.

Harris, M.I., R. Klein, T.A. Welborn and M.W. Knuiman. 1992. Onset of NIDDM occurs at least 4–7 years before clinical diagnosis. Diabetes Care 15: 815–819.

Heinz, W.F., and J.H. Hoh. 1999. Spatially Resolved Force Spectroscopy of Biological Surfaces using the Atomic Force Microscope. Tibtech. 17: 143–150.

Hekele, C., C.G. Goesselsberger and I.C. Gebeshuber. 2008. Nanodiagnostics performed on human red blood cells with atomic force microscopy. Mater. Sci. Technol. 24: 1162–1165.

Hovorka, R. 2005. Continuous glucose monitoring and closed-loop systems. Diabetes Medicine 23: 1–12.

Imanishi, Y., and Y. Ito. 1995. Glucose-sensitive insulin-releasing molecular systems. Pure & Appl. Chem. 67: 2015–2021.

Jansen, R., W. Dzwolak and R. Winter. 2005. Amyloidogenic self-assembly of insulin aggregates probed by high resolution atomic force microscopy. Biophys. J. 88: 1344–1353.

Jarvis, S.P., J.E. Sader and T. Fukuma. Frequency Modulation Atomic Force Microscopy in Liquids. pp. 315–351. In: B. Bharat and H. Fuchs. [eds.] 2008. Applied Scanning Probe Methods VIII. Springer-Verlag, Berlin, Germany.

Jin, H., X. Xing, H. Zhao, Y. Chen, X. Huang, S. Ma, H. Ye and J. Cai. 2010. Detection of erythrocytes influenced by aging and type 2 diabetes using atomic force microscope. Biochem. Biophys. Res. Commun. 391: 1698–1702.

Johnston, A.P.R., C. Cortez, A.S. Angelatos and F. Caruso. 2006. Layer-by-layer engineered capsules and their applications. Curr. Opin. Coll. Int. Sci. 11: 203–209.

Kalinin, S.V., B.J. Rodriguez, S. Jesse, K. Seal, R. Proksch, S. Hohlbauch, I. Revenko, G.L. Thompson and A.A. Vertegel. 2007. Towards local electromechanical probing of cellular and biomolecular systems in a liquid environment. Nanotechnology 18: 424020.

Kizilel, S., M. Garfinkel and E. Opara. 2005. The bioartificial pancreas: progress and challenges. Diabetes Technology & Therapeutics 7: 968–985.

Kodera, N., D. Yamamoto, R. Ishikawa and T. Ando. 2010. Video imaging of walking myosin V by high-speed atomic force microscopy. Nature 468: 72–76.

Krol, S., S. del Guerra, M. Grupillo, A. Diaspro, A. Gliozzi and P. Marchetti. 2006. Multilayer nanoencapsulation. New approach for immune protection of human pancreatic islets. Nano Lett. 6: 1933–1939.

Kueng A., C. Kranz, A. Lugstein, E. Bertagnolli and B. Mizaikoff. 2005. AFM-tip-integrated amperometric microbiosensors: high-resolution imaging of membrane transport. Angew. Chem. 117: 3485–3488.

Lal, R., and M.F. Arnsdorf. 2010. Multidimensional atomic force microscopy for drug discovery: a versatile tool for defining targets, designing therapeutics and monitoring their efficacy. Life Sciences 86: 545–562.

Lin, S., Y.-M. Wang, L.-S. Huang, C.-W. Lin, S.-M. Hsu and C.-K. Lee. 2007. Dynamic response of glucagon/anti-glucagon pairs to pulling velocity and pH studied by atomic force microscopy. Biosensors and Bioelectronics 22: 1013–1019.

Lorenzo, A., B. Razzaboni, G.C. Weir and B.A. Yankner. 1994. Pancreatic islet cell toxicity of amylin associated with type-2 diabetes mellitus. Nature 368: 756–760.

Mangenot, S., N. Buzhynskyy, J.-F. Girmens and S. Scheuring. 2009. Malformation of junctional microdomains in cataract lens membranes from a type II diabetes patient. Pflugers. Arch. Eur. J. Physiol. 457: 1265–1274.

Mauro, M., E.F. Craparo, A. Podestà, D. Bulone, R. Carrotta, V. Martorana, G. Tiana and P.L. San Biagio. 2007. Kinetics of different processes in human insulin amyloid formation. J. Mol. Biol. 366: 258–274.

Müller, D.J., and Y.F. Dufrêne. 2008. Atomic force microscopy as a multifunctional molecular toolbox in nanobiotechnology. Nat. Nanotech. 3: 261–269.

Nikiforov, M.P., G.L. Thompson, V.V. Reukov, S. Jesse, S. Guo, B.J. Rodriguez, K. Seal, A.A. Vertegel and S.V. Kalinin. 2010. Double-layer mediated electromechanical response of amyloid fibrils in liquid environment. ACS Nano 4: 689–698.

Odetti, P., I. Argano, R. Rolandi, S. Garibaldi, S. Valentini, L. Cosso, N. Traverso, D. Cottalasso, M.A. Pronzato and U.M. Marinari. 2000. Scanning force microscopy reveals structural alterations in diabetic rat collagen fibrils: role of protein glycation. Diabetes Metab. Res. Rev. 16: 74–81.

Odetti, P., I. Argano, S. Garibaldi, S. Valentini, M.A. Pronzato and R. Rolandi. 1998. Role of advanced glycation end products in aging collagen. Gerontology 44: 187–191.

Oh, B.K., M.E. Robbins, B.J. Nablo and M.H. Schoenfisch. 2005. Miniturized glucose biosensor modified with a nitric oxide-releasing xerogel microarray. Biosensors and Bioelectronics 21: 749–757.

Oliva, M., C. Caramella, I. Díez-Pérez, P. Gorostiza, C.-F. Lastra, I. Oliva and E.L. Mariño. 2002. Sequential atomic force microscopy imaging of a spontaneous nanoencapsulation process. Int. J. Pharm. 242: 291–294.

Oliva, M., I. Díez-Pérez, P. Gorostiza, C.F. Lastra, I. Oliva, C. Caramella and E.L. Mariño. 2003. Self-assembly of drug-polymer complexes: a spontaneous nanoencapsulation process monitored by atomic force microscopy. J. Pharm. Sci. 92: 77–83.

Opara E.C., S.H. Mirmalek-Sani, O. Khanna, M.L. Moya and E.M. Brey. 2010. Design of a bioartificial pancreas. J. Investig. Med. 58: 831–837.

Pickup, J.C., F. Hussain, N.D. Evans and N. Sachedina. 2005. In vivo glucose monitoring: the clinical reality and the promise. Biosensors and Bioelectronics 20: 1897–1902.

Pickup, J.C., Z.L. Zhi, F. Khan, T. Saxl and D.J.S. Birch. 2008. Nanomedicine and its potential in diabetes research and practice. Diabetes Metab. Res. Rev. 24: 604–610.

Podestà, A., G. Tiana, P. Milani and M. Manno. 2006. Early events in insulin fibrillization studied by time-lapse atomic force microscopy. Biophys. J. 90: 589–597.

Proksch, R. Multi-frequency atomic force microscopy. pp. 125–151. In: S.V. Kalinin and A. Gruverman. [eds.] 2011. Scanning Probe Microscopy of Functional Materials. Part 2. Springer Science+Business Media, New York, USA.

Putman, C.A.J., K.O. Van der Werf, B.G. De Grooth, N.F. Van Hulst and J. Greve. 1994. Tapping mode atomic force microscopy in liquid. Appl. Phys. Lett. 64: 2454–2456.

Rademacher, M., R.W. Tillman, M. Fritz and H.E. Gaub. 1992. From molecules to cells- imaging soft samples with the AFM. Science 257: 1900–1905.

Rask-Madsen, C., and G.L. King. 2010. Podocytes lose their footing. Nature. 468: 42–44.

Riehemann, K., S.W. Schneider, T.A. Luger, B. Godin, M. Ferrari and H. Fuchs. 2009. Nanomedicine—challenges and perspectives. Angew. Chem. Int. Ed. 48: 872–897.

Rohrer, H. 1994. The nanometer age: challenge and chance. Microsc. Microanal. Microstruct. 5: 237–246.

Schneider, S.W., K.C. Sritharan, J.P. eibel, H. Oberleithner and B.P. Jena. 1997. Surface dynamics in living acinar cells imaged by atomic force microscopy: identification of plasma membrane structures involved in exocytosis. Proc. Natl. Acad. Sci. USA 94: 316–321.

Sedman, V.L., S. Allen, W.C. Chan, M.C. Davies, C.J. Roberts, S.J.B. Tendler and P.M. Williams. 2005. Atomic force microscopy study of human amylin (20–29) fibrils. Prot. Pept. Lett. 12: 79–83.

Sitterberg, J., A. Özcetin, C. Ehrhardt and U. Bakowsky. 2010. Utilising atomic force microscopy for the characterization of nanoscale drug delivery systems. Eur. J. Pharm. Biopharm. 74: 2–13.

Slezak, L.A., A.S. Quinn, K.C. Sritharan, J.P. Wang, G. Aspelund, D.J. Taatjes and D.K. Andersen. 1999. Binding forces of hepatic microsomal and plasma membrane proteins in normal and pancreatitic rats: an AFM force spectroscopy study. Microsc. Res. Tech. 44: 363–367.

Sokolov, I., V. Subba-Rao and L.A. Luck. 2006. Change in rigidity in the activated form of the glucose/galactose receptor from Escherichia coli: a phenomenon that will be key to the development of biosensors. Biophys. J. 90: 1055–1063.

Suresh, S., J. Spatz, J.P. Mills, A. Micoulet, M. Dao, C.T. Lim, M. Beil and T. Seufferlein. 2005. Connections between single-cell biomechanics and human disease states: gastrointestinal cancer and malaria. Acta Biomaterialia 1: 15–30.

Ushiki, T., M. Shigeno and K. Abe. 1994. Atomic force microscopy of embedment-free sections of cells and tissues. Arch. Histol. Cytol. 57: 427–432.

Waizumi, K., M. Plomp and W. van Enckvort. 2003. Atomic force microscopy studies on growing surfaces of bovine insulin crystals. Colloids and Surfaces B: Biointerfaces 30: 73–86.

Wang, H., B.E. Layton and A.M. Sastry. 2003. Nerve collages from diabetic and nondiabetic Sprague-Dawley and biobreeding rats: an atomic force microscopy study. Diabetes Metab. Res. Rev. 19: 288–298.

Wang, J. 2008. Electrochemical glucose biosensors. Chem. Rev. 108: 814–825.

Welsh, G.I., L.J. Hale, V. Eremina, M. Jeansson, Y. Maezawa, R. Lennon, D.A. Pons, R.J. Owen, S.C. Satchell, M.J. Miles, C.J. Caunt, C.A. McArdle, H. Pavenstädt, J.M. Tavaré, A.M. Herzenberg, C.R. Kahn, P.W. Mathieson, S.E. Quaggin, M.A. Saleem and R.J.M. Coward. 2010. Insulin Signaling to the Glomerular Podocyte Is Critical for Normal Kidney Function. Cell Metab. 12: 329–340.

Xu, K., D.M. Hercules, I. Lacik and T.G. Wang. 1998. Atomic force microscopy used for the surface characterization of microcapsule immunoisolation devices. J. Biomed. Mater. Res. 41: 461–467.

Ye. S., C. Wang. X. Liu, Z. Tong, B. Ren and F. Zeng. 2006. New loading process and release properties of insulin from polysaccharide microcapsules fabricated through layer-by-layer assembly. J. Contr. Rel. 112: 79–87.

Yip, C.M., and M.D. Ward. 1996. Atomic force microscopy of insulin single crystals: direct visualization of molecules and crystal growth. Biophys. J. 71: 1071–1078.

Yip, C.M., M.L. Brader, M.R. DeFelippis and M.D. Ward. 1998. Atomic force microscopy of crystalline insulins: the influence of sequence variation on crystallization and interfacial structure. Biophys. J. 74: 2199–2209.

Yip, C.M., M.R. DeFelippis, B.H. Frank, M.L. Brader and M.D. Ward. 1998. Structural and morphological characterization of ultralente insulin crystals by atomic force microscopy: evidence of hydrophobically driven assembly. Biophys. J. 75: 1172–1179.

Yip, C.M., C.C. Yip and M.D. Ward. 1998. Direct force measurements of insulin monomer-monomer interactions. Biochemistry 37: 5439–5449.

Yip, C.M., M.L. Brader, B.H. Frank, M.R. DeFelippis and M.D. Ward. 2000. Structural studies of a crystalline insulin analog complex with protamine by atomic force microscopy. Biophys. J. 78: 466–473.

Zhao, W., J.-J. Xu and H.-Y. Chen. 2005. Extended-range glucose biosensor via layer-by-layer assembly incorporating gold nanoparticles. Frontiers in Bioscience 10: 1060–1069.

Zhong, Q., D. Inniss, K. Kjoller and V.B. Elings. 1993. Fractured polymer/silica fiber surface studied by tapping mode atomic force microscopy. Surf. Sci. Lett. 290: L688 (1993).

Zhou, Q., J. Brown, A. Kanarek, J. Rajagopal and D.A. Melton. 2008. *In vivo* reprogramming of adult pancreatic exocrine cells to β-cells. Nature 455: 627–632.

4

Nanocarriers of Antisense Oligonucleotides in Diabetes

Novella Rapini,[1,a] *Nunzio Bottini*[1,b] *and Massimo Bottini*[2,*]

ABSTRACT

Antisense oligonucleotides are short strands of deoxyribonucleotides that are complementary to specific encoding mRNA sequences and can block gene expression. Their potential use as therapeutic and gene validation tools has elicited great interest. However, poor intracellular delivery into several tissues in living animals currently limits the range of *in vivo* applications of antisense oligonucleotides. Innovative solutions to this problem are coming from nanomedicine, a discipline recently born from the marriage of nanotechnology and medicine. Through the use of nanotechnology-derived composites (nanocarriers), nanomedicine helps traditional drugs to avoid the body's defenses that the drug encounters following its systemic administration. In this chapter we will describe diabetes-relevant applications of nanocarriers as

---

[1]La Jolla Institute for Allergy & Immunology, 9420 Athena Circle, La Jolla, CA, 92037, USA.
[a]E-mail: nrapini@liai.org
[b]E-mail: nunzio@liai.org
[2]Sanford-Burnham Medical Research Institute, 10901 North Torrey Pines Road, La Jolla, CA, 92037, USA; E-mail: mbottini@sanfordburnham.org
*Corresponding author

List of abbreviations after the text.

delivery systems for antisense oligonucleotides. We divided the chapter in two sections. In the first section we will 1) describe the modes of action of antisense oligonucleotides and the chemical modifications that have been developed to ameliorate resistance to nucleases, enhance affinity and potency, and reduce toxicity; 2) the biological barriers encountered by the antisense oligonucleotides following their systemic administration; 3) the arsenal of solutions that nanomedicine can offer to enable the safe delivery of antisense oligonucleotides to the target site. In the second section, we will review the antisense oligonucleotides and delivery systems that have been developed as novel therapeutic weapons against diabetes. This includes research from our laboratory focusing on achieving targeted delivery of antisense oligonucleotides into T cells for therapy of Type 1 Diabetes.

1. INTRODUCTION

Antisense oligomers (ASOs) are short strands (typically 20bp in length) of deoxyribonucleotides that, upon intracellular delivery, hybridize to specific mRNAs and cause reduced expression of the encoded proteins (Chan et al. 2006). ASOs represent a powerful strategy for gene validation and for the cure of diseases associated to dysregulated protein expression, such as autoimmune diseases and cancer. Several chemical modifications of ASOs have been developed which dramatically improved their resistance to nucleases, circulation half-life, affinity for mRNA and silencing potency compared to first-generation oligos. However, a persisting limitation to the use of ASOs *in vivo* is their low spontaneous cell-permeability, especially in certain cell types. Several approaches based on physical methods (e.g., electroporation) or the use of delivery agents (e.g., viral vectors and cationic liposomes) are available to ensure efficient introduction of ASOs into cells in culture. However, comparable systems for *in vivo* application are still in the early stages of development, and the vast majority of ASOs undergoing clinical trials for FDA-approval are topically or systemically administered in "free" form. There is a critical need for novel delivery systems to enable the safe delivery of ASOs to targeted cells and tissues after systemic administration. Innovative solutions in this field are coming from nanomedicine, a discipline recently born from the marriage of nanotechnology and medicine. Through the use of nanotechnology-derived composites (*nanocarriers*), nanomedicine helps traditional drugs avoid the biological and biophysical barriers the drug encounters following its systemic administration. In this chapter we will describe diabetes-relevant applications of nanocarriers as delivery systems for ASOs. We divided the chapter in two sections. In the first section we will summarize the modes of action of ASOs and the chemical modifications that have been developed to ameliorate the resistance to nucleases, enhance affinity and potency, and reduce toxicity (§ 2.1.), the biological barriers encountered by ASOs following their systemic administration (§ 2.2. and § 2.3.) and the range of solutions that nanomedicine can offer to enable the safe delivery of ASOs to the target site (§ 2.4. and § 2.5). In the second section, we will review ASOs and delivery systems that have been developed for applications in the diabetes field. This includes research from our laboratory focused on achieving targeted delivery of ASOs into T cells for therapy of Type 1 Diabetes.

2. ASOs AND NANOMEDICINE

2.1 Nanodrugs for ASO Delivery & Targeting

Oligonucleotide-based therapy can be based on the use of ASOs, small interference RNA (siRNA) or aptamers. Each type of drug blocks the expression of proteins by means of different intracellular mechanisms. In this chapter we will focus our attention on ASOs as a weapon against diabetes and on nanotechnology-based approaches that have been developed or are under development to improve their potency and delivery.

ASOs are short (typically 20bp in length) single-stranded deoxyribonucleotide analogues that hybridizes with the complementary mRNA via Watson-Crick base pairing. ASOs are excellent tools for target validation and gene function studies, but also highly appealing as therapeutic strategy to selectively suppress the expression of disease-related proteins. It is worthy of notice that the single strand structure of ASOs makes them particularly suited for *in vivo* immunological applications since at least some types of ASOs are believed to be less prone to elicit "stress" immune responses through receptors on innate immune cells (such as toll-like receptor 3, TLR-3, which binds double helical nucleic acids). Formation of the ASO-mRNA heteroduplex can result in inhibition of the target gene expression by various mechanisms, depending on the chemical make-up (generation) of the ASO (Fig. 1). Despite usually showing much lower potency than siRNAs, unmodified DNA ASOs can operate in a way similar to siRNA by activating endogenous ribonuclease H (RNase H) which recognizes the DNA-RNA heteroduplex and cleaves the RNA strand leaving intact the ASO. However, the use of unmodified ASOs has been largely unsuccessful because of their susceptibility to degradation by nucleases and their inability to penetrate mammalian

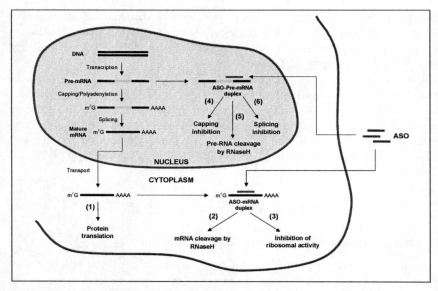

Fig. 1. Mechanism and location of action of ASOs. In absence of ASOs, mRNAs are normally translated into functional proteins (1). ASOs that are delivered into cells can hybridize with either the mRNA (in the cytoplasm) or the Pre-mRNA (in the nucleus). In the cytoplasm, the formation of the ASO-mRNA hetero-duplex can lead either to activation of RNaseH (2) or to inhibition of ribosomal activity (3). Alternatively, the ASO can migrate into the nucleus, and block the maturation of mRNA at different level of the process (4–6). The mechanism and location of action of an ASO depend on its chemical make-up and on the technique used to deliver it into the cells. Independently of mechanism and location, the action of an ASO will finally result in protein expression knock-down.

cells in culture. Various chemical modifications of the phosphodiester bond and/or the ribose sugar have been developed to decrease nuclease cleavage and increase the biostability and potency of the ASOs (Fig. 2). For instance, phosphorothioate-modified (first generation) ASOs have one of the non-bridging oxygen atoms in the phosphodiester bond replaced by a sulphur atom, whereas phosphorodiamidate morpholino-modified (third generation) ASOs are non-charged oligomers in which the phosphodiester bond and the ribose sugar are replaced by a phosphorodiamidate linkage and a six-membered morpholino ring, respectively. Morpholino ASOs are extremely suitable for *in vivo* applications because they are highly resistant to nucleases and proteases in biological fluids.

Despite these improvements in chemical make-up, the use of ASOs for *in vivo* knock-down of gene expression is still significantly limited by their poor intracellular delivery. Novel solutions to enhance ASO delivery are needed in order to fully exploit the potential of ASOs for diagnosis and therapy of human diseases. Some of these solutions are emerging from the

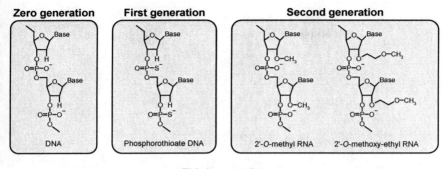

Fig. 2. Chemical modifications of antisense oligonucletides. Chemical modifications (generations) have been developed to decrease nuclease cleavage and increase the biostability and potency of the ASOs.

field of nanotechnology. Nanotechnology is the science of manipulating the matter at the atomic and molecular level to obtain materials with specifically enhanced chemical and physical properties. The application of nanotechnology in medicine (*nanomedicine*) aims at developing multifunctional nanoparticles (*nanodrugs*) for delivering pharmaceutical, therapeutic, and diagnostic agents to the target sites with high efficacy and specificity. Compared to traditional approaches, the ultimate promise of nanodrugs is to achieve therapeutic and diagnostic goal, using less amounts of drugs and with less or no side effects. A nanodrug can be thought as a tripartite structure composed by a nanotechnology-derived nanoparticle (*nanocarrier*), loaded with therapeutic and diagnostic payloads (*cargos*, for instance the ASOs) and with "helper" agents (*enhancers*) which enable the nanodrug to resist and/or penetrate the biological barriers encountered following *in vivo* administration (Fig. 3).

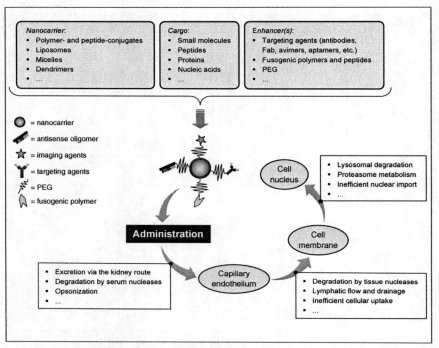

Fig. 3. The nanotechnology approach. Multifunctional nanoparticles (*nanodrugs*) are tripartite structures composed by a *nanocarrier* loaded with *cargos* (diagnostic and therapeutic payloads) and *enhancers* (targeting, permeation and imaging agents). Enhancers allow the nanoparticle to reach the site of interest avoiding the biological and biophysical barriers encountered following the systemic administration.

2.2 The Biological Barriers: RES, Extravasation & Diffusion

To address the challenge of delivery ASOs to the site of action (for instance the nucleus of a specific cell subpopulation) *in vivo* with nanotechnology, it is necessary to combine the rational design of a nanodrug with the fundamental understanding of the several biological and physical barriers, which the nanodrug will encounter after injection in a living organism (Fig. 3).

Following their administration, nanodrugs with molecular weights smaller than 5kDa (such as "free" ASOs, and polymer- and peptide-conjugates) may be rapidly excreted via the kidney route (Brenner et al. 1976). Larger nanodrugs are not filtered as efficiently through the kidney glomerular system and they are often significantly bound by opsonins in the bloodstream such as immunoglobulins, complement components and other serum proteins. This can be followed by internalization by professional phagocytic cells of the reticulo endothelial system (RES), and delivery into lysosomal compartments where they are subjected to degradation (Mosser and Edwards 2008). Cells of RES include circulating monocytes and tissue macrophages, including specialized ones such as the Kupffer cells in the liver. Liver and spleen are enriched with RES cells and characterized by a highly fenestrated vascular endothelium with gaps that allow the transit of particles up to 200 nm in diameter (Scherphof 1991). Thus it is not surprising that those organs are the major accumulation sites of nanodrugs. Strategies to avoid uptake by the RES are mainly based on passivation of the nanodrug surface with polyethylene glycol (PEG) chains (Juliano et al. 2008).

While circulating in the blood stream, the nanodrugs must pass the endothelial cells lining the vascular lumen in order to target specific tissue parenchymal cells. In general, nanodrugs with lateral dimension smaller than approximately 5 nm readily egress the capillary endothelium via paracellular routes involving imperfections in the cell junctions, whereas larger nanodrugs are able to pass across the capillary vasculature only in certain specialized tissues (for instance spleen and liver) and in defective tissues (such as tumor vessels and pancreatic islet blood vessels in diabetic patients). The tumor microenvironment is also characterized by a scarce lymphatic drainage, which, along with the leakiness of the vascular system, results in the so-called enhanced permeability and retention (EPR) effect. Several studies have been shown that larger nanodrugs (nanoparticles) accumulate in tumor tissues due to the EPR effect in higher amount than smaller nanodrugs ("free" ASOs and conjugates). However, the latter may be a better therapeutic choice if used in combination with anti-angiogenic agents, which are intended to "normalize" the tumor tissue vasculature.

Once extravasated from the blood vessels, the nanodrugs must diffuse into the extracellular matrix which is composed by a dense meshwork of proteins that might bind the nanodrugs and tends to block their diffusion (Juliano 2007). Once in the extracellular matrix, the nanoparticle must target some or all tissue cells, be internalized, and either diffuse in the cytoplasm or reach the nucleus. The process of cellular internalization of a nanoparticle is usually energy-dependent and follows one of the following pathways: 1) clathrin-dependent endocytosis, 2) the lipid raft pathway, 3) the caveolar pathway, 4) phagocytosis, and 5) pinocytosis (Fig. 4). Even if more rarely, nanoparticles may also diffuse through the plasma membrane and be internalized by an energy-independent process; an example of this is given by certain kind of functionalized carbon nanotubes that behave as nano-needles and migrate into the cytoplasm by puncturing the plasma

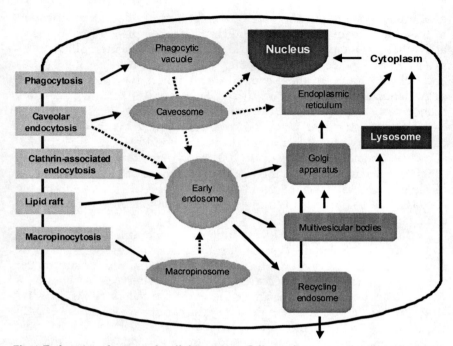

Fig. 4. Endocytic pathways and trafficking routes. Cells uptake macromolecules and particles by a process called endocytosis. *Phagocytosis* involves the ingestion of large substances by "professional" cells (phagocytes). The substances are ingested in the phagocytic vacuoles and transported in the lysosomes. *Macropinocytosis* is a clathrin-independent mechanism through membrane ruffling and responsible for the intracellular delivery of large amount of solutes and fluids through large vesicles of heterogeneous dimensions (micropinosomes). *Clathrin-mediated endocytosis* is generated by the invagination and pinching off of regions of the plasma membranes (coated pits) rich of a fibrous protein (clathrin) which together with a smaller polypeptide form a basketlike network of hexagon and pentagons on the surface of pits and intracellular vesicles. The endocytic vesicles subsequently lose their coats and deliver their content to the sorting endosomes. Three distinct pathways have been identified that sort the endosomes and bring their content to either the late endosomes/lysosomes for degradation, or the plasma membrane through recycling endosomes, or the endoplasmic reticulum through the trans-Golgi network and the Golgi apparatus (retrograde transport). Advances in the design of specific molecular inhibitors enabled to elucidate clathrin-independent pathways stimulated via certain microdomains of the plasma membrane, *caveolae* and *lipid rafts*, enriched in cholesterol and glycosphingolipids. Dark arrows represent established intracellular routes, whereas those represented by dotted arrows are supported by more preliminary evidence.

membrane without triggering any of the energy-dependent pathways described in Fig. 4 (Kostarelos et al. 2007). Following cellular uptake, the nanodrug is trafficked into a variety of intracellular compartments and, finally, either degraded in the lysosomes, or transported to a specific site.

2.3 Cutting Edge Area of Research: Active Targeting

Traditional drugs passively accumulate in tissues by mechanisms that exploit physical and/or chemical gradients or fenestrations in blood vessels. Even if those mechanisms could target particular tissues with high efficiency, they cannot be used to carry traditional drugs (or nanodrugs) to specific sites (active targeting). Cell-specific targeting is a form of active targeting where particles are transported into a specific cell subpopulation through receptor-mediated endocytosis (RME). Here, the drug is functionalized with agents (ligands) able to recognize specific targets on the surface of cells (antigens and receptors). Efficient cell-specific targeting is based on high target specificity; the antigen/receptor must be expressed only on the cell population of interest, and on the selectivity and efficacy of the targeting agent. The ligand could bind the target with high affinity and/or avidity—a way to enhance targeting efficacy through simultaneous interactions with low affinity. Traditional drugs are usually conjugated to a single ligand, monovalent structure, whereas nanodrugs are able to carry multiple ligands thereby enhancing targeting avidity.

A recent report showed that gold and silver spherical nanoparticles functionalized with Herceptin and having a diameter falling in a 25 nm-centered narrow range of values can be internalized by breast cancer cells through HerB2-mediated endocytosis. Nanoparticles with diameters not in that range of values were not internalized (Jiang et al. 2008). Since HerB2 is a normally internalization-impaired receptor, these results suggest that modulation of the physico-chemical properties of the nanocarrier can induce RME after engagement of surface receptors even the ones that are naturally less prone to post-engagement internalization.

2.4 The Arsenal of Nanocarriers

An extremely wide range of nanocarriers have been used as backbone for the fabrication of nanodrugs. Nanocarriers can be classified by their dimension (for instance into conjugates and nanoparticles), their chemical composition (polymeric, amphiphile-based, organic, inorganic), their shape and geometry (spherical, cylindrical, toroidal, shell, tree-like, worm-like), or their physical properties (magnetic, luminescent, flexible, soft, degradable). Here we will give a quick overview of the most commonly used nanocarriers, without following a specific classification.

Polymers are among the most widely investigated materials for the fabrication of novel and high-performance nanocarriers. In the simplest application, the nanocarrier is a polymeric chain directly conjugated to the ASO to ameliorate its solubility and/or help its release from the endolysosomal compartments. Otherwise, biodegradable polymers and block co-polymers have been used to fabricate nanoparticles encapsulating (by electrostatic interaction) the ASOs. The oligo is then released by diffusion or erosion of the polymeric lattice in a predictable manner (Khan et al. 2004).

Amphiphile-based nanoparticles are the most widely used nanocarriers and, in function of their architecture, they are classified into liposomes and micelles. Liposomes are FDA-approved spherical structures composed by several concentric lipid bilayers with a water core (water-in-water), and are used to deliver a wide range of drugs. Cationic liposomes are the most commonly used delivery systems for first and second generation ASOs. Micelles are self-assembled spherical lipid monolayers with a hydrophobic core and a hydrophilic outer shell (oil-in-water) and are under clinical evaluation as carriers for several water insoluble drugs (including phosphorodiamidate morpholino-modified ASOs) (DeLong et al. 1999).

Organic nanoparticles such as dendrimers, peptide- and/or protein-conjugates, and carbon nanotubes have been also used to deliver ASOs. Dendrimers are tree-like structures whose synthesis is based on the growth of branched synthetic polymers (Yoo and Juliano 2000), whereas peptide- and protein-nanoparticles are fabricated by conjugating the ASO directly to peptides or proteins in order to address solubility problems and/or facilitate their intracellular internalization and trafficking (Kang et al. 2008). Carbon nanotubes (NTs) are hollow graphitic tubes, about 1 nm in diameters and hundreds of nanometers long, which are able to emit light in the near infrared region and are characterized by a high surface-to-volume ratio. Our research group has recently used NTs to deliver ASOs into T cells *in vitro* (Delogu et al. 2009).

Inorganic nanoparticles have been extensively studied as nanocarriers because of their peculiar physico-chemical properties which can be exploited to fabricate multi-functional nanoparticles. Among the inorganic nanoparticles reported as nanocarriers for ASOs, gold nanoparticles are often preferred because of their easy synthesis and uniform size (Rosi et al. 2006; Rink et al. 2010). Quantum dots are inorganic semiconducting nanocrystals that are able to emit in a narrow range of wavelengths which can be controlled since it depends on the composition and dimension of the particle (Jia et al. 2007).

3. ANTISENSE OLIGONUCLEOTIDE THERAPY IN DIABETES

ASOs are under development for therapeutic applications in the type 1 and type 2 diabetes (T1D and T2D). However use of nanocarriers for the delivery of those ASOs is at its infancy and few studies has been reported (Table 1).

Table 1. ASOs for therapeutic applications in diabetes. Published studies about the use of ASOs for therapeutic applications applications in the type 1 and type 2 diabetes.

ASO & Generation	Target	Route	Nanocarrier	Reference
iCo-007 second	c-Raf kinase	intravitreous injection in mice	none	Hnik et al. 2009
ISIS113715 second	PTP-1B	intraperitoneal injection in mice	none	Waring et al. 2003
ISIS113715 third (LNA)	PTP-1B	intraperitoneal injection in mice	none	Koizumi et al. 2007
ISIS113715 second	PTP-1B	interscapular injection in monkeys	none	Swarbrick et al. 2009
-*	11βhydroxysteroid dehydrogenase	intraperitoneal injection in mice	none	Bhat et al. 2008
second	glucocorticoid receptor	intraperitoneal injection in mice	none	Liang et al. 2005
third (LNA)	PPARγ	in vitro	Lipofectamine 2000	Tachibana et al. 2007
second	CD80, CD86, CD40	subcutaneous injection in mice	PROMAXX microspheres	Phillips et al. 2008
-*	eGFP	in vitro	gold nanoparticles	Rink et al. 2010
second	PTPN22	in vitro	carbon nanotubes	Delogu et al. 2009

*ASO's generation was not specified in the *Materials and Methods* section.

3.1 iCo-007 for Diabetic Retinopathy and Diabetic Macular Edema

Hyperglycemia in diabetic patients alters hemodynamics of the retinal vasculature and causes chronic hypoxia, which in turns results in the upregulation of growth factors and in pathologies such as neovascularization, vascular permeability, macular edema, etc. The inhibition of the kinase c-Raf is envisaged as an effective therapeutic strategy for diabetic macular edema (DME) and diabetic retinopathy (DR) because c-Raf is a downstream signaling mediator for multiple growth factors responsible for DME and DR such as the vascular endothelial growth factor (VEGF), the insulin-like growth factor (IGF), the basic fibroblast growth factor (bFGF), and others (Grillone and Henry 2008). iCo-007 is a second generation ASO targeting c-Raf kinase that has been discovered by ISIS Pharmaceuticals Inc. (Carlsbad, California, USA). In pig models of branch vein occlusion and mouse models of laser induced choroidal neovascularization the ASO caused inhibition of c-Raf expression and improvement in the neovascularization severity score (Danis et al. 2003). iCo-007 was licensed to iCo Therapeutics Inc. (Vancouver, British Columbia, Canada) in 2005 and since then it has been in an open label, dose escalating Phase I clinical trial in patients with DME. In a study, the drug was administered as a single intravitreous injection to patients with DME which did not respond to other treatments. The drug did not show any serious adverse events and lead to reduction of macular edema in some patients (Hnik et al. 2009). Those results suggest that iCo-007 might be an effective treatment for clinical diseases associated with neovascularization and leakage (other than DME and DR), viral infections, intraocular inflammation and glaucoma.

3.2 Two Locked Nucleic Acid-modified ASOs for the Cure of T2D

The peroxisome proliferator-activated receptor (PPAR) γ and the protein-tyrosine phosphatase (PTP)-1B are validated targets for T2D and third-generation ASOs have been developed against them. PPARγ is a component of the nuclear hormone receptor superfamily and plays an important role in inflammation, cell cycle regulation and differentiation, and glucose homeostasis. Treatment of wild-type mice with a PPARγ antagonist has been reported to prevent high fat diet-induced obesity, insulin resistance, and T2D. Tachibana et al. designed a third-generation locked nucleic acid targeting the translation initiation site of human PPARγ mRNA and showed its inhibitory effect on two human cell lines (THP-1 and HCT116) (Tachibana et al. 2007). PTP-1B negatively regulates insulin receptor in insulin-responsive tissues (such as adipose, liver and muscle). ISIS Pharmaceuticals Inc. has been developing a second-generation ASOs (ISIS113715) to inhibit expression of PTP-1B. In preclinical studies ISIS113715 was able to normalize the plasma glucose levels in treated *ob/ob* and *db/db* mice and improved insulin sensitivity and adiponectin concentration in primates (Swarbrick et al. 2009). Recent phase II data showed that T2D patients well tolerated the ASOs and showed statistically significant reduction in several indices of blood glucose control. Furthermore, Koizumi et al. developed a third-generation locked nucleic acid isosequencial of ISIS113715 and shown to have enhanced *in vivo* efficacy in treated *db/db* mice respect to ISIS113715 (Koizumi et al. 2007).

3.3 Oligonucleotide-loaded PROMAXX® Microspheres, a Diabetes Suppressive Vaccine

PROMAXX® microspheres technology has been developed by Epic Therapeutics Inc. The microspheres are fabricated through a phase-separation process that takes place in an aqueous environment at mild temperatures and enables the encapsulization of biomacromolecules. The size of microspheres can be tuned by adjusting the conditions of the process (ionic strength, pH, rate of cooling, etc.) and the release of entrapped molecules can be programmed over times spanning from hours to weeks, thus drastically reducing the need for injections. Epic Therapeutics Inc. has characterized PROMAXX® formulations for pulmonary delivery of several drugs such as insulin, alpha-1 anti-trypsin, human growth hormone, etc. Phillips et al. used PROMAXX® microspheres to deliver ASOs into dendritic cells (DC) to down-regulate the co-stimulatory receptors CD80/CD86 and CD40 (immature state). Several reports have shown that the exogenous administration of DC in an immature state efficiently prevents and suppresses diabetes in NOD mice. However this technology has serious logistical limitations due to the fact that the clinic (where the patients are treated) is often far away from the facilities where the DC are engineered. In an effort to avoid those limitations, Phillips et al. use PROMAXX® microspheres to elicit immature state in DC directly *in vivo*. The microspheres were safe and neutral on DC maturation and, when loaded with ASOs against co-stimulatory receptors, able to prevent T1D and reverse hyperglycemia (Phillips et al. 2008). These results suggest that the ASO-microsphere approach is feasible and can be easily applied for other autoimmune diseases.

3.4 Oligonucleotide-conjugated Gold Nanoparticles, Novel Antisense Agents for T1D

Gold nanoparticles (GNs) are widely used as nanocarriers because they are easy to synthesize in uniform and tunable size and can be subjected to a variety of chemical functionalizations. GNs can be densely functionalized with thiol-modified oligonucleotides, to yield particles that are stable under physiological conditions and can bind the complementary nucleic acids with very high affinity. Recently Rosi et al. investigated GNs functionalized with ASOs (GN-ASOs) (Rosi et al. 2006) and reported that they were readily uptaken by various cell types (RAW 264.7, HeLa, NIH-3T3 and MDCK), did not affect cellular morphology and viability, and were not digested by intracellular nucleases up to 48 hours after internalization. GN-ASOs outperformed commercially available cationic liposomal delivery systems in a cell transfection system and their performance could be enhanced by increasing the number of ASOs decorating the GNs. Rink et al. successfully used GN-ASOs *in vivo* to enhance islet resistance after transplantation, and graft survival by targeting specific cytokine and chemokine receptors, pro apoptosis genes and second messenger signaling pathway molecules in β cells. In a recently published work they showed that GN-ASOs were able to fully penetrate into all the cells of purified human islet (Rink et al. 2010). This was a remarkable achievement since pancreatic islets are compact three dimensional collections of cells whose core is usually unreachable by transfection agents such as viral vectors and cationic liposomes. The same group also observed that GN-ASOs did not negatively impact the viability and functionality of β cells, since 1) conjugate-treated islets were able to secrete the same amount of insulin in response to glucose as untreated islets and 2) diabetic mice transplanted with treated islets were able to achieve euglycemia comparable to mice transplanted with untreated islets. Importantly the authors also showed that GNs conjugated with anti-eGFP ASOs were able to reduce the expression of eGFP in islets from transgenic MIP-eGFP mice that express eGFP under the control of the insulin promoter. These results suggest that GN-ASOs are a promising platform for the fabrication of ASO-based tools for modulating gene expression *in vitro* and *in vivo*.

3.5 PEGylated Carbon Nanotubes as Cell-specific Delivery Systems for ASOs

Our laboratory has been experimenting with carbon nanotubes (NTs) functionalized with polyethylene glycol (PEG) in the effort to generate an efficient targeted delivery systems for ASOs. We and others functionalize NTs with PEG chains (Liu et al. 2009; Delogu et al. 2009) by adsorption of phospholipids terminated with activated PEG chains (DSPE-PEG) onto NTs. This is a non-covalent procedure based on ultrasonication, stepwise ultrafiltration and purification by ultrafiltration (PEGylated carbon nanotubes or PNTs) (Fig. 5). PNTs are excellent nanocarriers due to their physicochemical properties (multivalency, nanoscopic dimension,

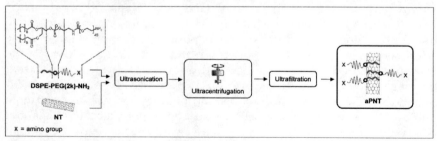

Fig. 5. Fabrication of functionalized PEGylated carbon nanotubes. Pristine (non-functionalized) carbon nanotubes (NTs) were ultrasonicated with phospholipids terminated with amino-functionalized 2-kDa mw PEG chains {1,2-distearoyl-*sn*-glycero-3-phosphoethanolamine-*N*-[amino(polyethylene glycol)$_{2000}$]} [DSPE-PEG(2k)-NH$_2$], fractionated by stepwise ultracentrifugation to isolate short hydrophilic amino-terminated PEGylated NTs and purified by ultrafiltration to remove free phospholipids (aPNT).

Color image of this figure appears in the color plate section at the end of the book.

and solubility under physiological conditions), low toxicity and excellent pharmacokinetic profile (Liu et al. 2009; Schipper et al. 2008). We focused on achieving PNT-based *in vivo* delivery of nanodrugs into T cells, which are key player in the pathogenesis of T1D. In a first report, we functionalized PNTs with monoclonal antibodies specific for receptors expressed on the T cell plasma membrane and a fusogenic polymer (PNT-mAb-p), and intravenously injected them in mice (Cato et al. 2008). The PNT-mAb-p were found to be delivered in the cytoplasm of T cells with high specificity, thus suggesting that the fusogenic polymer aided the nanoparticles to escape from the lysosomal compartments. Recently, we used PNTs to deliver ASOs in Jurkat T cells and achieve knock-down of the protein tyrosine phosphate PTPN22, a pathogenic factor and drug target for T1D (Delogu et al. 2009; Bottini et al. 2006). PNTs were conjugated with anti-PTPN22 phosphorothioate-modified ASOs through a disulfide bond (PNT-ASO) and incubated with T cells in culture. The disulfide bond was intended to be cleaved once the nanoassemblies were into the lysosomal compartments and enable the release of the ASOs from the PNTs. An order of magnitude higher reduction of PTPN22 expression than that produced by electroporation was observed. Our results suggest that PNTs-conjugated ASOs could be exploited to achieve gene knock-down in T cell *in vivo*, a goal which is relevant for therapy of T1D and other T cell mediated human diseases.

4. Applications to Areas of Health and Diseases

Target-specific ASOs are potentially much easier to optimize and handle than chemical inhibitors. These features of ASOs make them appealing as a tool for scientists involved in studies of mechanism of disease and as potential drugs in the therapeutic arena. ASOs have promising applications in the diabetes field. A few ASOs are already performing well in T2D clinical trials. Unfortunately poor spontaneous cell penetration has been a major drawback of ASOs in biomedical studies and therapeutic experimentation. Nanotechnology-based approaches can significantly boost ASO cell permeability. The use of nanocarriers will further expand the range of the *in vivo* applications of ASOs to include for example immune modulation and beta cell protection-regeneration in the T1D field. Nanotech also can further enhance the performance/safety profile of ASOs by enabling tissue targeting (for example to metabolically-relevant organs) or conferring to the ASO the ability to selectively affect specific immune cell subpopulations which play positive or negative role in the pathogenesis of T1D.

5. Key Facts of Drug Delivery

- ASOs represent a powerful therapeutic weapon for human diseases. However one of the major limitations of ASOs is their poor cell-permeability.
- The use of nanocarriers can improve the accumulation of ASOs in organs and tissues relevant to diabetes and their delivery into specific cell subpopulations.
- The multivalent structure of some nanocarriers can be exploited to improve the potency of ASOs by increasing the number of triggered receptors and/or of delivered ASOs per internalization event.
- Nanocarriers with finely tuned size and shape may trigger cell-mediated endocytosis through normally internalization-impaired receptors and therefore expand the spectrum of the receptors that can be exploited for cell internalization.

Dictionary

c-Raf kinase: c-Raf kinase is a mitogen-activated protein (MAP) kinase kinase kinase (MAP3K), encoded in humans by the *RAF1* gene and involved in cellular processes regulating differentiation, proliferation and apoptosis. c-Raf functions downstream of the Ras sub-family of membrane associated GTPases and regulates the activation of the dual specificity protein kinases MEK1 and MEK2, which, in turn, activate the serine/threonine specific protein kinases ERK1 and ERK2. The latter are important player in the control of the expression of genes involved in cell division, differentiation, migration and apoptosis.

Euglycemia: Euglycemia, also called normoglycemia, is the condition of having a normal level of sugar in the blood.

GFP: The green fluorescent protein (GFP) is a small protein (238 aminoacids, approximately 30kDa) that emits in the green range of the visible light if excited by blue light. It was first isolated from the jellyfish *Victoria Aequorea*. The GFP gene has been successfully introduced in cells, bacteria, fungi and in living animals to label specific proteins whose biological processes can be then followed by simple fluorescent microscopy. Several mutants of GFP have been engineered to change its spectral properties. In particular, the enhanced GFP (eGFP) is a one-point mutant that exhibits increased fluorescence and photostability respect to GFP.

Kinase: A kinase is an enzyme that transfers a phosphate group from a donor molecule (for instance ATP) to its substrate. This process is called phosphorylation. The inverse process, the removal of a phosphate group, is carried out by enzymes called phophatases.

PTPN22: The protein tyrosine phosphatase (PTP) N22 is expressed only in white blood cells, where it negatively regulates the T cell receptor (TCR) signalling by cleaving activating phosphate groups from several critical TCR signalling mediators. A genetically encoded variant of PTPN22 carrying a single Arg to Trp aminoacid substitution in position 620 is a strong risk factor for T1D and other autoimmune diseases. Recent data showed that the Trp620 mutant of PTPN22 weaken signalling through the TCR. Small reductions of TCR signalling negatively affect immune tolerance through a variety of mechanisms. Given the increased activity of the autoimmune-predisposing PTPN22 variant, selective inhibition of PTPN22 in carriers of the mutant has been envisaged as an effective etiological therapy for autoimmune diseases in carriers of the Trp620 variant.

Summary

- Antisense oligonucleotides are short (typically 20bp in length) single-stranded deoxyribonucleotide analogues that hybridizes with the complementary mRNA via Watson-Crick base pairing and can block gene expression.
- Antisense oligonucleotides represent a powerful therapeutic weapon for diseases associated with dysregulated protein expression.
- Several chemical modifications of the phosphodiester bond and/or the ribose sugar have been identified which improve the pharmacokinetic profile and increase the potency of the antisense oligonucleotides.
- All the antisense oligonucleotide-based drugs under clinical investigations are administered in "free" form or encapsulated into cationic liposomes.
- A major issue for antisense oligonucleotide-based therapeutics is their poor cell-permeability.
- Nanotechnology-derived particles (nanocarriers) offer the possibility to target delivery of antisense oligonucleotides with high efficiency to their site of action.
- The use of nanocarriers for the targeted delivery of antisense oligonucleotides is still a young field and the choice of the optimal system can be complicated.

Abbreviations

ASOs	:	antisense oligonucleotides
bFGF	:	basic fibroblast growth factor
DC	:	dendritic cell
DTT	:	Dithiothreitol
eGFP	:	enhanced green fluorescent protein

EPR	:	enhanced permeability and retention effect
DME	:	diabetic macular edema
DNA	:	deoxyribonucleic acid
DR	:	diabetic retinopathy
FDA	:	food and drug administration
GN	:	gold nanoparticle
IGF	:	insulin-like growth factor
MAP	:	mitogen-activated protein
mRNA	:	messenger RNA
NTs	:	carbon nanotubes
PEG	:	polyethylene glycol
PNTs	:	PEGylated carbon nanotubes
PPAR	:	peroxisome proliferator-activated receptor
pre-mRNA	:	precursor mRNA
PTPN22	:	protein tyrosine phosphate N22
RES	:	reticulo endothelial system
RME	:	receptor mediated endocytosis
RNA	:	ribonucleic acid
RNase	:	ribonuclease
RT-PCR	:	real time polymerase chain reaction
siRNA	:	small interference RNA
TCR	:	T cell receptor
TLR	:	toll like receptor
T1D	:	type 1 diabetes
T2D	:	type 2 diabetes
VEGF	:	vascular endothelial growth factor

References

Bhat, B.G., H. Younis, J. Herrera, K. Palacio, B. Pascual, G. Hur, B. Jessen, K.M. Ogilvie and P.A. Rejto. 2007. Antisense inhibition of 11betahydroxysteroid dehydrogenase type 1 improves diabetes in a novel cortisone-induced diabetic KK mouse model. Biochem. Biophys. Res. Commun. 365: 740–5.

Bottini, N., T. Vang, F. Cucca and T. Mustelin. 2006. Role of PTPN22 in type 1 diabetes and other autoimmune diseases. Semin. Immunol. 18: 207–213.

Brenner, B.M., W.M. Deen and C.R. Robertson. 1976. Determinants of glomerular filtration rate. Annu. Rev. Physiol. 38: 11–19.

Cato, M.H., F. D'Annibale, D.M. Mills, F. Cerignoli, M.I. Dawson, E. Bergamaschi, N. Bottini, A. Magrini, A. Bergamaschi, N. Rosato, R.C. Rickert, T. Mustelin and M. Bottini. 2008. Cell-type specific and cytoplasmic targeting of PEGylated carbon nanotube-based nanoassemblies. J. Nanosci. Nanotechnol. 8: 2259–69.

Chan, J.H.P., S. Lim and W.S.F. Wong. 2006. Antisense oligonucleotides: from design to therapeutic application. Clin. Exp. Pharmacol. Physiol. 33: 533–40.

Danis, R., M. Criswell, F. Orge, E. Wancewicz, K. Stecker, S. Henry and B. Monia. 2003. Intravitreous anti-raf-1 kinase antisense oligonucleotide as an angioinhibitory agent in porcine preretinal neovascularization. Curr. Eye Res. 26: 45–54.

Delogu, L.G., A. Magrini, A. Bergamaschi, N. Rosato, M.I. Dawson, N. Bottini and M. Bottini. 2009. Conjugation of antisense oligonucleotides to PEGylated carbon nanotubes enables efficient knockdown of PTPN22 in T lymphocytes. Bioconjug. Chem. 20: 427–31.

DeLong, R.K., H. Yoo, S.K. Alahari, M. Fisher, S.M. Short, S.H. Kang, R. Kole, V. Janout, S.L. Regan and R.L. Juliano. 1999. Novel cationic amphiphiles as delivery agents for antisense oligonucleotides. Nucleic Acids Res. 27: 3334–41.

Grillone, L.R., and S.P. Henry. Potential therapeutics: applications of antisense oligonucleotides in ophthalmology. pp. 585–600. *In*: S.T. Crooke [eds.] 2008. Antisense drug technology: principles, strategies, and applications. CRC Press Taylor & Francis Group, New York, USA.

Hnik, P., D.S. Boyer, L.R. Grillone, J.G. Clement, S.P. Henry and E.A. Green. 2009. Antisense oligonucleotide therapy in diabetic retinopathy. J. Diabetes Sci. Technol. 3: 924–30.

Jia, N., Q. Lian, H. Shen, C. Wang, X. Li and Z. Yang. 2007. Intracellular delivery of quantum dots tagged antisense oligodeoxynucleotides by functionalized multiwalled carbon nanotubes. Nano Lett. 7: 2976–80.

Jiang, W., B.Y.S. Kim, J.Y. Rutka and W.C.W. Chan. 2008. Nanoparticle-Mediated Cellular Response Is Size-Dependent. Nat. Nanotechnol. 3: 145–150.

Juliano, R.L. Biological barriers to nanocarrier-mediated delivery of therapeutic and imaging agents. pp. 263–278. *In*: C.M. Niemeyer, and C.A. Mirkin. [eds.] 2007. Nanobiotechnology II. Wiley-VCH, Weinheim, Germany.

Juliano, R.L., M.R. Alam, V. Dixit and H. Kang. 2008. Mechanisms and strategies for effective delivery of antisense and siRNA oligonucleotides. Nucleic Acids Res. 36: 4158–71.

Kang, H., M.R. Alam, V. Dixit, M. Fisher and R.L. Juliano. 2008. Cellular delivery and biological activity of antisense oligonucleotides conjugated to a targeted protein carrier. Bioconjug. Chem. 19: 2182–8.

Khan, A., M. Benboubetra, P.Z. Sayyed, K.W. Ng, S. Fox, G. Beck, I.F. Benter and S. Akhtar. 2004. Sustained polymeric delivery of gene silencing antisense ODNs, siRNA, DNAzymes and ribozymes: *in vitro* and *in vivo* studies. Drug Target 12: 393–404.

Koizumi, M., M. Takagi-Sato, R. Okuyama, K. Araki, W. Sun and D. Nakai. 2007. *In vivo* antisense activity of ENA oligonucleotides targeting PTP1B mRNA in comparison of that of 2′-MOE-modified oligonucleotides. Nucleic Acids Symp. Ser. 51: 111–2.

Kostarelos, K., L. Lacerda, G. Pastorin, W. Wu, S. Wieckowski, J. Luangsivilay, S. Godefroy, D. Pantarotto, J.P. Briand, S. Muller, M. Prato and A. Bianco. 2007. Cellular uptake of functionalized carbon nanotubes is independent of functional group and cell type. Nat. Nanotechnol. 2: 108–13.

Liang, Y., M.C. Osborne, B.P. Monia, S. Bhanot, L.M. Watts, P. She, S.O. DeCarlo, X. Chen and K. Demarest. 2005. Antisense oligonucleotides targeted against glucocorticoid receptor reduce hepatic glucose production and ameliorate hyperglycemia in diabetic mice. Metabolism 54: 848–55.

Liu, Z., S. Tabakman, K. Welsher and H. Dai. 2009. Carbon Nanotubes in Biology and Medicine: *In vitro* and *in vivo* Detection, Imaging and Drug Delivery. Nano Res. 2: 85–120.

Mosser, D.M., and J.P. Edwards. 2008. Exploring the full spectrum of macrophage activation. Nat. Rev. Immunol. 8: 958–69.

Phillips, B., K. Nylander, J. Harnaha, J. Machen, R. Lakomy, A. Styche, K. Gillis, L. Brown, D. Lafreniere, M. Gallo, J. Knox, K. Hogeland, M. Trucco and N. Giannoukakis. 2008. A microsphere-based vaccine prevents and reverses new-onset autoimmune diabetes. Diabetes 57: 1544–55.

Rink, J.S., K.M. McMahon, X. Chen, C.A. Mirkin, C.S. Thaxton and D.B. Kaufman. 2010. Transfection of pancreatic islets using polyvalent DNA-functionalized gold nanoparticles. Surgery 148: 335–45.

Rosi, N.L., D.A. Giljohann, C.S. Thaxton, A.K. Lytton-Jean, M.S. Han and C.A. Mirkin. 2006. Oligonucleotide-modified gold nanoparticles for intracellular gene regulation. Science 312: 1027–30.

Scherphof, G.L. *In vivo* behavior of liposomes: Interactions with the mononuclear phagocyte system and implications for drug targeting. pp. 285–300. *In*: R.L. Juliano. [eds.] 1991. Targeted Drug Delivery. Springer, New York, USA.

Schipper, M.L., N. Nakayama-Ratchford, C.R. Davis, N.W. Kam, P. Chu, Z. Liu, X. Sun, H. Dai and S.S. Gambhir. 2008. A pilot toxicology study of single-walled carbon nanotubes in a small sample of mice. Nat. Nanotechnol. 3: 216–21.

Swarbrick, M.M., P.J. Havel, A.A. Levin, A.A. Bremer, K.L. Stanhope, M. Butler, S.L. Booten, J.L. Graham, R.A. McKay, S.F. Murray, L.M. Watts, B.P. Monia and S. Bhanot. 2009. Inhibition of protein tyrosine phosphatase-1B with antisense oligonucleotides improves insulin sensitivity and increases adiponectin concentrations in monkeys. Endocrinology 150: 1670–9.

Tachibana, K., T. Katayama, C. Ueda, M. Sumitomo, M. Tagami, K. Ishimoto, D. Yamasaki, T. Tanaka, T. Hamakubo, J. Sakai, T. Kodama, S. Obika, T. Imanishi and T. Doi. 2007. Antisense activity of 2',4'-BNA targeted to PPAR gamma in THP-1 and HCT116 cells. Nucleic Acids Symp. Ser. 51: 441–2.

Yoo, H., and R.L. Juliano. 2000. Enhanced delivery of antisense oligonucleotides with fluorophore-conjugated PAMAM dendrimers. Nucleic Acids Res. 28: 4225–31.

Waring, J.F., R. Ciurlionis, J.E. Clampit, S. Morgan, R.J. Gum, R.A. Jolly, P. Kroeger, L. Frost, J. Trevillyan, B.A. Zinker, M. Jirousek, R.G. Ulrich and C.M. Rondinone. 2003. PTP1B antisense-treated mice show regulation of genes involved in lipogenesis in liver and fat. Mol. Cell Endocrinol. 203: 155–68.

Section 2: Glucose

5

Carbon Nanotubes Coupled to siRNA Generates Efficient Transfection and is a Tool for Examining Glucose Uptake in Skeletal Muscle

Johanna T. Lanner[1,a,]* and Håkan Westerblad[1,b]

ABSTRACT

Insulin resistance and type 2 diabetes reaches epidemic proportions all around the world today. Multiple genes together with sedentary lifestyle and obesity are factors that participate in the development of type 2 diabetes. Better understanding of the pathological processes in obesity, insulin resistance and type 2 diabetes together with improved pharmacological treatments are of major importance. Carbon nanotubes (CNT) consist of pure carbon and belong to the family of fullerenes. They are described as hollow cylinders formed by rolling single (SWNT) or multiple

---

[1]Karolinska Institutet, Department of Physiology and Pharmacology, 171 77 Sweden.
[a]E-mail: johanna.lanner@gmail.com
[b]E-mail: hakan.westerblad@ki.se
*Corresponding author

List of abbreviations after the text.

(MWNT) layers of graphene layers into cylinders. SWNTs have shown to be a unique tool for effectively delivery of small interfering RNA (siRNA), DNA, small peptides and protein into living cells both *in vivo* and *in vitro* with no evidence of nonspecific effects or cell damage. siRNA induces RNA interference to inhibit expression of target proteins in a highly specific manner, which makes siRNA a powerful therapeutical tool for treatment of a range of diseases, including genetic diseases, virus infections and cancer. Here we discuss how CNTs coupled to siRNA can be used as a tool to manipulate the expression of ion channels and hence Ca^{2+} fluxes important for glucose uptake and its potential use in therapeutic treatment.

INTRODUCTION

We are currently witnessing a world wide epidemic of people with type 2 diabetes (Zimmet et al. 2001; Wild et al. 2004; Cheng 2005; Hossain et al. 2007; Longo-Mbenza et al. 2010). Multiple genes together with environmental and behavioral factors, such as sedentary lifestyle and obesity, are etiological factors in type 2 diabetes (Zimmet et al. 2001; Cheng 2005). One major component in the progression and pathogenesis of type 2 diabetes is insulin resistance (Gu et al. 1998; Shulman 2000). Even with new therapies, type 2 diabetes is still associated with high morbidity and mortality, and heart disease is the leading cause of death (Abdul-Ghani & DeFronzo 2010). As a result, this causes a significant financial burden on the healthcare system (Jönsson 1998; ADA 2008). Thus, a better understanding of the pathological processes in obesity and type 2 diabetes is of major importance. Here we discuss how carbon nanotubes can be used as a tool to manipulate the expression of ion channels and hence Ca^{2+} fluxes important for glucose uptake and its potential use in therapeutic treatment.

Insulin-mediated Glucose Uptake in Skeletal Muscle

Skeletal muscle is the major site for insulin-mediated glucose disposal (DeFronzo et al. 1985). Both insulin and exercise (contractions) are able to stimulate glucose transporter (GLUT) 4 recruitment to the plasma membrane in skeletal muscle, although this occurs via different signaling pathways (Thong et al. 2005; Lanner et al. 2008). The insulin signaling pathway starts with activation of the insulin receptor (IR), which belongs to the tyrosine kinase receptor family. Upon activation, IRs undergo autophosphorylation and starts a cascade of intracellular signaling events that eventually results in glucose uptake (White 2003; Huang & Czech 2007; Lanner et al. 2008; Bryant & Gould 2011). GLUT4 is stored in intracellular vesicles under basal conditions and upon stimulation these vesicles translocate to the plasma membrane, where GLUT4 inserts in the membrane and glucose uptake is facilitated (Thong et al. 2005; Huang & Czech 2007; Lanner et al. 2008; Bryant & Gould 2011). The insulin-stimulated docking and fusion process of GLUT4-containing vesicles also involves a group of proteins known as SNARE proteins (Martin et al. 1996; Cheatham 2000; Foster et al. 2000; Khan et al. 2001; Bao et al. 2008; Schwenk et al. 2010; Bryant & Gould 2011) (Fig. 1). This process is reminiscent to the Ca^{2+}-dependent vesicle fusion in synaptic terminals (Burgoyne & Morgan 1993; Lou et al. 2005; Schneggenburger & Neher 2005). Documented members of this

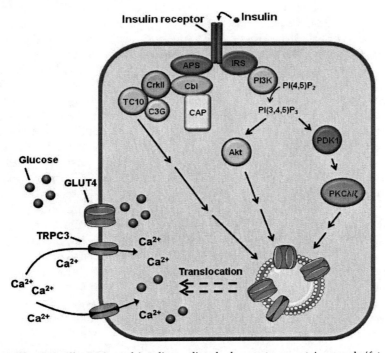

Fig. 1. Schematic illustration of insulin-mediated glucose transport in muscle/fat cells. Following insulin binding to its receptor, a series of signaling reactions initiate translocation of GLUT4-containing vesicles to the cell surface. At the same time, extracellular Ca^{2+} enters the cell via TRPC3, resulting in localized increases that facilitate docking/fusion/insertion of GLUT4 in the membrane, where it accelerates glucose transport. Single arrow reflects a direct effect. Multiple arrows in a pathway reflect two or more steps. For the sake of simplicity several established steps have been omitted. (Permission to publish; adapted from Lanner et al. Curr Opin Pharmacol 2008).

Color image of this figure appears in the color plate section at the end of the book.

large SNARE complex are; VAMP-2 (also called synaptobrevin), Munc-18, Synip, and Syntaxin-4 (Martin et al. 1996; Khan et al. 2001; Okada et al. 2007; Bao et al. 2008; Bryant & Gould 2011). Canonical transient receptor potential 3 (TRPC3) channels also translocates to the plasma membrane in an agonist-induced manner (Singh et al. 2004; Smyth et al. 2006) and are found in larger protein complexes that contain proteins that are also found in GLUT4-containing vesicles (e.g., VAMP2 and synaptophysin) (Singh et al. 2004). Accordingly, TRPC3 has been observed to translocate in response to insulin in cardiomyocytes of normal but not of insulin resistant mice, hence similar to the translocation of GLUT4 (Fauconnier et al. 2007).

Insulin, TRPC-mediated Ca^{2+}-influx and Targeted Exocytosis of GLUT4

A role of Ca^{2+} in insulin-mediated glucose uptake is not widely recognized (Klip et al. 1984; Cheung et al. 1987; Klip & Ramlal 1987; Wallberg-Henriksson et al. 1988). However, in line with earlier research (Schudt et al. 1976; Clausen 1980; Bihler et al. 1986; Bruton et al. 1999), we have shown that Ca^{2+} influx plays an important role in insulin-mediated glucose uptake in skeletal muscle (Lanner et al. 2006; Lanner et al. 2009). By manipulating the Ca^{2+} influx, we were able to alter the insulin-mediated glucose uptake: decreased Ca^{2+} influx using pharmacological agents (2-APB, MDL, or low extracellular $[Ca^{2+}]$) resulted in decreased insulin-stimulated glucose uptake in adult skeletal muscle fibers; increased Ca^{2+} influx stimulated with OAG or after removal of 2-APB resulted in increased insulin-stimulated glucose uptake (Lanner et al. 2006; Lanner et al. 2009). Furthermore, we were able to identify TRPC3 as being responsible for the insulin-mediated Ca^{2+}-influx by using the novel technique of carbon nanotubes (CNT) coupled to siRNA (CNT-siRNA) to transfect skeletal muscle fibers (Lanner et al. 2009). Transfection for 48 hours with CNT-siRNA resulted in 40% decrease in TRPC3 protein expression compared to control. This was accompanied by an ~80% decrease in insulin-mediated glucose uptake in transfected fibers. TRPC3 also co-immunoprecipitate and co-localize with GLUT4 in the transverse tubular system of muscle fiber, which is considered the major site of glucose uptake (Marette et al. 1992; Munoz et al. 1995; Lauritzen et al. 2006).Taken together, we propose that Ca^{2+} influx via TRPC3 modulates a late step of the insulin-dependent recruitment of GLU4-containing vesicles to the plasma membrane, which is crucial for optimal insulin-mediated glucose uptake in skeletal muscles (Lanner et al. 2009). Thus, TRPC3 is a potential therapeutic target for treatment of type 2 diabetes and mapping TRPC3 had not been possible without the efficiency of CNT-siRNA.

Carbon Nanotube Structure and Properties

CNTs consist of pure carbon and belong to the family of fullerenes. They are described as hollow cylinders formed by rolling single (single-walled CNT (SWNT), diameter <1nm) or multiple (multiple-walled CNT (MWNT), total outer diameter <10nm) layers of graphene into cylinders (Dresselhaus et al. 2004; Foldvari & Bagonluri 2008). In our skeletal muscle only SWNTs have been used (Fig. 2) (Lanner et al. 2009). SWNT has been shown to effectively transport biological molecules (including siRNA, DNA, small peptides and protein) into living cells both *in vivo* and *in vitro* (Kam & Dai 2005; Kam et al. 2005a; Kam et al. 2005b; Dumortier et al. 2006; Klumpp et al. 2006; Kostarelos et al. 2007; Wu et al. 2008). CNTs appear to be taken up by cells via an energy-dependent endocytosis mechanism since they are taken up by the cells at 37°C but not at 4°C (Mukherjee et al. 1997; Kam & Dai 2005; Kam et al. 2005a). The cytotoxicity effect of CNT uptake into cells has been thoroughly investigated and functionalized

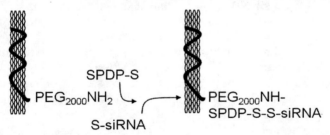

Fig. 2. Illustration of carbon nanotubes and siRNA conjugation. CNT-PEG2000 complex reacts with SPDP-S. Thiol-modified siRNA is then conjugated to the CNT-PEG2000-SPDP-S complex. (Permission from the publisher to reproduce image; Lanner et al. FASEB Journal, 2009).

SWNT appears to have very modest deleterious effects (Dumortier et al. 2006; Singh et al. 2006; Pulskamp et al. 2007; Schipper et al. 2008; Yang et al. 2008). However, if SWNT are not functionalized, they can cause toxicity effects and inflammatory responses, which in part is a consequence of SWNT forming aggregate (agglomeration) (Lam et al. 2004; Yokoyama et al. 2004; Lam et al. 2006; Singh et al. 2006; Belyanskaya et al. 2009). Since CNTs are hydrophobic by nature, they are practically insoluble and hence can accumulate in cells and organs with dangerous effects (Lam et al. 2004; Yokoyama et al. 2004; Lam et al. 2006; Singh et al. 2006; Belyanskaya et al. 2009). Therefore several methods to overcome hydrophobicity have been developed in order to functionalize CNTs. These include covalent binding of organic molecules to CNTs and non-covalent wrapping of CNTs by surfactants, synthetic molecules or biopolymers (Zheng et al. 2003; Tasis et al. 2006; Wu et al. 2006; Krajcik et al. 2008; Schipper et al. 2008; Bekyarova et al. 2010). MWNTs are more prone to undergo agglomeratation and hence have shown to have more deleterious effects, e.g., trigger apoptosis in human T cells *in vitro* and cause inflammation responses in mice *in vivo* (Bottini et al. 2006; Zhu et al. 2007; Qu et al. 2009).

siRNA and Transportation into Cells

RNAi is a post-translational gene-silencing process that targets messenger RNA resulting in inhibition of protein expression (Fire et al. 1998; Elbashir et al. 2001a; Elbashir et al. 2001b). siRNA consists of 21–25 nucleotides that induce RNA interference to inhibit expression of target proteins in a highly specific manner (Fire et al. 1998; Caplen et al. 2001; Elbashir et al. 2001a). These characteristics have made siRNA a potential powerful therapeutic tool for treatment of a range of diseases, including genetic diseases, virus infections and cancer (Wilson & Richardson 2006; Yang et al. 2006; Zhang et al. 2006; Smith et al. 2007; Wang et al. 2009; Phalon et al. 2010). We used siRNA to show that it is Ca^{2+} influx via TRPC3 that is important for the insulin-mediated glucose uptake in skeletal muscle fibers (Lanner et al. 2009). One major challenge with siRNA treatment has been poor efficiency due to limited cellular uptake *in vitro* and *in vivo* (Wilson & Richardson 2006; Wang et al. 2009; Phalon et al. 2010). However, SWNT coupled to siRNA has been highly successful for delivery of siRNA into cells with no evidence of nonspecific effects or cell damage (Kam et al. 2005a; Yang et al. 2006; Zhang et al. 2006; Wang et al. 2008; Lanner et al. 2009; Ladeira et al. 2010). Relative to other techniques, SWNT-siRNA is easier to use (Sapru et al. 2002) and has less biosafety concerns (Schipper et al. 2008; Yang et al. 2008) than viral vectors, and is more effective than lipofectamine (Pramfalk et al. 2004; Kam et al. 2005b). CNT efficiency is most evident in hard-to-transfect cell types, e.g., adult neurons, cardiac and skeletal muscle cells. Our rational for using adult mature skeletal muscle cells instead of immature muscle cells (myotubes) or cell lines is that Ca^{2+} handling is markedly different between these cell types and hence also the physiological responses. For example, intracellular Ca^{2+} transients in response to electrical stimulation have a half-duration of ~5 ms in adult mouse skeletal muscle cells, which is about 50 times faster than the rate obtained in myotubes (Baylor & Hollingworth 2003; Kubis et al. 2003). We reported that transfection by SWNT-siRNA is highly effective in adult muscle fibers, achieving a 40% reduction in TRPC3 protein expression with no evidence of nonspecific effects or cell damage (Fig. 3) (Lanner et al. 2009). Parallel to our findings in muscle fibers (Lanner et al. 2009), Ladeira and colleagues reported 96% silencing of the InsP3R-II gene by SWNT transfection in cardiomyocytes with no apparent toxicity or nonspecific gene silencing, which was 5-10-fold greater than a lipid-based

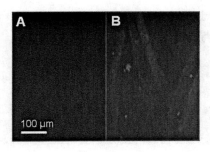

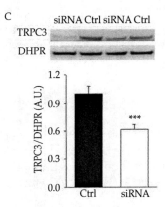

Fig. 3. Carbon nanotubes successfully transfected adult skeletal muscle fibers. (A) Non-transfected FDB fibers exposed to only SWNT demonstrating little autofluorescence. (B) Fibers transfected with SWNT-siRNA constructs labelled with Cy5; all fibers in the field of view show fluorescence indicating successful entry of the labelled SWNT-siRNA. Images were taken 48 hours post transfection. (C) Representative western blots of TRPC3 expression obtained after 48h incubation without (Ctrl) or with TRPC3-siRNA; the voltage-gated Ca^{2+} channel (dihydropyridin receptor, DHPR) was used as loading control. (Permission from the publisher to reproduce image; Lanner et al. FASEB Journal 2009).

Color image of this figure appears in the color plate section at the end of the book.

gene transfer system (Ladeira et al. 2010). Thus, SWNT-siRNA appears to possess the critical characteristics for successful target delivery into cells *in vitro* and *in vivo*, which makes it a high impact combination for future therapeutical interventions.

Practical example of transfection of skeletal muscle by functionalized SWNT coupled to siRNA

Here follows a description of how we successfully transfected isolated muscle fibers with functionalized SWNT coupled to siRNA (Lanner et al. 2009). HiPco® single walled CNTs (Unidym) were sonicated for 90 min in a 0.6% solution of 1,2-distearoyl-sn-glycero-3-phosphoethanolamine-

N-[amino(polyethylene glycol)2000] (PEG$_{2000}$, Avanti Polar Lipids). The suspension was then filtered through a 0.1 μm Millipore filter and resuspended in a 50 mM phosphate salt buffer. This was centrifuged at 25,000 g for 5 hours to sediment large nanotube bundles. The supernatant was collected and centrifuged again at 25,000 g for a further 5 hours. CNTs were incubated for 1 hour with 2.5 mg sulfosuccinimidyl 6-(3'-[2-pyridyldithio]-propionamido)hexanoate (Pierce) to put reactive sulfhydryl cross-linkers on the CNT surface. Thereafter, siRNA with an extra thiol group attached to the 3' end was added to the CNT and left overnight to allow chemical conjugation to occur. Intact adult *flexor digitorium brevis* (FDB) fibers were isolated and plated on 35 mm laminin-coated glass-bottomed Petri dishes and transfected with siRNA by adding 5 μl of the conjugated CNT-siRNA to the Petri dishes containing 3 ml minimum essential medium (MEM). In control experiments, 5 μl of CNT were added to dishes containing 3 ml MEM. It has previously been shown that once the CNT enters the cell, the disulfide bond is broken making the siRNA freely available (Kam et al. 2005b). The siRNA target, in our case TRPC3, was a double strand with a sense sequence and an antisense sequence (Lanner et al. 2009). The specificity of the siRNA sequences was checked against all known mammalian nucleotide sequences using a BLAST search. A randomly scrambled siRNA sequence (nonsense) (Lanner et al. 2009) was used to check for specificity of TRPC3 siRNA. To check that the constructs entered the cells were siRNAs labeled with a fluorescent tag (Cy3 or Cy5) attached to the 5' end. Efficacy of uptake of the conjugated CNT-siRNA was checked by measuring the Cy3 or Cy5 fluorescence inside the cells 48 hours post transfection and western blots were used to determine the levels of TRPC3 protein expression 48 hours post transfection with or without CNT-siRNA targeted against TRPC3 (Lanner et al. 2009).

Applications to Areas of Health and Disease

The high efficiency of SWNT uptake into cells with no evidence of nonspecific effects or cell damage makes it a potential powerful tool for delivery of pharmaceutical agents in disease treatment (Kam et al. 2005a; Yang et al. 2006; Zhang et al. 2006; Wang et al. 2008; Lanner et al. 2009; Ladeira et al. 2010). SWNT coupled to siRNA appears to possess the critical characteristics for successful target delivery into cells *in vitro* and *in vivo* and has already been tested on a range of diseases with beneficial effects, including genetic diseases, virus infections and cancer (Wilson & Richardson, 2006; Yang et al. 2006; Zhang et al. 2006; Smith et al. 2007; Wang et al. 2009; Phalon et al. 2010). We used SWNT-siRNA in skeletal muscle fibers to show that it is Ca^{2+} influx via TRPC3 is important for the insulin-mediated glucose uptake (Lanner et al. 2009). Thus, SWNT coupled to an agent that modulates TRPC3- mediated Ca^{2+} influx could be a future therapeutical intervention for treatment of insulin resistance and type 2 diabetes.

Key Facts

Glucose

- Glucose is a simple sugar (monosaccharide) and serves as a major energy source for the body. Continuous supply of glucose is essential for proper cell function and in humans the glucose level is maintained within a narrow range close to 5 mM. Altered glucose homeostasis leads to severe complications, including cardiac and renal failure, and damage to the nervous system, all observed in diabetes mellitus.

Type II diabetes

- The term diabetes mellitus describes a metabolic disorder characterized by chronic hyperglycemia with disturbances in carbohydrate, fat and protein metabolism, which result from defects in insulin secretion, insulin action, or both (Alberti & Zimmet 1998). The most common form of diabetes is type 2 diabetes, which accounts for ~ 90% of the global cases (Zimmet et al. 2001). Multiple genes together with environmental and behavioral factors, such as sedentary lifestyle and obesity, are etiological factors in type 2 diabetes (Zimmet et al. 2001).

Insulin resistance

- Insulin resistance means that cells show an inability to respond appropriately to normal circulating levels of insulin, which results in impaired glucose uptake and elevated plasma glucose concentration (Shulman 2000; Abdul-Ghani & DeFronzo 2010). To sustain insulin action and physiological blood glucose levels in insulin resistance, the β-cells have to produce and secrete more insulin (DeFronzo 1997). However, this compensatory process eventually fails, which results in an inability to maintain normal glucose levels (DeFronzo 1997; Porte & Kahn 2001).

Glucose transporters

- The transport of glucose into cells over the plasma membrane is mediated by members of the glucose transporter (GLUT) family and 13 members of this family are known today (Huang & Czech 2007). Basal glucose uptake occurs via the constitutively active GLUT1, which is expressed in many cell types, including muscle (Mueckler

1994). GLUT4 is expressed exclusively in striated muscle (i.e., skeletal and cardiac muscle) and adipose tissue (Mueckler 1994; Huang & Czech 2007; Bryant & Gould 2011). The translocation and insertion of GLUT4 into the plasma membrane is impaired in insulin-resistant conditions and type 2 diabetes, which results in impaired glucose uptake (Zisman et al. 2000; Bryant & Gould 2011).

Summary Points

- CNTs are a unique tool for effectively delivery of siRNA, DNA, small peptides and protein into living cells both *in vivo* and *in vitro*. This feature is extra beneficial when working with cell types that are known to be hard to transfect, e.g., adult skeletal muscle fibers, neurons.
- CNTs consist of pure carbon and belong to the family of fullerenes. They are described as hollow cylinders formed by rolling single or multiple layers of graphene into cylinders.
- The usage of siRNA to decrease and/or inhibit the expression of target proteins in a highly specific manner is of great interest because of its potential in treatment of a wide variety of severe diseases and disorders, e.g., cancer and virus diseases such as hepatitis.
- A major challenge with siRNA for therapeutical purposes has been to find methods for efficient delivery into cells with low toxicity. Here we refer to several reports showing that this problem might be overcome by using SWNT coupled to siRNA, which appears to be a proficient method to transfect cells with moderate side-effects.
- SWNT-siRNA is also a tremendous tool in finding new therapeutical targets; we used SWNT-siRNA when mapping Ca^{2+} influx via TRPC3 to play an important role in insulin-mediated glucose uptake in skeletal muscle and hence TRPC3 provides a new target for treatment of type 2 diabetes.

Definitions

Ca^{2+}: Ca^{2+} is a multifaceted messenger and apart from being crucial for muscle contraction, it also regulates other cellular events such as fertilization, synaptic transmission, gene transcription, hormonal signaling, and cell death. Duration, amplitude, and spatial distribution of Ca^{2+} play a significant role in facilitating these diverse effects. The Ca^{2+}-signaling time scale spans from milliseconds, e.g., in exocytosis of neurotransmitters and in skeletal muscle contractions, up to minutes and hours as observed in fertilization and immune responses.

TRPC: The TRPC channels are a subfamily of seven ion channels (designated 1–7) that belong to the large TRP super family. TRPC channels are all Ca^{2+} permeable non-selective cation channels.

Fullerene: A fullerene is a molecule composed solely of carbon in form of a hollow tube, sphere, or ellipsoid.

RNA interference (RNAi): RNAi is a biological process where the presence of certain fragments of double-stranded RNA interferes with the expression of a particular gene that shares a homologous sequence with the double-stranded RNA. This results in that a particular gene is not expressed and consequently the expression of the corresponding protein is inhibited.

siRNA: Small interfering RNA (siRNA), sometimes called short interfering RNA or silencing RNA, is a class of double-stranded RNA that consists of 21–25 nucleotides that induce RNA interference to inhibit expression of target proteins in a highly specific manner.

Abbreviations

GLUT	:	glucose transporter
TRP	:	transient receptor potential
TRPC	:	canonical transient receptor potential
CNT	:	carbon nanotube
SWNT	:	single-walled carbon nanotube
MWNT	:	multi-walled carbon nanotube
RNAi	:	RNA interference
siRNA	:	small interfering RNA
MEM	:	minimum essential medium
FDB	:	flexor digitorium brevis
2-APB	:	2-aminoethoxydiphenyl borate
MDL	:	cis-N-(-2-phenykcyclopentyl)azacyclotridec-1-en-2-amine
OAG	:	1-oleyl-2-acetyl-sn-glycerol
IRS	:	insulin receptor substrate
PI3K	:	phosphoinositide 3-kinase
$PI(4,5)P_2$	:	phosphatidylinositol 4,5-bisphosphate
$PI(3,4,5)P_3$	:	phosphatidylinositol 3,4,5-trisphosphate
Akt/PKB	:	protein kinase B
PDK	:	phosphoinositide-dependent kinase
PKC	:	protein kinase C
APS	:	adaptor protein containing pleckstrin homology and SH2 domains
CAP	:	c-Cbl-associated protein

CrkII   :   CT10-related kinase II
C3G   :   guanylnucleotide exchange factor

Acknowledgement

JT Lanner was supported by a postdoc fellowship from The Swedish Research Council.

Reference

Abdul-Ghani, M.A., and R.A. DeFronzo. 2010. Pathogenesis of insulin resistance in skeletal muscle. J. Biomed Biotechnol. 2010, 19.

Alberti, K.G., and P.Z. Zimmet. 1998. Definition, diagnosis and classification of diabetes mellitus and its complications. Part 1: diagnosis and classification of diabetes mellitus provisional report of a WHO consultation. Diabet. Med. 15: 539–553.

Association Diabetes Association. 2008. Economic costs of diabetes in the U.S. in 2007. Diabetes Care 31: 596–615.

Bao, Y., J.A. Lopez, D.E. James and W. Hunziker. 2008. Snapin interacts with the Exo70 subunit of the exocyst and modulates GLUT4 trafficking. J. Biol. Chem. 283: 324–331.

Baylor, S.M., and S. Hollingworth. 2003. Sarcoplasmic reticulum calcium release compared in slow-twitch and fast-twitch fibres of mouse muscle. J. Physiol. 551: 125–138.

Bekyarova, E., I. Kalinina, X. Sun, T. Shastry, K. Worsley, X. Chi, M.E. Itkis and R.C. Haddon. 2010. Chemically engineered single-walled carbon nanotube materials for the electronic detection of hydrogen chloride. Adv. Mater. 22: 848–852.

Belyanskaya, L., S. Weigel, C. Hirsch, U. Tobler, H.F. Krug and P. Wick. 2009. Effects of carbon nanotubes on primary neurons and glial cells. Neurotoxicology 30: 702–711.

Bihler, I., P. Charles and P.C. Sawh. 1986. Effects of strontium on calcium-dependent hexose transport in muscle. Can. J. Physiol. Pharmacol. 64: 176–179.

Bottini, M., S. Bruckner, K. Nika, N. Bottini, S. Bellucci, A. Magrini, A. Bergamaschi and T. Mustelin. 2006. Multi-walled carbon nanotubes induce T lymphocyte apoptosis. Toxicology Letters 160: 121–126.

Bruton, J.D., A. Katz and H. Westerblad. 1999. Insulin increases near-membrane but not global Ca^{2+} in isolated skeletal muscle. PNAS 96: 3281–3286.

Bryant, N.J. and G.W. Gould. 2011. SNARE proteins underpin insulin-regulated GLUT4 traffic. Traffic. [Epub ahead of print].

Burgoyne, R.D., and A. Morgan. 1993. Regulated exocytosis. Biochem. J. 293: 305–316.

Caplen, N.J., S. Parrish, F. Imani, A. Fire and R.A. Morgan. 2001. Specific inhibition of gene expression by small double-stranded RNAs in invertebrate and vertebrate systems. PNAS 98: 9742–9747.

Cheatham, B. 2000. GLUT4 and company: SNAREing roles in insulin-regulated glucose uptake. Trends Endocrinol. Metab. 11: 356–361.

Cheng, D. 2005. Prevalence, predisposition and prevention of type II diabetes. Nutr Metab. 2: 29.
Cheung, J.Y., J.M. Constantine and J.V. Bonventre. 1987. Cytosolic free calcium concentration and glucose transport in isolated cardiac myocytes. Am. J. Physiol. 252, C163–C172.
Clausen, T. 1980. The role of calcium in the activation of the glucose transport system. Cell Calcium 1: 311–325.
DeFronzo, R.A. 1997. Pathogenesis of type 2 diabetes mellitus: metabolic and molecular implications for identifying diabetes genes. Diabetes Rev. 5: 117–269.
DeFronzo, R.A., R. Gunnarsson, O. Björkman, M. Olsson and J. Wahren. 1985. Effects of insulin on peripheral and splanchnic glucose metabolism in noninsulin-dependent (type II) diabetes mellitus. J. Clin. Invest. 76: 149–155.
Dresselhaus, M.S., G. Dresselhaus, J.C. Charlier and E. Hernandez. 2004. Electronic, thermal and mechanical properties of carbon nanotubes. Philos Transact A Math Phys. Eng. Sci. 362: 2065–2098.
Dumortier, H., S. Lacotte, G. Pastorin, R. Marega, W. Wu, D. Bonifazi, J.P. Briand, M. Prato, S. Muller and A. Bianco. 2006. Functionalized carbon nanotubes are non-cytotoxic and preserve the functionality of primary immune cells. Nano Lett 6: 1522–1528.
Elbashir, S.M., J. Harborth, W. Lendeckel, A. Yalcin, K. Weber and T. Tuschl. 2001a. Duplexes of 21-nucleotide RNAs mediate RNA interference in cultured mammalian cells. Nature 411: 494–498.
Elbashir, S.M., W. Lendeckel and T. Tuschl. 2001b. RNA interference is mediated by 21- and 22-nucleotide RNAs. Genes Dev. 15: 188–200.
Fauconnier, J., J.T. Lanner, A. Sultan, S.J. Zhang, A. Katz, J.D. Bruton and H. Westerblad. 2007. Insulin potentiates TRPC3-mediated cation currents in normal but not in insulin-resistant mouse cardiomyocytes. Cardiovasc. Res. 73: 376–385.
Fire, A., S. Xu, M.K. Montgomery, S.A. Kostas, S.E. Driver and C.C. Mello. 1998. Potent and specific genetic interference by double-stranded RNA in Caenorhabditis elegans. Nature 391: 806–811.
Foldvari, M., and M. Bagonluri. 2008. Carbon nanotubes as functional excipients for nanomedicines: I. pharmaceutical properties. Nanomedicine 4: 173–182.
Foster, L.J., M.L. Weir, D.Y. Lim, Z. Liu, W.S. Trimble and A. Klip 2000. A functional role for VAP-33 in insulin-stimulated GLUT4 traffic. Traffic 1: 512–521.
Gu, K., C.C. Cowie and M.I. Harris. 1998. Mortality in adults with and without diabetes in a national cohort of the U.S. population, 1971–1993. Diabetes Care 21: 1138–1145.
Hossain, P., E.l. Kawar and M. Nahas. 2007. Obesity and diabetes in the developing world—a growing challenge. N. Engl. J. Med. 356: 213–215.
Huang S., and M.P. Czech. 2007. The GLUT4 glucose transporter. Cell Metab. 5: 237–252.
Jönsson, B. 1998. The economic impact of diabetes. Diabetes Care 21: C7-10.
Kam, N.W., and Dai H. 2005. Carbon nanotubes as intracellular protein transporters: generality and biological functionality. J. Am. Chem. Soc. 127: 6021–6026.

Kam, N.W.S., M. O'Connell, J.A. Wisdom and H. Dai. 2005a. Carbon nanotubes as multifunctional biological transporters and near-infrared agents for selective cancer cell destruction. PNAS 102: 11600–11605.

Kam, N.W.S., Z. Liu and H. Dai. 2005b. Functionalization of carbon nanotubes via cleavable disulfide bonds for efficient intracellular delivery of siRNA and potent gene silencing. J. Am. Chem. Soc. 127: 12492–12493.

Khan, A.H., D.C. Thurmond, C. Yang, B.P. Ceresa, C.D. Sigmund and J.E. Pessin. 2001. Munc18c regulates insulin-stimulated glut4 translocation to the transverse tubules in skeletal muscle. J. Biol. Chem. 276: 4063–4069.

Klip, A., G. Li and W.J. Logan. 1984. Role of calcium ions in insulin action on hexose transport in L6 muscle cells. Am. J. Physiol. 247: E297–E304.

Klip, A., and T. Ramlal. 1987. Cytoplasmic Ca^{2+} during differentiation of 3T3-L1 adipocytes. Effect of insulin and relation to glucose transport. J. Biol. Chem. 262: 9141–9146.

Klumpp, C., K. Kostarelos, M. Prato and A. Bianco. 2006. Functionalized carbon nanotubes as emerging nanovectors for the delivery of therapeutics. Biochim. Biophys. Acta 1758: 404–412.

Kostarelos, K., L. Lacerda, G. Pastorin, W. Wu, WieckowskiSebastien, J. Luangsivilay, S. Godefroy, D. Pantarotto, J.P. Briand, S. Muller, M. Prato and A. Bianco. 2007. Cellular uptake of functionalized carbon nanotubes is independent of functional group and cell type. Nat. Nano 2: 108–113.

Krajcik, R., A. Jung, A. Hirsch, W. Neuhuber and O. Zolk. 2008. Functionalization of carbon nanotubes enables non-covalent binding and intracellular delivery of small interfering RNA for efficient knock-down of genes. Biochem. Biophy. Res. Commun. 369: 595–602.

Kubis, H.P., N. Hanke, R.J. Scheibe, J.D. Meissner and G. Gros. 2003. Ca^{2+} transients activate calcineurin/NFATc1 and initiate fast-to-slow transformation in a primary skeletal muscle culture. Am. J. Physiol. Cell Physiol. 285: C56-C63.

Ladeira, M.S., V.A. Andrade, E.R.M. Gomes, C.J. Aguiar, E.R. Moraes, J.S. Soares, E.E. Silva, R.G. Lacerda, L.O. Ladeira, A. Jorio, P. Lima, M. Fatima Leite, R.R. Resende and S. Guatimosim. 2010. Highly efficient siRNA delivery system into human and murine cells using single-wall carbon nanotubes. Nanotechnology 21: 385101.

Lam, C-W., J.T. James, R. McCluskey and R.L. Hunter. 2004. Pulmonary toxicity of single-wall carbon nanotubes in mice 7 and 90 days after intratracheal instillation. Toxicol. Sci. 77: 126–134.

Lam, C.W., J.T. James, R. McCluskey, S. Arepalli and R.L. Hunter. 2006. A review of carbon nanotube toxicity and assessment of potential occupational and environmental health risks. Crit. Rev. Toxicol. 36: 189–217.

Lanner, J.T., J.D. Bruton, Y. Assefaw-Redda, Z. Andronache, S.-J. Zhang, D. Severa, Z-B. Zhang, W. Melzer, S-L. Zhang, A. Katz and H. Westerblad. 2009. Knockdown of TRPC3 with siRNA coupled to carbon nanotubes results in decreased insulin-mediated glucose uptake in adult skeletal muscle cells. FASEB Journal 23: 1728–1738.

Lanner, J.T., J.D. Bruton, A. Katz and H. Westerblad. 2008. Ca^{2+} and insulin-mediated glucose uptake. Curr. Opin. Pharmacol. 8: 339–345.

Lanner, J.T., A. Katz, P. Tavi, M.E. Sandström, S.-J. Zhang, C. Wretman, S. James, J. Fauconnier, J. Lännergren, J.D. Bruton and H. Westerblad. 2006. The role of Ca^{2+} influx for insulin-mediated glucose uptake in skeletal muscle. Diabetes 55: 2077–2083.
Lauritzen, H.P., T. Ploug, C. Prats, J.M. Tavare and H. Galbo. 2006. Imaging of insulin signaling in skeletal muscle of living mice shows major role of T-tubules. Diabetes 55: 1300–1306.
Longo-Mbenza, B., J.B. Kasiam Lasi On'kin, A. Nge Okwe, N. Kangola Kabangu and S. Mbungu Fuele. 2010. Metabolic syndrome, aging, physical inactivity, and incidence of type 2 diabetes in general African population. Diab. Vasc. Dis. Res. 7: 28–39.
Lou, X., V. Scheuss and R. Schneggenburger. 2005. Allosteric modulation of the presynaptic Ca^{2+} sensor for vesicle fusion. Nature 435: 497–501.
Marette, A., E. Burdett, A. Douen, M. Vranic and A. Klip. 1992. Insulin induces the translocation of GLUT4 from a unique intracellular organelle to transverse tubules in rat skeletal muscle. Diabetes 41: 1562–1569.
Martin, S., J. Tellam, C. Livingstone, J.W. Slot, G.W. Gould and D.E. James. 1996. The glucose transporter (GLUT-4) and vesicle-associated membrane protein-2 (VAMP-2) are segregated from recycling endosomes in insulin-sensitive cells. J. Cell Biol. 134: 625–635.
Mueckler, M. 1994. Facilitative glucose transporters. Eur. J. Biochem. 219: 713–725.
Mukherjee, S., R.N. Ghosh and F.R. Maxfield. 1997. Endocytosis. Physiol. Rev. 77: 759–803.
Munoz, P., M. Rosemblatt, X. Testar, M. Palacin, G. Thoidis, P.F. Pilch and A. Zorzano. 1995. The T-tubule is a cell-surface target for insulin-regulated recycling of membrane proteins in skeletal muscle. Biochem. J. 312: 393–400.
Okada, S., K. Ohshima, Y. Uehara, H. Shimizu, K. Hashimoto, M. Yamada and M. Mori. 2007. Synip phosphorylation is required for insulin-stimulated Glut4 translocation. Biochem. Biophys. Res. Commun. 356: 102–106.
Phalon, C., D.D. Rao and J. Nemunaitis. 2010. Potential use of RNA interference in cancer therapy. Expert Rev. Mol. Med. 12: e26.
Porte, D., Jr. and S.E. Kahn. 2001. Beta-cell dysfunction and failure in type 2 diabetes: potential mechanisms. Diabetes 50: S160–S163.
Pramfalk, C., J. Lanner, M. Andersson, E. Danielsson, C. Kaiser, I-M. Renstrom, M. Warolen and S. James. 2004. Insulin receptor activation and down-regulation by cationic lipid transfection reagents. BMC Cell Biol. 5: 7.
Pulskamp, K., S. Diabaté and H.F. Krug. 2007. Carbon nanotubes show no sign of acute toxicity but induce intracellular reactive oxygen species in dependence on contaminants. Toxicol. Lett. 168: 58–74.
Qu, G., Y. Bai, Y. Zhang, Q. Jia, W. Zhang and B. Yan. 2009. The effect of multiwalled carbon nanotube agglomeration on their accumulation in and damage to organs in mice. Carbon 47: 2060–2069.
Sapru, M.K., K.M. McCormick and B. Thimmapaya. 2002. High-efficiency adenovirus-mediated in vivo gene transfer into neonatal and adult rodent skeletal muscle. J. Neurosci. Methods 114: 99–106.

Schipper, M.L., N. Nakayama-Ratchford, C.R. Davis, N.W.S. Kam, P. Chu, Z. Liu, X. Sun, H. Dai and S.S. Gambhir. 2008. A pilot toxicology study of single-walled carbon nanotubes in a small sample of mice. Nat. Nano 3: 216–221.

Schneggenburger, R., and E. Neher. 2005. Presynaptic calcium and control of vesicle fusion. Curr. Opin. Neurobiol. 15: 266–274.

Schudt, C., U. Gaertner and D. Pette. 1976. Insulin action on glucose transport and calcium fluxes in developing muscle cells *in vitro*. Eur. J. Biochem. 68: 103–111.

Schwenk, R., E. Dirkx, W. Coumans, A. Bonen, A. Klip, J. Glatz and J. Luiken. 2010. Requirement for distinct vesicle-associated membrane proteins in insulin- and AMP-activated protein kinase (AMPK)-induced translocation of GLUT4 and CD36 in cultured cardiomyocytes. Diabetologia 53: 2209–2219.

Shulman, G.I. 2000. Cellular mechanisms of insulin resistance. J. Clin. Invest. 106: 171–176.

Singh, B.B., T.P. Lockwich, B.C. Bandyopadhyay, X. Liu, S. Bollimuntha, S.C. Brazer, C. Combs, S. Das, A.G. Leenders, Z.H. Sheng, M.A. Knepper, S.V. Ambudkar and I.S. Ambudkar. 2004. VAMP2-dependent exocytosis regulates plasma membrane insertion of TRPC3 channels and contributes to agonist-stimulated Ca^{2+} influx. Mol. Cell 15: 635–646.

Singh, R., D. Pantarotto, L. Lacerda, G. Pastorin, C. Klumpp, M. Prato, A. Bianco and K. Kostarelos. 2006. Tissue biodistribution and blood clearance rates of intravenously administered carbon nanotube radiotracers. PNAS 103: 3357–3362.

Smith, F.J.D., R.P. Hickerson, J.M. Sayers, R.E. Reeves, C.H. Contag, D. Leake, R.L. Kaspar and W.H.I. McLean. 2007. Development of therapeutic siRNAs for pachyonychia congenita. J. Invest. Dermatol. 128: 50–58.

Smyth, J.T., L. Lemonnier, G. Vazquez, G.S. Bird and J.W. Putney, Jr. 2006. Dissociation of regulated trafficking of TRPC3 channels to the plasma membrane from their activation by phospholipase C. J. Biol. Chem. 281: 11712–11720.

Tasis, D., N. Tagmatarchis, A. Bianco and M. Prato. 2006. Chemistry of carbon nanotubes. Chemical Reviews 106: 1105–1136.

Thong, F.S., C.B. Dugani and A. Klip. 2005. Turning signals on and off: GLUT4 traffic in the insulin-signaling highway. Physiology (Bethesda) 20: 271–284.

Wallberg-Henriksson, H., B.N. Campaigne and J. Henriksson. 1988. *In vitro* reversal of insulin resistance in diabetic skeletal muscle is independent of extracellular Ca^{2+} and Mg^{2+}. Acta. Physiol. Scand. 133: 125–126.

Wang, S.-L., H-H. Yao and Z.-H. Qin. 2009. Strategies for short hairpin RNA delivery in cancer gene therapy. Expert Opin. Biol. Ther. 9: 1357–1368.

Wang, X., J. Ren and X. Qu. 2008. Targeted RNA interference of cyclin A2 mediated by functionalized single-walled carbon nanotubes induces proliferation arrest and apoptosis in chronic myelogenous leukemia K562 cells. Chem. Med. Chem. 3: 940–945.

White, M.F. 2003. Insulin signaling in health and disease. Science 302: 1710–1711.

Wild, S., G. Roglic, A. Green, R. Sicree and H. King. 2004. Global prevalence of diabetes. Diabetes Care 27: 1047–1053.

Wilson, J.A., and C.D. Richardson. 2006. Future promise of siRNA and other nucleic acid based therapeutics for the treatment of chronic HCV. Infect. Disord. Drug Targets 6: 43–56.

Wu, Y., J.S. Hudson, Q. Lu, J.M. Moore, A.S. Mount, A.M. Rao, E. Alexov and P.C. Ke. 2006. Coating single-walled carbon nanotubes with phospholipids. J. Phys. Chem. 110: 2475–2478.

Wu, Y., J.A. Phillips, H. Liu, R. Yang and W. Tan. 2008. Carbon nanotubes protect DNA strands during cellular delivery. ACS Nano 2: 2023–2028.

Yang, R., X. Yang, Z. Zhang, Y. Zhang, S. Wang, Z. Cai, Y. Jia, Y. Ma, C. Zheng, Y. Lu, R. Roden and Y. Chen. 2006. Single-walled carbon nanotubes-mediated *in vivo* and in vitro delivery of siRNA into antigen-presenting cells. Gene Therapy 13: 1714–1723.

Yang, S-T., X. Wang, G. Jia, Y. Gu, T. Wang, H. Nie, C. Ge, H. Wang and Y. Liu. 2008. Long-term accumulation and low toxicity of single-walled carbon nanotubes in intravenously exposed mice. Toxicol. Lett 181: 182–189.

Yokoyama, A., Y. Sato, Y. Nodasaka, S. Yamamoto, T. Kawasaki, M. Shindoh, T. Kohgo, T. Akasaka, M. Uo, F. Watari and K. Tohji. 2004. Biological behavior of hat-stacked carbon nanofibers in the subcutaneous tissue in rats. Nano Letters 5: 157–161.

Zhang, Z., X. Yang, Y. Zhang, B. Zeng, S. Wang, T. Zhu, R.B.S. Roden, Y. Chen and R. Yang. 2006. Delivery of telomerase reverse transcriptase small interfering RNA in complex with positively charged single-walled carbon nanotubes suppresses tumor growth. Clin. Cancer Res. 12: 4933–4939.

Zheng, M., A. Jagota, E.D. Semke, B.A. Diner, R.S. McLean, S.R. Lustig, R.E. Richardson and N.G. Tassi. 2003. DNA-assisted dispersion and separation of carbon nanotubes. Nat. Mater. 2: 338–342.

Zhu, L., D.W. Chang, L. Dai and Y. Hong. 2007. DNA damage induced by multiwalled carbon nanotubes in mouse embryonic stem cells. Nano Letters 7: 3592–3597.

Zimmet, P., K.G.M.M. Alberti and J. Shaw. 2001. Global and societal implications of the diabetes epidemic. Nature 414: 782–787.

Zisman, A., P.D. Odile, D.E. Abel, M.M. Dodson, F. Mauvais-Jarvis, L.B. Bradford, J.F.P. Wojtaszewski, M.F. Hirshman, A. Virkamaki, L.J. Goodyear, R.C. Kahn and B.B. Kahn. 2000. Targeted disruption of the glucose transporter 4 selectively in muscle causes insulin resistance and glucose intolerance. Nat. Med. 6: 924–928.

6

Glucose Sensors Based on Functional Nanocomposites

Kwang Pill Lee,[1,a,]* *Anantha Iyengar Gopalan*[1,b] *and Shanmugasundaram Komathi*[1,c]

ABSTRACT

Diabetes mellitus (DM) is a growing public health problem throughout the world that needs to be addressed with great care. DM is a heterogeneous group of metabolic disorders characterized by chronic hyperglycemia with disturbances of carbohydrate, fat and protein metabolism resulting from defects in insulin secretion, action or both. In simple sense, diabetes is characterised by disordered metabolism and inappropriate blood glucose levels resulting from either lower presence of the harmone insulin or from abnormal resistance to insulin's effects. It affects the heart, kidney, eyes, and other organs including the gastrointestinal tract in human beings. The complications of DM include retinopathy, nephropathy, neuropathy, and increased risk for atherosclerotic vascular disease. DM is the leading cause of blindness in young

---

[1]Department of Chemistry Education, Kyungpook National University, Daegu-702-701, Republic of Korea.
[a]E-mail: kplee@knu.ac.kr
[b]E-mail: algopal_99@yahoo.com
[c]E-mail: s_komathi83@yahoo.com
*Corresponding author

List of abbreviations after the text.

people and is comparable with mascular degeneration as a cause of blindness in older adults. DM is the leading cause of end-stage renal disease requiring dialysis or transplantation. DM is the leading cause of nontraumatic amputations of the lower extremity, a result of peripheral neuropathy and peripheral vascular disease. DM is associated with a twofold to fivefold increased risk for coronary heart disease. Glycemic control is associated with a reduced risk for the microvascular and neuropathic complications in diabetic patients. Regular aerobic exercise and periodic blood glucose level monitoring reduces high risk factors and complications in diabetic patients. Glucose oxidase (GOx) is an ideal enzyme that has been extensively used in electrochemical glucose sensors to monitor the blood glucose levels.

Electrochemical glucose biosensor integrates a biological element like an enzyme with a physiochemical transducer to produce an electronic signal proportional to glucose concentration which is then conveyed to a detector. Development of highly sensitive, low cost and reliable enzyme based glucose sensors has been the subject of research for few decades. For the electrochemical biosensor applications, the electrode modifying material is expected to possess several characteristics such as good electron transduction, physical or chemical environment for the enzyme, bioactivity, easy accessibility towards the analyte and large surface area. However, it is rare to have all the above mentioned characteristics in a single component material. There is a need to develop multifunctional materials. Nanomaterials provide almost unlimited combinations of sizes, dimensions, shapes and chemical compositions, which can be tailored in order to develop new functional nanoprobes for glucose detection. Nanocomponents such as carbon nanotubes (CNTs), conducting polymers (CPs), metal/metal oxide nanoparticles and sol-gel nanomaterials silica have been employed as electrode modifiers either as an individual component or in binary/ternary/multiple combinations. This chapter focuses on the development of functional/multi component nanomaterials based electrochemical glucose biosensors and the different methodologies that have been adopted to fabricate such GOx based electrochemical glucose biosensors. The effective immobilization methodologies of GOx into these functional components and strategies for the construction of glucose biosensors to enhance the performances have also been detailed. The *in-vivo* implantations of these functional bioprobes for self powered blood glucose monitoring over a period of time are underway. Substantial work needs to be done in future for the application of these functional nanomaterials towards monitoring of glucose concentration from tears, sweat and saliva and also for closed loop insulin delivery.

1. INTRODUCTION

Diabetes is a wide spread disease causing health problem throughout the world. Currently, nearly 285 million people are affected by diabetes across the world and the value is expected to increase more than 450 million by the year 2030. Diabetes means "siphon" (liquefaction of the flesh and bones into urine) as coined by Arateus in his work "Acute and Chronic Diseases". By definition, diabetes mellitus is a heterogeneous group of metabolic disorders characterized by chronic hyperglycemia with disturbances of carbohydrate, fat and protein metabolism resulting from defects in insulin secretion, action or both. In simple sense, diabetes is characterised by disordered metabolism and inappropriate blood glucose levels resulting from either lower presence of the harmone insulin or from abnormal resistance to insulin's effects. Although there is no impossible cure for diabetes, tight glucose control can substantially reduce morbidity and mortality. The following are the classification of diabetes: (i) Type 1 (insulin-dependent diabetes mellitus); (ii) Type 2 (non-insulin-dependent diabetes mellitus); (iii) Gestational diabetes mellitus; (iv) Impaired glucose tolerance; (v) Impaired fasting glucose; (vi) other types (associated with pancreatic diseases, removal endocrinopathies, genetic syndromes). The effects of diabetes mellitus include long-term damage, dysfunction and failure of various organs especially the eyes, kidneys, heart and blood vessels. Frequent or continuous testing of glucose levels is crucial for the confirmation of diabetes and subsequent treatment. Therefore, development of highly sensitive, low cost and reliable glucose sensors has been the subject of research for few decades. A variety of glucose monitoring devices are now available that give a digital readout of the blood glucose concentration. The devices continue to be improved and the time required for the test to be completed is now as short as 5 s. In addition, the size of the blood samples has decreased and many meters use a direct activation system. Some of the glucose meters include mechanical lancet to obtain blood and computerized memory to record the blood glucose levels. There are many approaches to monitor glucose concentration, such as optical technique, surface plasma resonance, capacitance detection, electrochemical luminacence, calorimetry etc. Among them electrochemical method is a promising one due to its simplicity, reliability, low detection limit, sensitivity, low cost and ease of use.

2. ELECTROCHEMICAL DETERMINATION OF GLUCOSE

Glucose oxidase (GOx), an ideal enzyme for use in bioelectrochemistry, has been extensively used to monitor the blood glucose levels in diabetics. GOx is a typical flavin enzyme with FAD (flavin adenine dinucleotide)/$FADH_2$ as the redox prosthetic group. The biological function of GOx is to catalize glucose to glucanolactone. When GOx catalyzes glucose oxidation, GOx-FAD is reduced to GOx-$FADH_2$ which can be oxidized by the electrode back to GOx-FAD (Liu et al. 2009). Generally, the electrochemical oxidation of GOx is difficult at the bare electrode. The active site of FAD/$FADH_2$ is located at a depth of 13Å within the insulative protein layer, apoenzyme. Hence, at the bare electrode, even if the enzyme is tightly immobilized, the distance between the either one of its two FAD/$FADH_2$ centers and electrode surface exceeds the critical tunneling distance. The insulative shell of GOx blocks, the electron transfer between FAD/$FADH_2$ redox center and electrode that results in much lowered electroactivity. Enzymatic glucose sensors underwent three major generations of transitions on the mode of electrochemical sensing. First generation glucose biosensors are dependent on the presence of oxygen as a co-substrate to ensure the catalytic regeneration of the FAD centre. In the second generation glucose sensors, electron transfer is being facilitated by synthetic, electron-accepting mediators such as ferrocene derivatives, ferricyanide, quinones and transition-metal complexes. Consequent re-oxidation of the mediator at the electrode results in a quantifiable amperometric current. Third generation enzymatic glucose sensors involve direct electron transfer (DET) between the enzyme and the electrode, without the need for natural or synthetic mediators. The emerging technologies and future prospects on glucose monitoring technology have been reviewed (Flanagan 2009; Reach and Choleau 2008). In 2008, two extensive reviews on enzymatic electrochemical glucose sensors were published (Heller and Feldman 2008, Wang 2008). Recently, an overview of glucose sensors with specific focus on non-enzymatic electrochemical glucose sensor was reported by Compton (Toghill and Compton 2010).

The ultimate goal of the glucose electrochemical biosensors lies on fine tuning the salient characteristics such as sensitivity, selectivity, response time, good linear concentration range, stability and reproducibility to their best practical values. Several approaches have been developed in the process of performance improvements which include development of efficient matrix for enzyme immobilization and preparation of electrocatalytic components. For the electrochemical biosensor applications, the electrode modifying material needs to possess several characteristics such as good electron transduction capability, physical or chemical environment for the stable immobilization of enzyme, bioactivity, easy accessibility towards the analyte and large surface area. It has been noticed that nanostructured material facilitates DET between GOx and electrode.

3. NANOMATERIALS IN ELECTROCHEMICAL GLUCOSE SENSORS

Recently, the potential advantages of nanoscale materials have been exploited for glucose sensor application. Nanomaterials provide almost unlimited combinations of sizes, dimensions, shapes and chemical compositions, which can be tailored to develop new functional nanoprobes for glucose detection. Nanotechnology has been incorporated into glucose sensors through two primary approaches. First, sensors can be designed using macro- or microscale components (such as electrodes, membranes and supporting hardware) but incorporate either a nanostructured surface or a nanomaterial into this design. The nanoscale properties of these modified systems have several advantages, including higher surface areas (yielding larger currents and faster responses) and improved catalytic activities. The nanostructured material based sensors would be implanted and used for continuous monitoring. Second, nanofabrication techniques can be used for glucose sensors.

In recent years, rapid advancement in the development of nano materials has grossly increased electrode modification strategies. Nanostructured materials penetrate the insulative protein layer and effectively shorten the tunneling distance through their quantum confined physical, electrical and chemical properties (Wang et al. 2009a). In DET based biosensors, DET from enzyme to electrode occurs and current corresponding to enzyme oxidation is directly observed without the complications of mediators and intrusion of electroactive interferences or dependence on dissolved oxygen. Literature reveals that all the important characteristics required to achieve DET cannot be obtained through a single nanomaterial. Hence, there is always a demand for the development of composite nanomaterials, comprising two or more components, to achieve adequate sensitivity and stability for the biosensors. Multiple nano components and multifunctional nanoprobes can be conveniently designed to improve the performance of glucose biosensors.

Nanocomponents such as carbon nanotubes (CNTs), metal/metal oxide nanoparticles and sol-gel nanomaterials (like silica) have been employed as electrode modifiers either as an individual component or in binary/ternary/multiple combinations. When these nanocomponents are judiciously combined to form matrix for GOx immobilization, glucose sensors exhibited much improved performance characteristics like sensitivity, selectivity and response time. This chapter focuses on the development of functional/multi component nanomaterials based glucose biosensors and the different methodologies that have been adopted to fabricate such glucose biosensors. The effective immobilization methodologies of GOx into these functional components and strategies for the construction of glucose biosensors to enhance the performances have also been detailed. Typically, plenty of reports on the individual use of CNTs, metal/metal oxide nanoparticles, conducting polymer (either as such or in nanostructured form), biocompatible polymers (like chitosan) and ionic liquid in conjunction are available in literature. Recently, functional nanomaterials involving multiple components were reported for the fabrication of glucose sensors. Due to space limitations, we provide only few examples for each of the strategy. Readers, who have more interest, may refer the comprehensive literature in this topic.

3.1. Carbon Nanotube—polymer Based Biosensors

CNTs have been one of the most studied electrode materials due to their unique electronic, electrochemical and mechanical properties. CNTs effectively shuttle the electron from the active site of enzymes to the electrode surface. Direct or enhanced electrochemistry of enzymes is therefore witnessed. CNTs have several advantages for electrochemical sensors: small size with large surface area, high sensitivity, chemical stability, fast response time, good electron transfer capability, ease of protein immobilization and retention of bioactivity. However, pristine CNTs are hydrophobic and have poor dispersability in organic/aqueous solvents. Hence, strategies were evolved for the modification of CNTs with ionomers, conducting polymers (CPs), non-conducting polymers and metal nanoparticles (MNPs) for biosensor application. The electrode modifying components were used either alone or in binary or ternary or multiple combinations to achieve efficient immobilization of GOx and good biosensor performances. And, in many cases DET from enzyme to electrode have also been reported.

CNTs were coupled with a perfluorinated polymer (Nafion), redox hydrogel (poly(vinylimidazole) complexed with Os(4,4'-dimethylbpy)(2)Cl), polyaniline (PANI)/prussian blue and polystyrene sulfonic acid towards fabrication of glucose sensors. CNTs/CP composites were generally configured as bilayers in the fabrication of modified electrodes. In such cases, the bilayers were generated through electrochemical deposition of a layer of a CP over the preformed layer of CNTs. In the bilayer configuration, the beneficial characteristics of CNTs could not be fully retrieved as the CP layer completely/partly masks the surface of CNTs. On the contrary, if chains of CP are covalently linked (grafted) to the surface of CNTs, the resulting materials (CP linked CNTs) could exhibit synergistic properties of CNTs and CP. Studies revealed the importance of grafting of PANI units onto the surface of CNTs (Lee et al. 2009). In the case of PANI based glucose biosensors, one should remember that PANI has good electroactivity only at lower pHs (< 3). This restricts the combination of CNT and PANI for application in neutral medium. However, sulfonated PANI (SPANs) exhibit electrochemical activity beyond pHs 4.0. Hence, CNT–SPAN nanocomposites are more effective in glucose sensors as they possess the combined advantages of CNT and SPAN. Also, if SPANs are present as polymer networks on the surface of CNTs, the interconnected polymer networks provide adequate microenvironment for the immobilization of enzyme. Keeping these aspects into consideration, interconnected SPAN based three dimensional (3D) nanonetworks (SPAN-NW) were formed on the surface of multi-walled carbon nanotube (MWNT) (Lee et al. 2010). The 3D SPAN network (MWNT-g–SPAN-NW) was generated onto MWNT surface through electrochemical co-polymerization of a mixture of diphenyl amine 4-sulfonic acid, 4-vinyl aniline and 2-acrylamido-2-methyl-1-propane sulfonic acid in the presence of amine functionalized MWNT. GOx was immobilized onto MWNT-g–SPAN-NW to fabricate MWNT-g–SPAN-NW/GOx glucose biosensor. The MWNT-g–SPAN-NW/GOx glucose biosensor exhibited DET with an electron transfer rate constant, (k_s) of 4.11 s^{-1}. The efficient DET implies that GOx is present in its native state and augments electron transfer to the electrode. The high value of k_s indicates that MWNT-g–SPAN-NW acts as an excellent promoter for the electron transfer between redox-active sites of GOx and electrode. MWNT-g–SPAN-NW/GOx biosensor exhibited a rapid and sensitive response to glucose in the wide concentration range, at an applied potential of -0.1 V. The sensitivity and the detection limit of MWNT-g–SPAN-NW/GOx electrode were 4.34 µA/mM and 0.11 µM (S/N = 3), respectively. The high sensitivity of MWNT-g–SPAN-NW/GOx biosensor has been attributed to the effective immobilization of GOx onto the porous polymer network. The porous polymer network provides high surface area and the interconnected 3D network acts as a molecular wire for efficient electron transfer from GOx to the electrode. The sensitivity of MWNT-g–SPAN-NW/GOx biosensor is comparatively higher than GOx immobilized onto pristine CNTs/nafion matrix (0.33 µA/mM). Studies have been performed on electrochemical polymerization of polypyrrole (PPY) onto SWNT, neutral red modified MWNT, polyvinylpyrrolidone (PVP) protected prussian blue nanoparticles (PBNPs)-PANI/MWNT hybrid composites towards immobilization of GOx.

3.2 Carbon Nanotube/Metal or Metal Oxide Based Biosensors

The emergence of nanotechnology opens up new horizons for the use of various nanoparticles in biosensors. MNPs have large surface area to volume ratios and are used in the fabrication of electrochemical biosensors. A facile strategy to prepare CNT loaded Pt nanoparticle (Pt-CNT) composites has been developed (Wen et al. 2009). The method involves the electrochemical reduction of Pt source in the pores of anodic alumina membranes under hydrothermal conditions. The Pt-CNT nanocomposites contain large amounts of oxygen-rich groups that are beneficial to improving its solubility in water and for retaining the bioactivity of GOx. The Pt-CNT-based glucose biosensor exhibited a wide linear concentration range for glucose (0.16–11.5 mM) and a low detection limit of 0.055 mM. Furthermore, the biosensor exhibits excellent characteristics such as high selectivity towards the detection of glucose. CNTs are used as substrates for initial Pt coating and subsequent GOx immobilization to detect glucose (Wang et al. 2008a).

A biosensor is prepared by cross-linking GOx with glutaraldehyde at the electrode surface modified with Au nanoparticles decorated MWNTs. The glucose sensor showed wide linear range from 2.0μM to 0.15mM and a fast response time within 5s. AuNPs provide more reactive sites and facilitate faster electron transfer from GOx to CNTs. A glucose biosensor is constructed based on the adsorption of GOx onto AuPt - MWNT - ionic liquid (i.e., 1-octyl-3-methylimidazolium hexafluorophosphate, [OMIM]PF6) composite (Zhang et al. 2010). The Au-Pt nanoparticles were electrodeposited onto MWNT. Owing to the synergistic action of Au-Pt nanoparticles, MWNT and [OMIM]PF$_6$, the biosensor shows good response to glucose, with wide linear range (0.01 to 9.49 mM), short response time (3 s), and high sensitivity (3.47 μA mM^{-1}).

GOx was immobilized onto CdS nanoparticles embedded in a chitosan (CS) matrix. The CS-GOx-CdS/ACNTs-Pt$_{nano}$ electrode shows sensitive response to the glucose with a detection limit of 46.8 μM (S/N = 3). Glucose biosensor has been fabricated through immobilization of GOx within mesoporous CNT—titania—Nafion composite film coated on a platinized glassy carbon electrode. The biosensor responded linearly to glucose for a wider concentration range (0.5μM to 5.0mM) with a good sensitivity of 154 mA.M^{-1}.cm^{-2}. Due to the mesoporous nature of CNT-titania—Nafion composite film, the biosensor showed a very fast response time (~2 s). An amperometric biosensor based on CNT, a nano-thin plasma-polymerized film immobilized with GOx was developed (Muguruma et al. 2008). To facilitate the electrochemical communication between the CNT layer and GOx, CNT was treated with nitrogen or oxygen plasma. The resulting device showed good current response to glucose because electron transfer rate was 4-16-fold larger as compared to pristine CNT. The glucose biosensor showed high sensitivity (42 $\mu A.mM^{-1}.cm^{-2}$), detection limit (6 μM), high selectivity (almost no interference by 0.5 mM ascorbic acid) for glucose quantification and rapid response (<4 s to reach 95% of maximum response). Additionally, the devices showed a small and stable background current (0.350 ± 013 μA). A GOx immobilized nanodendritic poly[meso-tetrakis(2-thienyl)porphyrinato]cobalt(II)-SWNT modified GCE (pCoTTP-SWNTs-Nafion-GOx/GCE) has been reported (Fig. 1) (Chen et al. 2010). GOx retained its activity with a fast k_s of 1.01 s^{-1}. A high sensitivity of 16.57 $\mu A.mM^{-1} cm^{-2}$ was achieved with a low detection limit of 5.33 μM (S/N = 3). The biosensor also shows excellent selectivity against 0.2 mM uric acid and ascorbic acid. Novel hybrid films based on CNTs/nickel hexacyanoferrate (NiHCF) nanocomposites were synthesized for electrochemical glucose sensing. GOx was immobilized onto CNTs/NiHCF nanocomposite. The high sensitivity is attributed to the high surface area of CNT and excellent electrocatalytic activity of the modifiers.

GOx was covalently immobilized onto MWNTs with potassium ferricyanide as the redox mediator (Chen et al. 2009). MWNTs were grown directly on a layered structure of Co/Ti/Cr on a SiO_2/Si substrate by microwave-heated chemical vapor deposition. The mediator helps to shuttle the electrons between GOx and electrode. The sensitivity of biosensors towards glucose was found to depend on the acid pretreatment and GOx reaction times. The biosensor exhibited high sensitivity of 20.6 $\mu A.mM^{-1}.cm^{-2}$, a linear range of up to 8 mM, and a response time of <5 s. The free-standing CNT film with ZnO has been utilized for the construction of GOx based glucose sensor. Using polydiallyl dimethyl amine (PDDA), nanostructured ZnO coated CNT film and GOx, a multi layer biosensor PDDA/GOx/ZnO/MWNTs is constructed (Bai et al. 2010). The amperometric response of the PDDA/GOx/ZnO/MWNTs bioelectrode towards the detection of glucose has been reported. Au metal clusters modified CNTs (Rakhi et al. 2009), Pt nanospheres modified CNTs (Claussen et al. 2010) are also reported.

3.3. Carbon Nanotube/bio Polymer/Metal or Metal Oxide Based Biosensors

CS has been used for enzyme immobilization due to its excellent film forming ability, high water permeability and biocompatibility. The enormous signal amplification associated with nanoparticle-biomolecule assemblies provides the basis for ultrasensitive optical and electrical

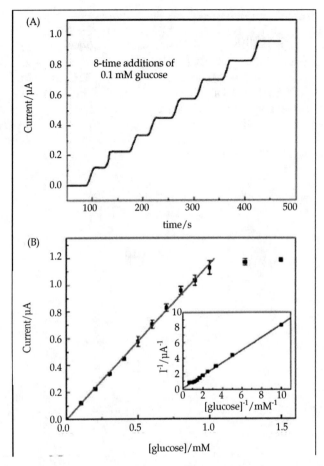

Fig. 1. (A) Amperometric response to successive addition of 0.1 mM glucose at the pCoTTP-SWNTs-Nafion-GOD/GCE after subtracting the baseline. (B) Calibration plot for glucose in air-saturated 0.01 M pH 7.4 PBS buffer at an applied potential of −0.2 V vs. Ag/AgCl. Inset presents the Lineweaver–Burk plot.

detection. Gold nanoparticles (AuNPs) attract researchers for its numerous bioanalytical applications. AuNPs provide adequate micro-environment to enhance DET between protein and electrode. High performance glucose biosensors were developed based on immobilization of GOx into CNTs/CS/CuMNPs, PANI coated magnetic(Fe_3O_4) CNTs composite (Liu et al. 2008) and PB@MWNT/hollow PtCo composite film (Che et al. 2010).

Considering the benefits of CS, ionic liquid, CNTs and AuNPs, a biosensor is fabricated to exploit their synergistic contributions. The fabrication of glucose biosensor consists of three steps: (i) thiol

(–SH) functionalization of MWNTs and dispersion into CS–IL matrix, (ii) deposition of AuNPs into MWNT(SH)–CS–IL matrix and (iii) immobilization of GOx to obtain, GOx/Au/CS–IL–MWNT(SH). Response time of GOx/Au/CS–IL–MWNT(SH) biosensor to glucose concentration is lower (6 s) than other glucose biosensors, CS/Glutaraldehyde/GOx (10 s), CS/MWNT/GOx (8 s), CS/GOx/AuNPs (7 s) and (MWNTs/GOx) (6.7 s). The biosensor exhibits electrochemical detection of glucose over a wide linear concentration range (1–10 mM) with a high sensitivity (4.10 $\mu A.mM^{-1}$) and a low response time (6 s) due to the combined catalytic influence of MWNT(SH) and AuNPs in CS–IL matrix (Fig. 2).

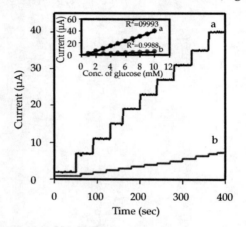

Fig. 2. (A) Chronoamperometric current responses of electrodes: (a) GOx/Au/CS–IL–MWNT(SH)/ITO and (b) GOx/MWNT(SH)/ITO to successive addition of 1 mM glucose. Inset: Calibration plot of concentration of glucose (1–10 mM) vs. current; (a) GOx/Au/CS–IL–MWNT(SH)/ITO and (b) GOx/MWNT(SH)/ITO.

Color image of this figure appears in the color plate section at the end of the book.

3.3.1 Layer by Layer approach in glucose sensors

One of the problems associated with electrochemical biosensors is the leaching of the enzyme from the substrate during the electrode operation. The leaching of enzyme could be controlled by either anchoring the enzyme into the matrix through molecular interactions or sandwiching them in between two stable matrices. It is possible to load a large amount of enzyme in a multilayer configuration. Layer-by-layer (LbL) deposition offers striking advantages such as precise control of composition, thickness of film, wide choice of materials, simplicity of procedure and DET between enzyme and underlying electrode. MWNTs were combined with AuNPs or an insulating polymer to form LbL assembly with GOx (Wu et al. 2007a). Also, the combination of GOx, AuNPs, an insulating polymer and

LbL assembly has been utilized for fabricating a glucose sensor (Wu et al. 2007b). The synergistic advantages of MWNTs, AuNPs, CPs and LbL assembly were exploited to achieve effective electron shuttling between the electrode and GOx. The three components, i.e., MWNT, Au NPs, PANI(SH), were integrated into a compact LbL film and a biosensor was fabricated with immobilization of GOx. A new strategy was adapted to form LbL assembly consisting of alternate layers of the nanocomposite and GOx (scheme 1) (Komathi et al. 2009).

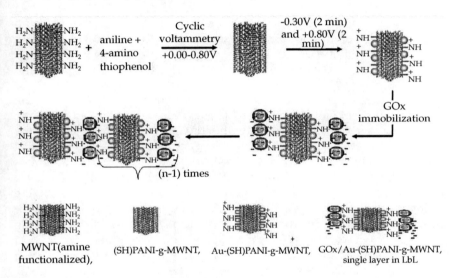

Scheme 1. Fabrication of LbL based biosensor comprising of multi-components, MWNT, PANI(SH), Au particles and GOx.

Color image of this figure appears in the color plate section at the end of the book.

Towards this purpose, thiol functionalized PANI (PANI(SH)) was grafted onto MWNTs and subsequently loaded with AuNPs to obtain the nanocomposite Au/(SH)PANI-g-MWNT. LbL film was successfully constructed by inducing electrostatic interactions between positively charged Au/(SH)PANI-g-MWNT and negatively charged GOx. The {GOx/Au-(SH)PANI-g-MWNT}$_n$ (n = number of stacks). The electrostatic interactions between positive charges of amine or imine groups in PANI(SH) and GOx as well as the effective binding of Au particles through –SH groups in PANI(SH) provides the compactness and stability to the

LbL assembly. There was no leaching of GOx to the electrolyte for a long period (10 days). The components in the nanocomposite provide adequate electron transfer path between GOx and the electrode. A high value for the rate constant of electron transfer process (27.84 s^{-1}) was observed at {GOx/Au-(SH)PANI-g-MWNT}$_n$/ITO electrode. The higher value of k_s informs that the components in {GOx/Au-(SH)PANI-g-MWNT}$_{10}$ synergistically contribute to the electron transfer path. The biosensor exhibited a lower detection limit (0.06 µM at S/N = 3) than reported for multilayer films of chitosan/MWNT/GOx (21 µM), MWNT/AuNPs/GOx (6.7 µM) and chitosan/AuNPs/GOx (7.0 µM). Thus, the inclusion of conducting polymer in the LbL assembly significantly enhances the k_s (27.84 s^{-1}), sensitivity 3.97 µA/mM (n = 10) and also lowers the detection limit compared to the LbL assemblies having MWNTs, AuNPs and an insulating polymer.

An amperometric glucose biosensor was constructed, based on the immobilization of GOx in CS on a GC electrode. The modified electrode was fabricated by LbL assembly based CNT/CS/AuNPs multilayer films (Wang et al. 2009b). The linear range for the electrochemical detection of glucose was 6 µM to 5 mM, with a detection limit of 3 µM. A multicomponent multilayer film was formed via LbL assembly. MWCNTs were introduced with carboxyl groups and used to interact with the amino groups of poly(allylamine) (PAA) and cysteamine via 1-ethyl-3-(3-dimethylaminopropyl)carbodiimide/N-hydroxysuccinimide crosslinking reaction (Wu et al. 2007a). A cleaned Pt electrode was immersed in PAA, MWCNTs, cysteamine and AuNPs, respectively followed by the adsorption of GOx. Anionic GOx was immobilized on the negatively charged CNTs surface. Alternative assembly of cationic Pt-DENs and an anionic GOx layers was developed. LbL film provides a favorable microenvironment to keep the bioactivity of GOx and prevents enzyme molecules from leakage. The excellent electro-catalytic activity of CNTs and Pt-DENs resulted in improved characteristics such as a low detection limit of 2.5 µM, a wide linear range, a short response time (within 5 s), and high sensitivity (30.64 µA.mM^{-1}.cm^{-2}) and stability (80% remains after 30 days).

GOx was immobilized onto the negatively charged CNT surface by alternative assembly of cationic poly(ethylenimine) (PEI) layer and GOx layers (Deng et al. 2010). DET of GOx was observed. The ultrathin {GOx/PEI}$_n$ film on the CNT surface provides a favorable microenvironment to keep the bioactivity of GOx. The PEI/{GOx/PEI}$_n$/CNT/GC biosensor exhibited a high sensitivity of 106.57 µA.mM^{-1}.cm^{-2}. GOX-SWNT conjugates can be assembled to fabricate biosensors with a poly[(vinylpyridine)Os(bipyridyl)$_2$Cl$^{2+/3+}$] redox polymer (PVP-Os) through a LbL self-assembly (Tsai et al. 2009). Incorporation of SWNT-enzyme conjugates into the LbL films resulted in high current densities

440 µA/cm². CNT coated PANI was encapsulated with Pt nanoparticles (Pt-DENs) (Xu et al. 2009). GOx was immobilized onto Pt-NPs/PANI/CNT composite film. The fabricated GOx/Pt-NPs/PANI/CNT biosensor exhibits excellent response to glucose, such as low detection limit (0.5 µM), wide linear range (1 µM–12 mM), short response time (5 s) and high sensitivity ($42\mu A.mM^{-1}.cm^{-2}$).

A glucose biosensor was constructed utilizing MWNTs and ZnO nanoparticles, PDDA and GOx (Wang et al. 2009c). GOx was electrostatically bound to ZnO nanoparticles at the electrode surface. A cationic PDDA layer was further coated onto the layer of GOx-containing ZnO NPs, which prevents leakage of GOx. This unique multi-layer structure (PDDA/GOx/ZnO/MWNTs) provides a favorable microenvironment to maintain the bioactivity of GOx, to result in rapid amperometric response toward glucose. By loading 0.5 U GOx at the sensor surface, a wide linear response range of 0.1-16 mM was witnessed. This nanomaterials-based glucose sensor showed high sensitivity and favorable stability over a relatively long-term storage (160 days).

3.4 Carbon Nanotube/sol-gel Silica Based Glucose Biosensors

The crux of bio-electrocatalysis, is the development of enzyme-immobilization techniques that provide effective electron transfer from enzyme to electrode, whilst maintaining high catalytic activity and enzyme stability. Different sol-gel matrices such as inorganic, organically modified (ormosils), hybrid sol-gels and interpenetrating polymer networks have been used for the immobilization of biomolecules. Organically modified sol-gels have tunable porosities and electrochemical activities. Sol-gel cage restricts the conformational change of macromolecules and enables the biomolecules to retain its bioactivity. The pores of sol-gel matrix can be tuned to allow unrestricted transport of molecules including buffer ions, substrates, products of the reaction, and analytes. Bio-electrocatalysis for glucose oxidation has been demonstrated by entrapping GOx into a silica/CNT composite (Vnitski et al. 2008). The immobilized enzyme is stable for a period of one month and retains catalytic activity for the oxidation of glucose.

3.4.1 Carbon nanotube/polymer/sol-gel silica based biosensors

To overcome the insulating property of the sol-gel matrices and slow diffusion of the electroactive species, sol-gel materials are incorporated with CNTs or AuNPs/rods or palladium or graphite nanostructures. Nafion, silica, MWNT, and CP have individually been known to improve the electrocatalytic activity, electron conduction path, sensitivity, stability of biosensors, respectively. Sol–gel silica, CNT, PANI and Nafion were used either alone or in binary combinations for the fabrication of glucose biosensor. An electrochemical biosensor-based on a multi-component composite comprising of MWNT, silica, Nafion and PANI has been developed (Gopalan et al. 2009). The four different components in the composite were judiciously integrated by employing the following strategy (Scheme 2).

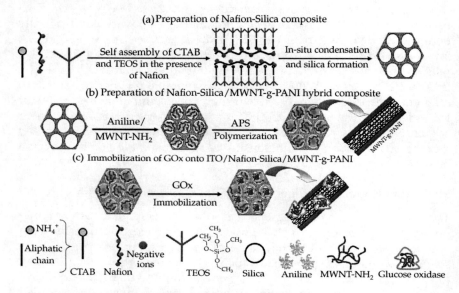

Scheme 2. Fabrication of Nafion-silica/MWNT-g-PANI/GOx biosensor.
Color image of this figure appears in the color plate section at the end of the book.

First, Nafion–silica composite was prepared by the condensation of silica precursors in a Nafion solution. In the subsequent step, MWNT-g-PANI was incorporated into silica–Nafion composite. Glucose biosensor was fabricated by the immobilization of GOx into Nafion–silica/MWNT-g-PANI composite. The sensitivity of the Nafion–silica/MWNT-g-PANI/GOx biosensor (5.01 µA/mm) is superior than reported for the glucose biosensor fabricated with silica or Nafion or CNT or CNT–Nafion composites or PANI-g-MWNT as matrix for GOx immobilization. Typically, Nafion–silica/MWNT-g-PANI/GOx biosensor shows superior sensitivity than Nafion/GOx/CNTs (0.33 µA/mM), Silica sol–gel/GOx/CNTs (0.196 µA/mM), Sol–gel/GOx/copolymer (0.6 µA/mM), and MWNT/AuNPs/GOx (2.527 µA/mM) biosensors. The superior sensitivity of Nafion–silica/MWNT-g-PANI/GOx arises from the judicious design for the fabrication. Nafion–silica/MWNT-g-PANI/GOx shows a lower response time of 6 s for the detection of glucose to reach a steady state current. The larger active surface area provided by the MWNT and augmented electron transduction at MWNT-g-PANI unit are the reasons for the effective electron mediation at Nafion–silica/MWNT-g-PANI/GOx biosensor to result a lower response time for glucose detection. Importantly, Nafion–silica/MWNT-g-PANI/GOx biosensor exhibited the combination of high sensitivity (5.01 µA/mm), low detection limit (0.1 µM) and lower response time (6 s) because of the combined presence of silica–Nafion and MWNT-g-PANI.

3.4.2 Carbon nanotube/metal nanoparticle/sol-gel silica based biosensors

For electrochemical biosensors, the immobilization of biological recognition elements onto the surface of the electrochemical transducer is an important process. The methods for enzyme immobilization into the electron transducer include adsorption, encapsulation, entrapment, cross-linking, and covalent bonding. Among these various enzyme immobilization protocols, adsorption is the simplest and involves minimal preparation time. The bioactivities of the immobilized enzyme can be retained well because adsorption does not require any chemical reagent. Many substances adsorb enzymes on their surface, for example, alumina, silica, Pt nanoparticles, CNTs, and glass. SWNT microarrays were patterned on SiO_2 substrates. Over the SWNT micropatterns, highly electroactive polycrystalline Pt nanoparticles (PtNPs) were electrochemically deposited (Zeng et al. 2010). GOx was further immobilized to obtain (GOx)/BSA/PtNP-SWCNT biosensor. The sensor shows the detection limit and sensitivity for glucose as 0.04 mM and 4.54 $µA.mM^{-1}.cm^{-2}$, respectively. The improved electrocatalytic activity originates from the PtNP-SWCNT micropatterns.

A glucose biosensor based on the adsorption of GOx onto Pt-MWNT-alumina-coated silica (Pt-MWCNT-ACS) nanocomposite was reported (Tsai et al. 2009). The MWCNT-ACS nanocomposite based glucose biosensor displayed a wide linear range of up to 10.5 mM, a high sensitivity of 113.13 $mA.M^{-1}.cm^{-2}$ and a response time of less than 5 s. The sensitivity for GOx-Pt-MWCNT-ACS nanocomposite is better as compared to GOx-Pt-CNT modified electrode. A biosensor based on electrodeposition of PtNPs, MWNTs, chitosan-SiO_2 sol-gel and GOx has been presented. The synergistic contributions of Pt, MWNTs and the biocompatibility of chitosan-SiO_2 sol-gel matrix resulted excellent electrocatalytic activity and high stability. The biosensor exhibited wide linear range from 1 µM to 23 mM and a low detection limit 1 µM to glucose. The biosensor also shows a much low response time (within 5 s) and a high sensitivity (58.9 $µAmM^{-1}.cm^{-2}$).

A new strategy was reported for fabricating a sensitivity-enhanced glucose biosensor based on MWNT, PtNPs and sol-gel of CS/silica organic-inorganic hybrid composite (Kang et al. 2008). PtNP-CS solution was synthesized through the reduction of $PtCl_6^{2-}$ by $NaBH_4$. PtNP was stabilized using CS. The CS/silica hybrid sol-gel was prepared by mixing methyltrimethoxysilane (MTOS) and CNT-PtNP-CS. The biosensor showed glucose response with linear range of 1.2μM to 6.0mM and a detection limit of 0.3 μM^{-1}. The results showed that the biosensor provided the high synergistic electrocatalytic action from the components. PtNPs-doped sol-gel solution is prepared. The incorporation of GOx within the Pt-CNT-silicate matrix resulted a Pt-CNT paste-based biosensor (Yang et al. 2006). The glucose sensor showed linear range from 1 to 25 mM with a response time <15 s, and the sensitivity is 0.98 $\mu A\ mM^{-1}\ cm^{-2}$ (Fig. 3). The sensitivity of the Pt-CNT paste-based biosensor is almost four times larger than that of the CNT-based biosensor (0.27 $\mu A.mM^{-1}.cm^{-2}$ at 0.1 V).

A glucose biosensor based on silicon dioxide coated magnetic nanoparticle decorated MWNT ($Fe_3O_4@SiO_2$/MWNTs) on a GCE has been developed (Baby et al. 2010). Functionalized MWNTs have been decorated with magnetic Fe_3O_4 nanoparticles and coated with biocompatible SiO_2. Amperometric biosensor has been fabricated by the deposition of GOx over Nafion-solubilized $Fe_3O_4@SiO_2$/MWNTs electrode. The bioelectrode retains its biocatalytic activity and offers fast and sensitive glucose quantification. The glucose biosensor exhibits a linear response from 1 μM to 30 mM with an excellent detection limit of 800 nM.

4. Summary

- Diabetes mellitus is a heterogeneous group of metabolic disorder and causes public health problem throughout the world.
- Nanomaterials differ from molecular or bulk species with their high surface areas and they possess unique properties.
- With the advent of nanomaterials as an electrode modifying material, the electrochemical glucose biosensor provoke the possibility of regulation of diabetes through close monitoring of glucose levels in blood.
- Intricate strategies are required for the fabrication of glucose biosensors involving binary/ternary and multiple combinations of nanomaterials like carbon nanotubes (CNTs), insulating/conducting polymers, metal or metal oxide nanoparticles and silica. Along with the above, effective immobilization of glucose oxidase into the sensor matrix determines the sensitivity of the sensors.

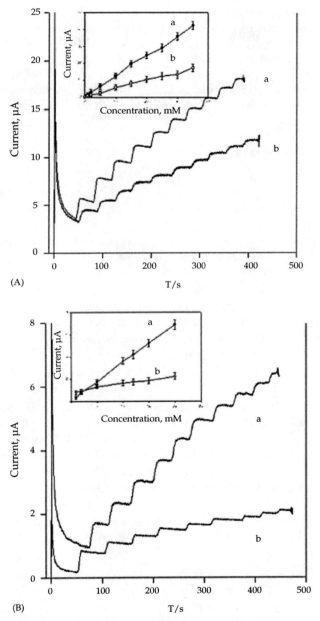

Fig. 3. Current–time recordings for successive addition of 2 mM glucose at the (a) Pt-CNTPE-GOx and (b) CNTPE-GOx biosensors measured at (A) 0.6 and (B) 0.1 V vs. SCE. Insert is the calibration curve. Error bars = ±S.D. and n = 4.

- Technological developments in recent periods are directed towards continuous glucose monitoring systems and artificial closed-loop systems that can control insulin release continuously and automatically as direct response to blood glucose levels.
- Multiple nano components and multifunctional nanoprobes can be conveniently designed for the self powered continuous blood glucose monitoring and determination of glucose from tears, sweat and saliva in the near future.

Dictionary of Key Terms for Novice

Diabetes mellitus: Diabetes mellitus is a group of metabolic diseases in which a person has high blood sugar, either because the body does not produce enough insulin, or because cells do not respond to the insulin that is produced.

Biomolecule: A biomolecule is a chemical compound that naturally occurs in living organisms. Biomolecules consist primarily of carbon and hydrogen, along with nitrogen, oxygen, phosphorus and sulfur (e.g., glucose).

Electrochemical glucose biosensor: Electrochemical glucose biosensor integrates a biological element with a physiochemical transducer to produce an electronic signal that is proportional to glucose concentration.

Carbon nanotube: A single sheet of graphene rolled up in the form of cylinder to form tube like structure.

Conductive polymers: Conductive polymers are organic polymers that conduct electricity.

Electrochemical methods: Methods for the qualitative and quantitative analysis based on electrochemical phenomena occurring within a medium or at the phase boundary that is related to changes in the structure, chemical composition, or concentration of the compound being analyzed.

Glucose oxidase: The glucose oxidase enzyme (GOx) is an oxido-reductase that catalyses the oxidation of glucose to hydrogen peroxide and D-glucono lactone.

Metal nanoparticles: Metal nanoparticles can be defined as any metal particle that has at least one dimension in the nanometre scale.

Abbreviations

GOx	:	Glucose oxidase
CNTs	:	Carbon nanotubes
CPs	:	Conducting polymers
FAD	:	Flavin adenine dinucleotide
MNPs	:	Metal nanoparticles

PANI	:	Polyaniline
SPANs	:	Sulfonated polyanilines
AAMs	:	Anodic alumina membranes
ACNTs	:	Aligned CNTs
CS	:	Chitosan
LbL	:	Layer-by-layer

References

Baby, T.T., and S. Ramaprabhu. 2010. SiO_2 coated Fe_3O_4 magnetic nanoparticle dispersed multiwalled carbon nanotubes based amperometric glucose biosensor. Talanta 80: 2016–2022.

Bai, D., Z. Zhang and K. Yu. 2010. Synthesis, field emission and glucose-sensing characteristics of nanostructural ZnO on free-standing carbon nanotubes films. Appl. Surf. Sci. 256: 2643–2648.

Che, X., R. Yuan, Y. Chai, J. Li, Z. Song and W. Li. 2010. Amperometric glucose biosensor based on prussian blue-multiwall carbon nanotubes composite and hollow PtCo nanochains. Electrochim. Acta. 55: 5420–5427.

Chen, Y.S., J.H. Huang and C.C. Chuang. 2009. Glucose biosensor based on multiwalled carbon nanotubes grown directly on Si. Carbon 47: 3106–3112.

Chen, W., Y. Ding, J. Akhigbe, C. Brueckner, C.M. Li and Y. Lei. 2010. Enhanced electrochemical oxygen reduction-based glucose sensing using glucose oxidase on nanodendritic poly[meso-tetrakis(2-thienyl)porphyrinato]cobalt(II)-SWNTs composite electrodes. Biosens. Bioelectron. 26: 504–510.

Claussen J.C., S.S. Kim, A.l. Haque, M.S. Artiles, D.M. Porterfield and T.S. Fisher. 2010. Electrochemical glucose biosensor of platinum nanospheres connected by carbon nanotubes. J. Diabetes Sci. Technol. 4: 312–319.

Deng, C.Y., J.H. Chen, Z. Nie and S.H. Si. 2010. A sensitive and stable biosensor based on the direct electrochemistry of glucose oxidase assembled layer-by-layer at the multiwall carbon nanotube-modified electrode. Biosens. Bioelectron. 26: 213–219.

Flanagan, D.E. 2009. The future of glucose sensors—using the technology within a clinical service. Diabetic Med. 26: 195–196.

Gopalan, A.I., K.P. Lee, D. Ragupathy, S.H. Lee and J.W. Lee. 2009. An electrochemical glucose biosensor exploiting a polyaniline grafted multiwalled carbon nanotube/perfluorosulfonate ionomer-silica nanocomposite. Biomaterials 30: 5999–6005.

Heller, A., and B. Feldman. 2008. Electrochemical glucose sensors and their applications in diabetes management. Chem. Rev. 108: 2482–2505.

Kang, X., Z. Mai, X. Zou, P. Cai and J. Mo. 2008. Glucose biosensors based on platinum nanoparticles-deposited carbon nanotubes in sol-gel chitosan/silica hybrid. Talanta 74: 879–886.

Komathi, S., A.I. Gopalan and K.P. Lee. 2009. Fabrication of a novel layer-by-layer film based glucose biosensor with compact arrangement of multi-components and glucose oxidase. Biosens. Bioelectron. 24: 3131–3134.

Lee, K.P., A.I. Gopalan and S. Komathi. 2009. Direct electrochemistry of cytochrome c and biosensing for hydrogen peroxide on polyaniline grafted multi-walled carbon nanotube electrode. Sensor Actuat. B-Chem. B141: 518–525.

Lee, K.P., S. Komathi, N.J. Nam and A.I. Gopalan. 2010. Sulfonated polyaniline network grafted multi-wall carbon nanotubes for enzyme immobilization, direct electrochemistry and biosensing of glucose. Microchem. J. 95: 74–79.

Liu, Z., J. Wang, D. Xie and G. Chen. 2008. Polyaniline-coated Fe_3O_4 nanoparticle-carbon-nanotube composite and its application in electrochemical biosensing. Small 4: 462–466.

Liu, J., C. Guo, C.M. Li, Y. Li, Q. Chi, X. Huang, L. Liao, and T. Yu. 2009. Carbon-decorated ZnO nanowire array: A novel platform for direct electrochemistry of enzymes and biosensing applications. Electrochem. Commun. 11: 202–205.

Muguruma, H., Y. Shibayama and Y. Matsui. 2008. An amperometric biosensor based on a composite of single-walled carbon nanotubes, plasma-polymerized thin film, and an enzyme. Biosens. Bioelectron. 23: 827–832.

Rakhi, R.B., K. Sethupathi and S. Ramaprabhu. 2009. A glucose biosensor based on deposition of glucose oxidase onto crystalline gold nanoparticle modified carbon nanotube electrode. J. Phys. Chem. B 113: 3190–3194.

Reach, G., and C. Choleau. 2008. Continuous glucose monitoring: Physiological and technological challenges. Curr. Diabetes Rev. 4: 175–180.

Tsai, M.C., and Y.C. Tsai. 2009. Adsorption of glucose oxidase at platinum-multiwalled carbon nanotube-alumina-coated silica nanocomposite for amperometric glucose biosensor. Sensor Actuat. B-Chem. B141: 592–598.

Tsai, T.W., G. Heckert, L.F. Neves, Y. Tan, D.Y. Kao, R.G. Harrison, D.E. Resasco and D.W. Schmidtke. 2009. Adsorption of glucose oxidase onto single-walled carbon nanotubes and its application in layer-by-layer biosensors. Anal. Chem. 81: 7917–7925.

Toghill, K.E., and R.G. Compton. 2010. Electrochemical non-enzymatic glucose sensors: a perspective and an evaluation. Int. J. Electrochem. Sc. 5: 1246–1301.

Vnitski, D., K. Artyushkova, R.A. Rincon, P. Atanassov, H.R. Luckarift and G.R. Johnson. 2008. Entrapment of enzymes and carbon nanotubes in biologically synthesized silica: glucose oxidase-catalyzed direct electron transfer. Small 4: 357–364.

Wang, J. 2008. Electrochemical glucose biosensors. Chem. Rev. 108: 814–825.

Wang, H., C. Zhou, L. Jiahua, H. Yu, F. Peng and J. Yang. 2008a. High sensitivity glucose biosensor based on Pt electrodeposition onto low-density aligned carbon nanotubes. Int. J. Electrochem. Sc. 3: 1258–1267.

Wang, Z., S. Liu, P. Wu and C. Cai. 2009a. Detection of glucose based on direct electron transfer reaction of glucose oxidase immobilized on highly ordered polyaniline nanotubes. Anal. Chem. 81: 1638–1645.

Wang, Y., W. Wei, X. Liu and X. Zeng. 2009b. Carbon nanotube/chitosan/gold nanoparticles-based glucose biosensor prepared by a layer-by-layer technique. Mater. Sci. Eng. C. 29: 50–54.

Wang, Y.T., L. Yu, Z.Q. Zhu, J. Zhang, J.Z. Zhu and C. Fan. 2009c. Improved enzyme immobilization for enhanced bioelectrocatalytic activity of glucose sensor. Sensor Actuat. B-Chem. B136: 332–337.

Wen, Z., S. Ci and J. Li. 2009. Pt nanoparticles inserting in carbon nanotube arrays: nanocomposites for glucose biosensors. J. Phys. Chem. C 113: 13482–13487.

Wu, B.Y., S.H. Hou, F. Yin, Z.X. Zhao, Y.Y. Wang, X.S. Wang and Q. Chen. 2007a. Amperometric glucose biosensor based on multilayer films via layer-by-layer self-assembly of multi-wall carbon nanotubes, gold nanoparticles and glucose oxidase on the Pt electrode. Biosens. Bioelectron. 22: 2854–60.

Wu, B.Y., S.H. Hou, F. Yin, J. Li, Z.X. Zhao, J.D. Huang and Q. Chen. 2007b. Amperometric glucose biosensor based on layer-by-layer assembly of multilayer films composed of chitosan, gold nanoparticles and glucose oxidase modified Pt electrode. Biosens Bioelectron. 22: 838–844.

Xu, L., Y. Zhu, X. Yang and C. Li. 2009. Amperometric biosensor based on carbon nanotubes coated with polyaniline/dendrimer-encapsulated Pt nanoparticles for glucose detection. Mater. Sci. Eng. C 29: 1306–1310.

Yang, M., Y. Yang, Y. Liu, G. Shen and R. Yu. 2006. Platinum nanoparticles-doped sol-gel/carbon nanotubes composite electrochemical sensors and biosensors. Biosens. Bioelectron. 21: 1125–1131.

Zhang, Y., G. Guo, F. Zhao, Z. Mo, F. Xiao and B. Zeng. 2010. A novel glucose biosensor based on glucose oxidase immobilized on aupt nanoparticle—carbon nanotube—ionic liquid hybrid coated electrode. Electroanal. 22: 223–228

Zeng, Z., X. Zhou, X. Huang, Z. Wang, Y. Yang, Q. Zhang, F. Boey and H. Zhang. 2010. Electrochemical deposition of Pt nanoparticles on carbon nanotube patterns for glucose detection. Analyst 135: 1726–1730.

Zinc Oxide Nanorods and Their Application to Intracellular Glucose Measurements

Muhammad H. Asif,[1,a,*] Magnus Willander,[1,b] Peter Strålfors[2] and Bengt Danielsson[3,#]

ABSTRACT

Nanostructure metal-oxides have opened innovative and exciting opportunities for exploring intracellular glucose biosensing applications. Zinc oxide (ZnO) is relatively a bio-safe, bio-compatible and large surface area nanostructure with polar surface that can be used for biomedical applications. ZnO nanostructures have been widely explored to develop biosensors with high sensitivity, fast response time and sufficient stability

---

[1]Department of Science and Technology, Campus Norrköping, Linköping University, SE-601 74 Norrköping, Sweden.
[a]E-mails: muhammad.asif@liu.se; mzmasif108@gmail.com
[b]E-mail: magnus.willander@liu.se
[2]Department of Clinical and Experimental Medicine, Division of Cell Biology, Linköping University, SE- 581 85 Linköping, Sweden; E-mail: peter.stralfors@liu.se
[3]Pure and Applied Biochemistry, Lund University, Box 124, SE-221 00 Lund, Sweden.
[#]*Present address:* Acromed Invest AB, Magistratsvägen 10, SE-22643 Lund, Sweden; E-mail: bengtd2001@yahoo.com
*Corresponding author

List of abbreviations after the text.

for the determination of intracellular glucose by electrochemical methods. In this chapter, we report on a functionalised ZnO-nanorod-based selective electrochemical sensor for intracellular glucose. To apply the sensor concept to intracellular glucose measurements, hexagonal ZnO nanorods were grown on the tip of a silver-covered borosilicate glass capillary (0.7 μm diameter) and coated with the enzyme GOD. The enzyme-coated ZnO nanorods exhibited a glucose-dependent electrochemical potential difference versus an Ag/AgCl reference micro-electrode. The potential difference was linear over a wide concentration range of interest. The effect of the hormone insulin, which increased the concentration of intracellular glucose, was also demonstrated. These results demonstrate the capability to perform biologically relevant measurements of glucose within living cells. The ZnO-nanorod glucose electrode thus holds promise for minimally invasive dynamic analyses of single cells. The measured glucose concentration in human adipocytes or frog oocytes using ZnO nanorod sensor was consistent with values of glucose concentration reported in the literature; furthermore, the sensor was able to show that insulin increased the intracellular glucose concentration. This microelectrode device represents a simple but powerful technique to measure intracellular glucose concentration which is of particular importance with regard to investigations into diabetes and its consequences.

1. INTRODUCTION

The application of nanotechnology to medicine, referred to as nanomedicine, has the potential to provide fundamental understanding of phenomena and materials that at the nanoscale have novel properties and functions. Nanotechnology and nanoscience have the capacity to make a new wave of medical techniques through the manufacturing of bioactive nanoscale structures. It can bring fundamental changes to the study and understanding of biological processes in health and disease, as well as enable novel diagnostics and interventions for treating diseases like diabetes. The size domains of the components involved with nanotechnology are similar to that of biological structures. For example, a quantum dot is about the same size as a protein (<10 nm) and drug-carrying nanostructures are the same size as some viruses (<100 nm). Because of this similarity in scale and certain functional properties, nanotechnology is a natural progression of research such as synthetic and hybrid nanostructures that can sense and repair biological lesions and damage, similarly to biological nanostructures. Nanotechnology offers sensing technologies that provide more accurate medical information more rapidly for diagnosing disease, and miniature devices that can manage treatment automatically (Arya et al. 2008). Nanotechnology has introduced new and exciting opportunities by using newly prepared nanostructure materials, for instance glucose biosensors. Because of its ubiquitous role in biology and medicine glucose is one of the most targeted analytes in bioanalysis whether it is in medicine or biotechnology. In biosensor based analysis glucose has attracted a clearly dominating interest due to the importance of glucose measurement for control of diabetes. The huge potential market for an artificial pancreas to optimally treat diabetes by continuously delivering the required concentration of insulin in the patient's blood has stimulated the development of a large variety of glucose biosensors since the presentation of the first glucose sensor by (Clarke and Lyons 1962) based on an amperometric oxygen electrode in combination with the enzyme GOD. Since then many different variations of glucose sensors have been proposed but amperometry has continued to be the dominating transducer technology and GOD the most commonly used enzyme. Enzymes are biological recognition molecules commonly employed in research and development because most chemical reactions in living systems are catalyzed by very specific enzymes. Earlier biosensors based on GOD measured the decrease in oxygen concentration but with time this concept was replaced by measuring the formation of hydrogen peroxide which provided more sensitive determinations. Similar amperometric electrode systems as for oxygen could be used but with reversed polarity. Among other glucose-specific enzymes, hexokinase has found widespread use especially in colorimetric analysis. GDH, such as NAD-dependent GDH and GDH-PQQ (with pyrroloquinoline quinone as cofactor that may be bound to the enzyme) have also attracted interest in order to avoid dependence on dissolved oxygen as for GOD. However, GOD is usually preferred since it is highly stable and specific and also cheap. It can also be used with other electron acceptors than oxygen, such as different quinones and ferrocene (Cass et al. 1984). The development of new amperometric glucose sensors have continued in different directions towards screen-printed devices, miniaturized and implantable devices and sensors incorporating nanomaterials (Santos et al. 2010). Nevertheless the long term stability of the sensors continues to be a serious limitation and in many cases sensors are fabricated only for single measurements and for a short period of time (hours). As a consequence, immobilization strategies for enzymes are important to preserve their biological activity (Wink et al. 1997). In contrast most other biosensors, thermal biosensors operating with a small flow-reactor with immobilized GOD/catalase can work with thousands of samples and year-long stability (Ramanathan and Danielsson 2001). There are many processes and methodologies developed for creating new glucose biosensors such as alternative electrochemical methods (Wang et al. 2008), colorimetry (Morikawa et al. 2002), conductometry (Miwa et al. 1994), optical methods (Mansouri et al. 1984), and fluorescent spectroscopy (Pickup et al. 2005). Among them, the electrochemical glucose sensors have attracted most attention over the last 40 years because of their excellent sensitivity and selectivity. Moreover, electrochemical techniques show lower detection limit, faster response time, better long term stability and inexpensiveness (Mahbubur Rahman et al. 2010). Nanostructure metal-oxides such as ZnO have been extensively investigated to develop biosensors with high sensitivity, fast response time, and stability for the determination of glucose by electrochemical oxidation. In this chapter we demonstrate a functionalized ZnO nanostructure-based electrochemical sensor for selective detection of glucose in single human adipocytes and frog oocytes. Hexagonal ZnO nanorods were grown on the tip of a silver-covered borosilicate glass capillary (diameter 0.7 µm) to make possible microinjection of specific reagents, which can interrupt or activate signal transmission to glucose, into the relatively large cells of adipocytes and oocytes (Asif et al. 2010).

2. Applications to Areas of Health and Disease

An overview of clinical applications of micro- and nanoscale biosensors (Morrison et al. 2008) point at promising developments of a variety of sensor concepts that may lead to improved sensors for the monitoring of pathogenic and physiologically relevant molecules including glucose. Some of the micro-sensor constructions presented in recent years may be able to perform on quite small samples (Xie et al. 1994) and are portable/implantable (Murday et al. 2009), but they are usually not possible to apply in intracellular measurements. In fact, there are not many intracellular biosensors reported in the literature. As one of few examples Abe et al. (Abe et al. 1992) described a sufficiently small glucose micro-sensor with a tip diameter of 2 µm that could provide reliable and accurate measurements of glucose during a few hours. As an alternative to direct measurements in cells, metabolite concentrations can be assessed by indirect, non-invasive measurements using various techniques, such as NMR that will give average concentrations on samples containing several cells (Cline et al. 1998). Other techniques, like those employing fluorescently labeled glucose molecules, may permit real time measurements in single living cells (Yamasda et al. 2000).

There are, however, situations when real-time monitoring in single cells with probes inserted in living cells is a preferred configuration. In our studies of (bio)sensors based on probes covered by ZnO nanorods we have found that it is possible to make probes small enough for glucose measurements inside large cells (Asif et al. 2010; Fulati et al. 2010). Besides glucose, other analytes have also been targeted such as Ca^{2+} (Asif et al. 2009) and pH (Safaa et al. 2007).

3. ZINC OXIDE NANOSTRUCTURE

Zinc oxide (ZnO) is a wide band gap material from group II-VI semiconductors with band gap energy of 3.37eV at room temperature and has a large excitonic binding energy of 60 meV. In addition, it is a piezoelectric, bio-safe and biocompatible material. Zinc oxide is a polar semiconductor with two crystallographic planes with opposite polarity and different surface relaxation energies. This leads to a higher growth rate along the c-axis. The crystal structures formed by ZnO are wurtzite, zinc blende, and rocksalt. ZnO is an important multifunctional material which has wide applications in telecommunications, chemical, and biochemical sensors and optical devices. In this chapter ZnO nanorods will be discussed in the context of electrochemical biosensors to detect glucose in extra- and intracellular measurements.

ZnO nanostructures have received extensive attention due to their eminent performance in electronics, optics and photonics. The synthesis of ZnO nanostructures has been an active field for the last fifteen years because of their applications as sensors, transducers and catalysts. In recent years, semiconducting nanostructures have been the focus of considerable research due to their unique properties that can be exploited in various functional nanodevices. Nanodevice functionality has been demonstrated with these nanostructure materials in the form of electric field-effect switching (Kim et al. 2000), single electron transistors (Stone and Ahmad 1998), biological and chemical sensing (Cui et al. 2001), and luminescence (Huang et al. 2001) for one dimensional semiconducting nanostructure.

Due to the small dimensions of nanowires/nanorods combined with very large contact surface and strong binding with biological and chemical reagents, nanowire/nanorods will have important applications in biological and biochemical research. The diameter of these nanostructures can be engineered comparable to the size of the biological and chemical species being sensed, which intuitively makes them represent excellent primary transducers for producing electrical signals. The structure of ZnO can be described by alternating planes composed of hexagonal with neutral surface to have equal possibility of contribution to the ions with ionicity of around 60%. A literature survey reveals that ZnO nanorods show n-type semiconducting property and their electrical transport is highly dependent on the adsorption/desorption nature of chemical species (Li et al. 2005). Among a variety of nanosensor systems, our nanostructure electrochemical probe is one that can offer high sensitivity and real-time detection. The detection sensitivity of the glucose sensor may be increased to the single molecule level of detection by monitoring the very small changes in electrochemical potential caused by binding of biomolecular species on the surface of the probe.

4. BIOLOGICAL FUNDAMENTALS OF INTRACELLULAR GLUCOSE

Glucose is in many instances a preferred energy substrate in the cell and during hypoxic conditions, as in hard working muscles, glucose conversion in glycolysis is the only means of generating cellular ATP. While most tissues and cells can oxidize fatty acids for their energy supply, the brain is critically dependent on a constant glucose supply. The brain is, moreover, sensitive to too high concentrations of glucose, why the circulating concentration of glucose in the blood is kept within very narrow limits. During a meal when large amounts of glucose is entering the blood from the intestines, insulin stimulates the uptake of glucose into skeletal and heart muscles and into the adipose tissue fat cells. All the cells in the body can take up and oxidize glucose via a family of GLUT proteins that facilitate the diffusion of glucose, down its concentration gradient, through the plasma membrane. Most cell types constitutively express GLUT1, but the insulin sensitive muscle and adipocytes additionally express the insulin regulated glucose transporter GLUT4. GLUT4 is normally residing in intracellularly localized vesicles. Insulin induces the translocation of these GLUT4 containing vesicles to the plasma membrane where the vesicles fuse with the membrane to insert the transporters in the plasma membrane to access extracellular glucose and thus allow glucose uptake to take place.

Inside the cells glucose is immediately phosphorylated by hexokinase, which maintains the downhill concentration gradient of glucose. Thus generated glucose-6-phosphate can then be channelled through glycolysis to complete oxidation by the citric acid cycle and electron transport chain in the mitochondria to generate ATP, CO_2 and H_2O. In many cells, especially in skeletal muscle and liver, a large part of glucose is channelled to storage as polymerized glucose in the form of glycogen.

A characteristic of type 2 diabetes is the failure to maintain normal glucose levels, such that after meal glucose concentrations remain very high for a prolonged time and also during fasting, glucose levels are too high in the blood. This results from a failure to induce glucose uptake by muscle and fat cells, but also from a failure to inhibit release of glucose from the liver. Type 2 diabetes is strongly coupled to obesity and the disease starts as reduced sensitivity of the expanding adipose tissue to insulin. This insulin resistance of the adipocytes is transmitted, perhaps through the excessive release of fatty acids, to skeletal muscle and liver. The pancreatic beta-cells can often for many years mask the insulin resistance by churning out more insulin, but eventually the beta-cells can no longer compensate and type 2 diabetes is manifested. The insulin resistance in skeletal muscle and adipocytes is due to dysfunctional response to insulin because of failure in the intracellular signal transduction from the activated insulin receptor. In the adipocytes the expression of GLUT4 is also down regulated in type 2 diabetes. In type 1-diabetes the beta-cells have been destroyed by an attack of the immune system and the disease is therefore characterized by a lack of insulin and a consequent lack of stimulus to cells and tissues to dispose of glucose. A number of monogenic forms of diabetes result from mutations and consequent dysfunctions of specific proteins in the beta-cells, which make them unable to properly gauge extracellular levels of glucose. MODY2 is due to such a mutation in the gene coding for glucokinase, which in the beta-cells catalyzes the initial phosphorylation of glucose.

Maintenance of a constant well-defined concentration of circulating glucose is thus a fundamental basis for energy homeostasis in the body. Malfunctioning results in disease with dire long term consequences. Presently the incidence of obesity and type 2 diabetes is dramatically increasing around the world. An accurate and rapid measurement of glucose is obviously critical to diagnosis and treatment of the different forms of diabetes, as well as in research to understand the diabetes diseases and to develop new therapies. It is desirable to be able to get continuous measurements not only of circulating glucose, but also of interstitial and of intracellular concentrations in the living human being and in living cells, respectively.

5. INTRACELLULAR POTENTIOMETRIC MEASUREMENTS

Potentiometry is usually used to measure glucose concentration greater than around 10^{-5} M, which is in the physiological range mostly. The potential difference between the reference electrode and the indicator electrode (ZnO nanorods) is measured at zero current flow. The mostly nonpolarizable reference electrode provides a constant potential, while the ZnO nanorods based indicator electrode shows an erratic potential depending on the concentration of glucose. The potential difference at the ZnO nanorod based electrode-electrolyte interface arising from unstable activities of species m in the electrolyte phase (α) and in the electrode phase (β) is related by the following Nernst equation:

$$E = E° + \frac{RT}{Z^m F} \ln \frac{a_m^\alpha}{a_m^\beta}$$

Where E° is the standard electrode potential, R the gas constant, T the absolute temperature, F the faraday constant, a_m the activity of species m, and Z_m the number of moles of electron involved. The selective intracellular glucose measurements were performed by a potentiometric method utilising two electrodes. A ZnO-nanorod-decorated electrode coated with enzyme served as the intracellular working electrode, and an Ag/AgCl electrode was used as the intracellular reference microelectrode. The electrochemical response of the glucose probe was measured with a Metrohm pH meter model 827 versus the Ag/AgCl reference microelectrode, which had been calibrated externally versus an Ag/AgCl bulk reference electrode. This calibration showed approximately constant potential difference using glucose solution with wide concentrations range such as from 0.5 μM to 1000 μM. Subsequently, the potentiometric response of the glucose probe was studied in glucose solutions within the same concentrations range. A very fast response time was noted over the entire concentration range. After the extracellular measurements, the probe was used to selectively measure the intracellular concentration of glucose in two types of cells: human adipocytes (fat cells) and frog oocytes (egg cells).

6. ZnO NANORODS AND REFERENCE MICRO-ELECTRODES

To prepare the sensor and reference electrodes, we affixed the aforementioned borosilicate glass capillaries inside a flat support of the vacuum chamber of an evaporation system (Evaporator Satis CR725) to uniformly deposit chromium and silver films on the outer surface of the capillary tips. After some optimisation, the reference electrode Ag/AgCl tip was electrochemically prepared described in (Asif et al. 2010). The outer end of the Ag/AgCl layer was connected to a copper wire and fixed by means of high-purity-silver conductive paint. To prepare the working electrode, hexagonal single crystals of ZnO nanorods were grown on another silver-coated capillary glass tip using a low-temperature method by ACG, which is a common and cost-effective low-temperature technique. The growth procedure is as follows: the ZnO nanorods were grown on Ag coated tips of borosilicate glass capillaries in a solution of zinc nitrate hexahydrate [$Zn(NO_3)_2 \cdot 6H_2O$, 99.9% purity] and hexamethylenetetramine

($C_6H_{12}N_4$, 99.9% purity). The concentrations of both were fixed at 0.025 M. All the aqueous solutions were prepared in distilled water and we restrict the results to glass tip substrate. The glass capillaries substrates were immersed into the solution and tilted against the wall of the beaker. After that, the beaker was put into an oven at around 93 °C for different length of time to get aligned ZnO nanostructure. Then the substrate was removed from the solution and cleaned with de-ionized water. The as-grown ZnO nanorods on glass tip have been studied by FESEM at different magnifications. The ZnO nanorods were also characterized with HRTEM using a Tecnai G2 UT instrument operated at 200 kV with 0.19 nm point resolution. The TEM specimen was made by scraping the nanorods onto a copper grid with carbon film. The ZnO nanorod layer covered a small part of the silver-coated film. The part of the capillaries covered with ZnO nanorods varied from 3 mm down to 10 μm. The electrical contact was made on the other end of the Ag film for obtaining an electrical signal during measurements.

Careful efforts were taken to ensure sufficiently small tip geometry. Intracellular electrodes must have extremely sharp tips (sub-micrometer dimensions) and must be >10 μm in length. These characteristics are necessary for effective bending and gentle penetration of the flexible cell membrane.

7. PREPARATIONS OF CELLS

Human adipocytes (fat cells) were isolated by collagenase digestion of pieces of subcutaneous adipose tissue obtained during elective surgery at the university hospital in Linköping, Sweden (all patients gave their informed consent, and procedures were approved by the local ethics committee). The adipocytes were incubated overnight before use as described by (Strålfors and Honnor 1989) and used in a Krebs-ringer solution buffered with 20 mM HEPES, pH 7.40 and with additives, as in (Danielsson et al. 2005). A glass slide substrate (5 cm in length, 4 cm in width, and 0.17 mm in thickness) with sparsely distributed fat cells was placed on a prewarmed microscope stage set at 37°C. The indicator electrode and reference electrode were mounted and micromanipulated into the adipocytes.

Female *Xenopus laevis* were anesthetised in a bath with tricaine (1.4 g/L, Sigma-Aldrich, Sweden), and ovarian lobes cut off through a small abdominal incision (procedure approved by the local ethics committee). Oocytes were manually dissected into smaller groups and defolliculated by enzymatic treatment with liberase (Roche Diagnostics, Sweden) for 2.5 hours. Stage-III and -VI oocytes (approximately 1 mm in diameter) without spots and with clear delimitation between the animal and vegetal pole were selected. The experimental procedures are described in more detail by (Börjesson et al. 2010). During measurements, oocytes were placed on a glass slide substrate and bathed in a PBS solution supplemented with 1 mM glucose. Measurements were carried out at room temperature (20–23°C). The indicator electrode and reference electrode were mounted and micromanipulated into the oocytes according to the procedure described for adipocytes.

8. ANALYSIS OF FUNCTIONALIZED ZnO NANORODS BASED INTRACELLULAR GLUCOSE MEASUREMENT

The morphology of the as-grown high-density and aligned ZnO nanorods was investigated by FESEM. Figure 1 shows typical FESEM images, at different magnifications, showing that the ZnO nanorods grown on the glass tip substrate have a rod-like shape with a hexagonal cross section and are primarily aligned along the surface perpendicular direction. HRTEM image and SAED pattern are shown in Fig. 1 (c) with insert revealing that the nanorod long axis is along the [0001] direction.

Fig. 1. SEM and TEM images of ZnO nanorods. Field emission scanning electron microscopy images of the ZnO nanorods grown on Ag-coated glass capillaries, at different magnifications, without (A), with (B, D) and (C) SAED pattern from typical ZnO nanorods using low temperature aqueous chemical solution growth.

The construction of a two-electrode electrochemical potential cell was as follows:

reference electrode | reference electrolyte || test electrolyte | indicator electrode

The electrochemical cell voltage (electromotive force) changed when the composition of the test electrolyte was modified. These changes can be related to the concentration of ions in the test electrolyte via a calibration procedure. The actual electrochemical potential cell can be described by the diagram below:

Ag | ZnO | buffer || Cl^- | AgCl | Ag

The experimental setup for the intracellular measurements is shown in Fig. 2. The response of the electrochemical potential difference of the ZnO nanorods to the changes in buffer electrolyte glucose was measured for the range of 500 nM to 1 mM and shows that this glucose dependence is linear and has sensitivity equal to 42.5 mV/decade at around 23°C (Fig. 3). This linear dependence implies that such sensor configuration can provide a large dynamic range.

The sensing mechanism of the electrochemical glucose sensors is based on an enzymatic reaction catalysed by GOD with β D -glucose, according to the following:

$$H_2O + O_2 + \text{D-Glucose} \xrightarrow{GOD} \text{D-gluconolactone} + H_2O_2 \quad (1)$$

As a result of this reaction, D-gluconolactone and hydrogen peroxide are produced. With H_2O availability in the reaction, gluconolactone is spontaneously converted to gluconic acid, which at neutral pH forms the charged products of gluconate$^-$ and a proton according to the following equation:

$$\text{D-Gluconolactone} \xrightarrow{\text{Spontaneous}} \text{Gluconate}^- + H^+ \quad (2)$$

This proteolytic reaction of D-gluconolactone to gluconic acid shown in equation (2), results in a decrease of the medium pH that can be used for determination of the glucose concentration (Dam et al. 2003). In our case, it is the resulting change in ionic distribution around the ZnO nanorods that causes a change of the overall potential of the ZnO-nanorod electrode. Depending on the sample properties different selective mechanisms may be required to avoid influence by other ions present or other reactions taking place during the measurements. At glucose determination in serum

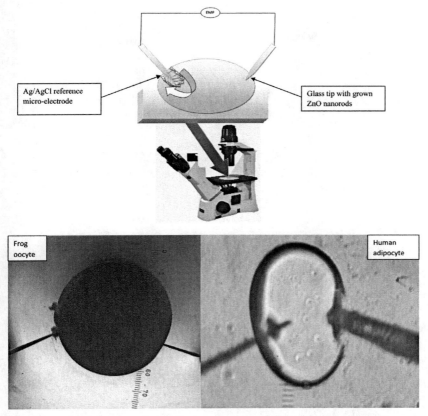

Fig. 2. Experimental setup of intracellular measurements. (a) Schematic diagram illustrating the selective intracellular glucose measurement setup. (b) Microscope images of a single frog (*Xenopus laevis*) oocyte and a single human fat cell (adipocyte) during measurements with a functionalised ZnO-nanorod coated probe as a working electrode and with an Ag/AgCl reference microelectrode.

Color image of this figure appears in the color plate section at the end of the book.

samples by amperometric GOD methods, ascorbic acid and uric acid are well known interferents. In earlier studies (Usman et al. 2010) it was shown that the proposed method was not affected by these compounds. On the other hand the same study showed that the performance of the sensor could be improved by membrane coatings with respect to stability and measuring range. In the measurements described in the actual work the sensors were not used for repeated measurements in the cells and the measurement conditions were more constant. Sensor performance and stability were quite acceptable without any protective membranes or other selective measures.

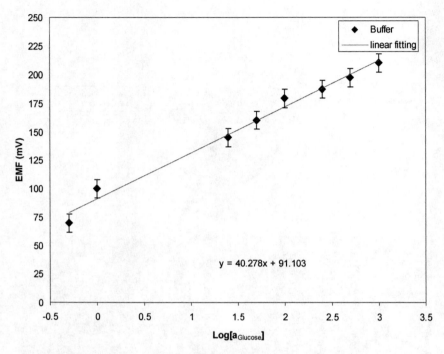

Fig. 3. Calibration curve between EMF and glucose concentration. A calibration curve showing the electrochemical potential difference versus the glucose concentration (0.5–1000 µM) using functionalised ZnO-nanorod-coated probe as a working electrode and an Ag/AgCl microelectrode reference microelectrode.

First, we used the nanosensor to measure the free concentration of intracellular glucose in a single human adipocyte. The glucose-selective nanoelectrode, mounted on a micromanipulator, was moved into position in the same plane as the cells. The ZnO nanoelectrode and the reference microelectrode were then gently micromanipulated a short distance into the cell (Fig. 2). Once the ZnO nanorod working electrode and the Ag/AgCl reference microelectrode were inside the cell, that is, isolated from the buffer solution surroundings, an electrochemical potential difference signal was detected and identified as the presence of glucose. The intracellular glucose concentration was estimated to be 50 ± 15 µM (n = 5). This can be compared with the 70 µM intracellular concentration determined by nuclear magnetic resonance spectroscopy in rat muscle tissue in the presence of a high, 10 mM, extracellular glucose concentration (Cline et al. 1998). Insulin stimulates glucose uptake by causing translocation of insulin-sensitive GLUT4 to the plasma membrane; GLUT4 then allows glucose to enter the cell along a concentration gradient. The role of insulin was investigated by adding 10 nM insulin after getting stable potential for intracellular measurement. About one and half minute after the insulin addition, the glucose concentration in the cell increased from 50 ± 15 µM to 125 ± 15 µM after few minutes (Fig. 4).

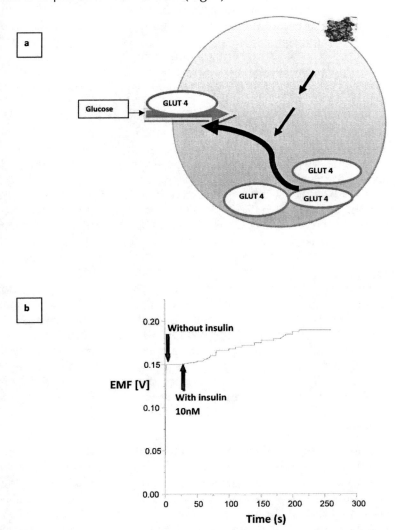

Fig. 4. Activation of glucose uptake and output response with insulin. (a) Intracellular mechanism for insulin-induced activation of glucose uptake. (b) Output response with respect to time for intracellularly positioned electrodes when insulin is applied to the extracellular solution.

Color image of this figure appears in the color plate section at the end of the book.

In another set of experiments, we used the nanosensor to measure intracellular glucose concentration in single frog oocytes. The intracellular glucose concentration was 125 ± 23 µM (n = 5). This is slightly higher than what has been reported earlier (<50 µM; Umbach et al. 1990). We do not know the reason for this difference, but one possibility is that the electrodes behave slightly differently inside the oocyte than outside, where they were calibrated. However, to test whether the electrode is measuring the glucose concentration inside the oocytes, we added 10 nM insulin to the cell medium to stimulate glucose uptake. Indeed, the glucose concentration in the frog oocytes increased from 125 ± 23 µM to 250 ± 19 µM.

The viability of the penetrated cells strongly depends on the size of the ZnO nanorods. By reducing the length of the ZnO nanorods, the total diameter of the tip will be reduced, which in turn increases the cell viability, and the sensitivity of the device is also expected to increase. The morphology of the functionalised ZnO intracellular sensor electrode was checked by scanning electron microscopy directly after measurements, shown in the images of Fig. 5. Obviously some components from the cell and the cell membrane adhere to the probe and possibly this contamination occurs mainly when the probe is pulled out from the cell. In any case the glucose response of the electrode does not seem to be affected, which is in line with what could be expected from a potentiometric device as long as the blockage of the active surface is only partial. If proper cleaning in deionised water is performed, the immobilised GOD will retain its enzymatic activity due to the strong electrostatic interaction between ZnO and GOD. We have attempted to clean the stuck cell components from the electrode after intracellular measurements. Figure 5b shows the immobilized electrode after cleaning. As clearly seen in the figure, the immobilized ZnO nanorods are still in good condition and that some residues form the cell components are still stuck to the electrode.

Conclusions

In conclusion, we have demonstrated that functionalised hexagonal ZnO nanorods grown on sub-micron silver-covered capillary glass tips works as a selective sensor for intracellular glucose concentration in single human adipocytes and frog oocytes, a technique offering new approaches in the investigation of the molecular basis of both type 1 diabetes and type 2 diabetes. The functionalised GOD retained its enzymatic activity due to excellent electrostatic interaction between ZnO and GOD. The proposed intracellular biosensor showed a fast response with a time constant of less than 1 s and had quite a wide linear range from 0.5 µM to 1000 µM. The performance regarding sensitivity, selectivity, and freedom

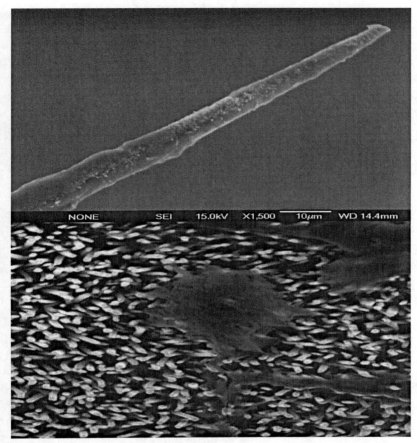

Fig. 5. SEM images after intracellular measurements. Scanning electron microscopy images showing the working electrode after intracellular measurements.

from interference when the sensor was exposed to intra- and extracellular glucose measurements were quite acceptable. The stability of the sensing ZnO layer was, however, limited and should be improved although the experiments described here have a short duration and could be performed without influence of this drawback. The effect of the hormone insulin, which increased the concentration of intracellular glucose, was also demonstrated. These results demonstrate the capability to perform biologically relevant measurements of glucose within living cells. The ZnO-nanorod glucose electrode thus holds promise for minimally invasive dynamic analyses of single cells. All of these advantageous features can make the proposed biosensor applicable in medical, food and other areas. Moreover, the fabrication method is simple and can be extended to immobilise other enzymes and other bioactive molecules with low isoelectric pH for a variety of biosensor designs. We also plan to take the same approach for other intracellular activity measurements.

Key Facts

- Zinc oxide is a semiconductor with a direct and wide band gap of 3.37 eV and a large excitation binding energy (60 meV). It can act as both a UV absorber and emitter and can be used to develop solid state devices.
- Relatively zinc oxide forms bio-safe and bio-compatible nanostructures with polar surface that can be used for biomedical applications.
- The extremely large surface area makes ZnO nanorods attractive for sensing.
- Zinc oxide can be easily fabricated on cheap substrates such as glass and into different kinds of nanostructures. The diameter of ZnO nanorods can be comparable with the size of biochemical species.
- Zinc oxide is a piezoelectric material which is a key property in building electromechanically coupled sensors and transducers.
- Glucose metabolism and insulin control of glucose metabolism in cells are key features to address in understanding the pathogenesis of diabetes as well as in developing new treatments for the disease.

Summary

- Current research trends include nanostructure applications to intracellular environment for glucose measurement, of particular interest in research on the molecular basis of type 1 and type 2 diabetes.
- ZnO nanostructures have been explored in many biological applications because ZnO is relatively a bio-safe and bio-compatible polar semiconductor material.
- The ability of solid and hollow ZnO nanorods to penetrate into the cells offers the potential for the delivery of drug molecules and nucleic acid.
- The structure of ZnO can be described by alternating planes with opposite polarity composed of hexagonal with neutral surface to have equal possibility of contribution to the ions with ionicity of around 60%.
- There is the possibility to make the diameter of these nanostructures comparable to the size of the biological and chemical species being sensed, which intuitively makes them represent excellent primary transducers for producing electrical signals.
- The innovative ZnO nanorod based materials are greatly promising for nanomedicine applications.

Definitions

Biosensor: A biosensor is an analytical device which converts a biological quantity into an electrical signal. The name biosensor signifies that the device is a combination of two parts: a bioelement and a sensor element in close proximity to each other.

Transducer: A transducer is a device that converts an observed change (physical or chemical) into a measureable signal. The word transducer is derived from the Latin verb *traduco*, which means a device that transfers energy from one system to another in the same or another form. Transducers can be optical, electrochemical, mass sensitive or mechanical thermal etc. The transduction mechanism relies on the interaction between the surface and the analyte directly or through mediators.

Electrochemical biosensors: Electrochemical biosensors respond to electron transfer, electron consumption, or electron generation during a chem/bio-interaction process. This class of sensors is of major importance and they are more flexible to miniaturization than most other biosensors. They are further divided into conductometric, potentiometric, and amperometric devices.

Potentiometric biosensors: In potentiometric sensors the potential change due to the accumulation of charge (electrons) on the working electrode is measured relative to a reference electrode when no current is flowing. The working electrode potential must depend on the concentration of the analyte in the solution. The reference electrode is needed to provide a defined reference potential.

Membrane for selectivity: Biosensors are usually covered with a thin membrane that has several functions, including diffusion control, reduction of interference and mechanical protection of the sensing probe. Commercially available polymers, such as polyvinyl chloride (PVC), polyethylene, polymethacrylate and polyurethane are commonly used for the preparation of the functionalization interface due to their suitable physical and chemical properties.

Biosafe and biocompatible material: Those materials do not irritate the surrounding structure in biosystems such as cells. Bio-safe materials do not provoke an abnormal inflammatory response and do not stimulate allergic or immunologic reactions.

Glucose: Also known as grape sugar or corn sugar, glucose is a fundamental carbohydrate in biology. Glucose is one of the main products of photosynthesis and serves as the human body's primary source of energy.

Diabetes: When energy homeostasis is not properly controlled, such as in obesity, glucose levels in the bloodstream are not properly controlled by insulin and diseases such as diabetes can develop.

Abbreviations

ZnO	:	Zinc Oxide
GOD	:	Glucose Oxidase
ATP	:	Adenosine triphosphate
GLUT	:	Glucose transporter
MODY2	:	Maturity onset diabetes of the young
GDH	:	Glucose dehydrogenases
NAD	:	Nicotineamide adenine dinucleotide
PQQ	:	Pyrroloquinoline quinone
PBS	:	Phosphate buffer saline
ACG	:	Aqueous chemical growth
FESEM	:	Field emission scanning electron microscopy
HRTEM	:	High resolution transmission electron microscopy
SAED	:	Selected area electron diffraction
NMR	:	Nuclear magnetic resonance
HEPES	:	4-(2-hydroxyethyl)-1-piperazineethanesulfonic acid

References:

Abe, T., Y.Y. Lau and A.G. Ewing. 1992. Characterization of glucose microsensors for intracellular measurements. Anal. Chem. 64: 2160–2163.

Al-Hilli S.M., M. Willander, A. Öst. and P. Strålfors. 2007. ZnO nanorods as an intracellular sensor for pH measurements. J .Appl. Phys. 102: 084304–5.

Asif, M.H., Syed M. Usman Ali, O. Nur, M. Willander, Cecilia Brännmark, Peter Strålfors, Ulrika Englund, Fredrik Elinder and B. Danielsson. 2010. Functionalised ZnO-nanorod-based selective electrochemical sensor for intracellular glucose. Biosensors Bioelectron. 25: 2205–2211.

Asif, M.H., A. Fulati, O. Nur, M. Willander, Cecilia Brännmark, Peter Strålfors, S.I. Börjesson and Fredrik Elinder. 2009. Functionalized zinc oxide nanorod with ionophore-membrane coating as an intracellular Ca^{2+} selective sensor. Appl. Phys. Lett. 95: 023703–5.

Arya, A.K., L. Kumar, D. Pukharia and K. Tripathi. 2008. Application of nanotechnology in diabetes. Digest journal of Nanomaterials and Biostructures 3: 221–225.

Börjesson, S.I., T. Parkkari, S. Hammarström and F. Elinder. 2010. Electrostatic tuning of cellular excitability. Biophys. J. 98: 396–403.

Cass, A.E.G., G. Davis, G.D. Francis, H.A.O. Hill, W.G. Aston, I.J. Higgins, E.V. Plotkin, L.D.L. Scott and A.P.F. Turner. 1984. Ferrocene mediated enzyme electrode for amperometric determination of glucose. Anal. Chem. 56: 667–671.

Clark, L.C. Jr., and C. Lyons. 1962. Electrode systems for continuous monitoring in cardiovascular surgery. Ann. N. Y. Acad. Sci. 102: 29–45.

Cline, G.W., B.M. Jucker, Z. Trajanoski, and A.J.M. Rennings and G.I. Shulman. 1998. A novel ^{13}C NMR method to assess intracellular glucose concentration in muscle, in vivo. Am. J. Physiol. 274: Endocrinol Metab. 37: E381–E389.

Cui, Y., Q. Wei, H. Park and C.M. Lieber. 2001. Nanowire Nanosensors for Highly Sensitive and Selective Detection of Biological and Chemical Species. Science 293: 1289–1292.

Dam, T.V., D. Pijanowska, W. Olthuis and P. Bergveld. 2003. Highly sensitive glucose sensor based on work function changes measured by an EMOSFET. Analyst 128(8): 1062–1066.

Danielsson, A., A. Öst, E. Lystedt, P. Kjolhede, J. Gustavsson, F.H. Nyström and P. Strålfors. 2005. Insulin resistance in human adipocytes occurs downstream of IRS1 after surgical cell isolation but at the level of phosphorylation of IRS1 in type 2 diabetes. FEBS J. 272: 141–151.

Fulati, A., S.M.U. Ali, M.H. Asif, N. Alvi, M. Willander, C. Brännmark, P. Strålfors, S.I Börjesson, F. Elinder and B. Danielsson. 2010. An intracellular glucose biosensor based on nanoflake ZnO. Sensors and Actuators B 150 (2010) 673–680.

Huang, M.H., S. Mao, H. Feick, H. Yan, Y. Wu, H. Kind, E. Weber, R. Russo and P. Yang. 2001. Room-Temperature Ultraviolet Nanowire Nanolasers. Science 292: 1897.

Kim, G.T., J. Muster, V. Krstic, J.G. Park, Y.W. Park, S. Roth and M. Burghard. 2000. Field-effect transistor made of individual V_2O_5 nanofibers. Appl. Phys. Lett. 76: 1875–77.

Li, Q.H, T. Gao, Y.G. Wang and T.H.Wang. 2005. Adsorption and desorption of oxygen probed from ZnO nanowire films by photocurrent measurements. Appl. Phys. Lett. 86: 123117–19.

Mahbubur Rahman, Md., and A.J. Saleh Ahmad, J.H Jin, S.J. Ahn and J.J. Lee. 2010. A Comprehensive Review of Glucose Biosensors Based on Nanostructured Metal-Oxides. Sensors 10: 4855–4886.

Mansouri, S., and J.S. Schultz. 1984. A Miniature Optical Glucose Sensor Based on Affinity Binding. Nature Biotech. 2: 885–890.

Miwa, Y., M. Nishizawa, T. Matsue and I. Uchida. 1994. A Conductometric Glucose Sensor Based on a Twin-Microband Electrode Coated with a Polyaniline Thin Film. Bull. Chem. Soc. J. p. 67: 2864–2866.

Morikawa, M., N. Kimizuka, M. Yoshihara and T. Endo. 2002. New Colorimetric Detection of Glucose by Means of Electron-Accepting Indicators. Chem. Eur. J. 8: 5580–5584.

Morrison, D.W.G, M.R. Dokmeci, U. Demirci and A. Khademhosseini. 2008. Clinical applications of micro- and nanoscale biosensors, in Biomedical Nanostructures (eds. K.E. Gonsalves, C.L. Laurencin, C.R. Halberstadt, L.S. Nair) John Wiley & Sons. pp. 433–454.

Murday, J.S., R.W. Siegel, J. Stein and J.F. Wright. 2009. Translational nanomedicine: status assessment and opportunities. Nanomedicine: Nanotechnology, Biology, and Medicine 5: 251–273.

Pichup, J.C., F. Hussain, N.D. Evans, O.J. Rolinski and D.J.S. Birch. 2005. Fluorescence-Based Glucose Sensors. Biosens. Bioelectron. 20: 2555–2565.

Ramanathan, K., and B. Danielsson. 2001. Principles and applications of thermal biosensors. Biosens. Bioelectron. 16: 417–423.

Santos, A.N, D.A.W. Soares and A.A.A. de Queiroz. 2010. Low potential stable glucose detection at dendrimers modified polyanailine nanotubes. Materials Research 13: 5–10.

Stone, N.J., and H. Ahmed. 1998. Silicon single electron memory cell. Appl. Phys. Lett. 73: 2134–36.

Strålfors, P., and R.C. Honnor. 1989. Insulin-induced dephosphorylation of hormone-sensitive lipase correlation with lipolysis and CAMP-dependent protein kinase activity. Eur. J. Biochem. 182: 379–385.

Usman, A.S.M., O. Nur, M. Willander and B. Danielsson. 2010. A fast and sensitive potentiometric glucose microsensor based on glucose oxidase coated ZnO nanowires grown on a thin silver wire. Sensors and Actuators B 145: 869–874.

Wang, Y., H. Xu, J. Zhang and G. Li. 2008. Electrochemical sensors for Clinic Analysis. Sensors 8: 2043–2081.

Wink, T., S.J. van Zuilen, A. Bult and W.P. van Bennekom. 1997. Self-assembled Monolayers for Biosensors. Analyst 122: 43R.

Xie, B., U. Harborn, M. Mecklenburg and B. Danielsson. 1994. Urea and lactate determined in 1-µL whole blood with a miniaturized thermal biosensor. Clin. Chem. 40: 2282–2287.

Yamada, K., M. Nakata, N. Horimoto, and M. Saito, H. Matsuoka and N. Inagaku. 2000. Measurement of glucose uptake and intracellular calcium concentration in single, living pancreatic β-cells. J. Biol. Chem. 275: 22278–22283.

Section 3: Insulin

8

Nanoprobes to Monitor Cell Processes in the Pancreas

Claire Billotey,[1] Caroline Aspord,[2] Florence Gazeau,[3] Pascal Perriat,[4] Olivier Tillement,[5] Charles Thivolet[6] and Marc Janier[7]

ABSTRACT

In vivo detection of specific cells implies the labeling of the cells with a physical agent locatable via imaging. Taking into account the sensitivity of scintigraphic methods, these should be the methods of choice, but their application is limited for exploration of the pancreas. Developments are oriented towards the utilization of nanoprobes detectable by MRI. Thus, a clinical study has recently

---

[1] E-mail: claire.billotey@univ-lyon1.fr
[2] E-mail: Caroline.Aspord@Efs.Sante.Fr
[3] E-mail: Florence.Gazeau@Univ-Paris-Diderot.Fr
[4] E-mail: Pascal.Perriat@Insa-Lyon.Fr
[5] E-mail: olivier.tillement@univ-lyon1.fr
[6] E-mail: Charles.Thivolet@Chu-Lyon.Fr
[7] E-mail: Janier@Univ-Lyon1.Fr

List of abbreviations after the text.

reported the possibility to detect using MRI macrophages (thanks to IONP labeling) which are involved insulitis in recent-onset T1D patients following. Certain limitations are linked to the iron-based nanoprobes, notably weak detection sensitivity and the absence of signal quantification, which explains the development of other types of nanoprobes, notably gadolinium-based nanoprobes. The preclinical validation tests must verify several aspects, one very essential being that the functional properties of the labeled cells are not altered.

Cellular imaging of the pancreas is in its infancy, and is based upon the development of nanoprobes for cellular imaging via MRI following *in vitro* labeling of lymphocytes or *in vivo* labeling of macrophages; cellular imaging of the pancreas also relies on radio-labels detectable by PET for the imaging of β-cells labeled *in vivo*.

INTRODUCTION

In vivo identification and quantification of the presence of specific cells implies the ability to label the cells with a physical agent inducing a signal which can be detected and measured by an external method. The cellular **imaging** performed here consists in creating/generating a mapping of the distribution of the tagged cells.

In routine clinical practice, this type of image has been obtained for more than 20 years by the use of radio-pharmaceuticals which enable, for example, the detection of an infection or digestive bleeding site.

Scintigraphy techniques cannot be used to monitor a cellular process for more that 2–3 days, due to the radioactivity decreasing. Furthermore, their use is limited for the exploration of the pancreas which is located in the vicinity of areas of non-specific high uptake, such as hepatic or splenic regions, with an intense spill-over phenomenon.

The labeling of cells with a stable physical agent inducing a signal detectable via MRI has been proposed since 1995, in order to combine the advantages of an imaging providing anatomical details and the possibility of monitoring the cells of interest over a long period. Since then, metallic agents have been used in nanoparticle form or in macro-molecular structures. Polysaccharide or lipid complexes, as well as an organic-sulfur treatment or dendrimers are the basis of these structures allowing, thanks to their physicochemical properties, the solubilisation of these complexes and their interaction with the cells of interest and therefore their cell internalization.

The purpose of this chapter is to describe:
1. The objectives and general principles of in *vivo* cellular imaging in human;
2. The tools and the different technological limits of cellular MRI;
3. The actual and in development tools of cellular MRI applied to the exploration of the pancreas.

MAIN TEXT

I. *In vivo* Cellular Imaging

1. *Purposes*

The general purposes of *in vivo* cellular imaging of the pancreas are:
- To detect at the earliest possible stage and to quantify a lymphocyte infiltration process of the islets of Langerhans, preceding the start of T1D;
- To quantify the mass of functional β cells.

2. *General principles of cellular imaging*

The identification, using *in vivo* imaging, of a specific cell population implies its tagging with a labeling agent which induces a physical signal which can be detected by an external means using different available methods of medical imaging. Those having the highest sensitivity, i.e., the scintigraphic methods (SPECT and PET), are the most appropriate for the detection of a signal of low intensity, except for the exploration of the pancreas. MRI, therefore, has its place among the cellular imaging of the pancreas.

a. *The different types of cells labeling*

Two different types of labeling can be used. The first type is performed by direct incubation of the cells of interest with the labeling agent *in vitro*; the labeled cells are then injected into the organism. This method is called **in vitro or direct labeling**. This implies that the autologous cells can be extracted from the organism, as this is typically the case for blood cells (erythrocytes, leukocytes, or platelets). In the second type of labeling, the cells of interest are labeled *in vivo*. Hence the labeling agent, which is injected *in vivo* needs, to be vectorized towards the cells in place. This method is called **in vivo or indirect labeling**.

i. Indirect or *in vivo* labeling

During indirect labeling, the target cells to be identified and located are the autologous cells "in place" *in vivo*. In order to label these cells, it is necessary to use a specific targeting probe of these cells which will lead the labeling agent to the targeted cells. Typically, one can use an antibody specifically recognized by membrane antigens, peptides, or other analogues of specific receptor ligands or membrane carriers of the targeted cells.

The first difficulty for the identification of the cells of interest by indirect labeling is the absence of absolute specificity of the cellular targets The basis of the location of the targeted cells lies on the over-expression of these cellular targets, and thus on the basis of a ratio of signal between the tissue where the cells to be targeted are located and the surrounding tissues.

Antibodies against the epitope of the carcino-embryonic antigen carried by the granulocytes radio-labeled with ^{99m}Tc are used for the detection of infection sites. However, as the liver is widely involved in the elimination of peptides and many other biological molecules, the liver activity is often very intense in SPECT examinations conducted with an indirect labeling, and an important spill-over phenomenon is the cause of great difficulties in the analysis of capture at the pancreatic level. Therefore, the development of radiopharmaceuticals recognizing specifically the lymphocytes such as the interleukine-2 labeled with ^{123}I (Signore et al. 1994) does not allow for efficient detection of insulitis with SPECT.

The better resolution and sensitivity of PET than SPECT allows researchers to overcome the spill-over problem, and PET tracers very specific of pancreatic β cells open a new prospect regarding the quantification of the β-cell mass.

The efficiency of the indirect labeling relies also on the target accessibility. Thus, following parenteral injection, the labeling agent will have to cross the capillary wall, except in case of the targeting of endothelial cells, and to be able to interact with the target as a natural ligand. The size and the steric effect of the complexes are parameters which have an important impact on the bioavailability of these agents.

Macrophage endocytosis of those macromolecular or particular complexes is the basis of the *in vivo* labeling of these cells. This can occur directly after the parenteral injection of the labeling agent, or by the internalization of the labeling agent released by the initially labeled cells after their lyses, providing a non-specific signal.However, this process can be used to demonstrate an invasive tumor in tissue rich in RES cells, such as the liver or the lymph nodes, or an inflammatory process such as insulitis, as will be shown below. For example, intravenous injection of

IONP permits a complete appraisal using MRI of all intrahepatic tumor lesions (areas in which Küpfer cells have disappeared, forming zones that do not uptake the labeled agent).

ii. Direct or *in vitro* labeling

This labeling is by nature **specific,** since only the cells of interest are labeled, even though the labeling agent does not have any specificity for a given type of cell. This technique, however, requires a procedure that isolates and labels the cells, resulting in significant costs for the clinical practice.

Typically, this type of labeling permits the detection of local accumulation sites of specific heterologous or autologous cells.

In this way, autologous leucocytes or erythrocytes are radio-labeled *in vitro* with a lipophilic molecule, such as ^{99m}Tc hexamethylpropylene-amine-oxime or Ox^{111}In. Then, following intravenous reinjection, SPECT permits the detection of leukocyte diapedesis as observed in the cases of an infected bone or vessel prosthesis, an intestinal inflammation or abscess, or an area of intestinal hemorrhage. Nonetheless, due to the very elevated level of activity in the liver and spleen (depending upon the type of radio-label used), this technique cannot be used to determine the infiltration of the pancreatic lymph nodes and the islets of Langerhans (insulitis) using autologous lymphocytes.

It is also possible to label cells *in vitro* by transducting cells to express a gene that will be the origin of a protein and will serve as an imaging signal source. These methods are particularly interesting, as they allow for evaluating the viability of the cells over time and for monitoring cells with a strong potential for multiplication; however, they are difficult to use in clinical studies.

b. Preferred indications for the different labeling methods

Indirect cellular labeling is used for the detection or *in situ* characterization of specific cells such as tumor cells, macrophages, or lymphocytes. Imaging provides a snapshot of cellular populations, and the use of iterative images allows for the monitoring of a pathological process or the evaluation of the impact of therapy (Table 1).

Direct labeling is most commonly used for the detection of infectious processes. This method can also be used to monitor physio-pathological processes such as the invasion of the pancreas by T cells, the onset and progression of T1D and T2D, or the clinical efficacy of cellular therapies. Due to proximity of the pancreas to the liver and the spill-over phenomenon, SPECT methods cannot be used in this area, which explains the development of MRI labels.

Table 1. Main characteristics and indications of the different types of labeling (direct and indirect) (as unpublished material of the author).

Characteristics	Indirect labeling	Direct labeling
Specificity	Depends on the specificity of the targeting molecule	+++
Stability/dilution	Absence of dilution The image reflects the status of the cells at one specific moment	Depends on the stability of the contrast agent and of the bond cell-contrast agent, and on the division grade of the labeled cells
Physiology	+++ (*In situ* labeling of cells)	++ (The cells of interest can be activated by the labeling process)
Preferential indications	*In situ* detection of specific cells	Longitudinal follow-up of cell migration, yet requiring stability of the contrast agent

3. Technological limits of cellular imaging

a. Sensitivity of detection

The sensitivity of cellular imaging can be defined as the number of cells detected per unit volume.

Regardless of whether direct or indirect labeling is used, the signal given off by the labeling agent will be weak. The type of imaging technique used must, therefore, have elevated intrinsic detection sensitivity. As the sensitivity of SPECT and PET is elevated, i.e., 10^3 and 10^4 higher than MRI, respectively, these are the methods of choice for molecular and functional imaging, allowing investigators to use very low radio-pharmaceutical doses ("trace" doses). Optical imaging techniques are more sensitive, but the existence of a significant; non-specific signal (auto-fluorescence) results in a contrast lower than that obtained with scintigraphic techniques. In addition, the elevated absorption of the fluorescence by biological tissues only permits imaging in the mouse or the exploration of superficial areas in humans.

The issue of sensitivity is more relevant to the methods of indirect labeling, related to the lower SBR and labeled cell number than in case of direct labeling. Therefore, only scintigraphic methods are pertinent and, in the case of the pancreas, only PET can be used.

b. Stability of the labeling

i. Stability of the bond between cells and the labeling agent

In order to deduce the signal detected in the presence of labeled cells, the bond between the cell and the labeling agent must be stable. Stability would be increased in the presence of internalization of the labeling agent, even though an efflux process is possible. As in the case of cell lysis, there can be a secondary internalization by macrophages.

ii. Stability of the labeling over time

This stability impacts the duration for studying cellular processes and depends on two parameters: the physical stability of the labeling agent and the absence of label dilution. Direct cellular labeling by PET tracers (^{18}FDG, ^{64}Cu, or ^{55}Co) is rather unstable due to an important efflux, particularly with ^{18}FDG (Prince et al. 2008), which makes them unsuitable for studying cellular processes on account of their very short physical period. Radiopharmaceuticals used in SPECT display short physical period, which limit their radio-biological effects. The use of ^{111}In prolongs *in vivo* exploration for 2 to 3 days. Metal oxides used as labeling agents in MRI have a large physical stability, which permits studies over several weeks (Billotey et al. 2005 & Fig. 1). Only the incorporation of a gene coding for the labeling agent within the genome prohibits the dilution of the label due to cellular divisions.

c. Absence of altered cellular functional properties

The labeling process must not alter the viability or the functionality of cells. An exhaustive series of *in vitro* tests must be performed, depending on the

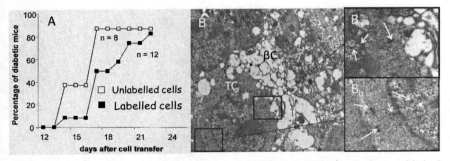

Fig. 1. *In vivo* evaluation of the function of *anionic IONP*–loaded CD3$^+$ cells (as unpublished material of the author). A plot illustrating the delay in diabetes occurrence in two male NOD mice populations following the transfer of 5x10^6, unlabeled (white squares) and labeled with anionic IONP (black squares), CD3$^+$-TC. The identical level of diabetes cases at Day 23 in the two populations supports the functional integrity of the CD3$^+$-TC labeled cells. B plot illustrating *TEM analysis* on pancreatic slices taken 24 days following transfer of CD3$^+$-TC labeled cells, revealing a necrotic β cell (βC) with the presence of several CD3$^+$-TC (TC) adhering to its cell membrane; magnification focalized on CD3$^+$-TC (B$_a$ and B$_b$) reveals the accumulation of multiple clusters of IONP (arrows), mostly in nucleus. These results allow to confirm the functional integrity of CD3$^+$-TC labeled cells, as they cause β-cell necrosis; and to support the stability of the *in vivo* labeling.

cell types. The results of these tests on labeled cells must be compared to those undertaken on unlabeled cells. The effect of a given labeling agent may differ from one type of cell to another and therefore, the tests must be performed for each particular type of cell.

For dendritic cells or macrophages, it is essential to check that there has been no activation triggered by the labeling process nor any altered "activability" of these cells.

Finally, altered similar behavior *in vivo* for labeled and unlabeled cells validates the labeling process (Fig. 1).

d. Absence of general toxicity of the labeling agent

The labeling agent must not exhibit overall toxic effects.

Taking into account the high sensitivity of scintigraphic methods, labeling agents for SPECT and PET are used at trace doses without any toxic effects. In this vein, $Ox^{111}In$ and ^{99m}Tc hexamethylpropylene-amine-oxime (CERETEC®) are accredited by the majority of national regulatory agencies for the leukocyte labeling.

It has been shown that after 2 to 4 weeks, the iron of IONP incorporated in macrophages is totally biodegraded, the iron being incorporated in hemoglobin (Weissleder et al. 1989). Several types of IONP have received regulatory approval for the imaging of hepatic tumors (Endorem®, Resovist®, and Teslacan®) and for lymphograhy (Sinerem® and Combidex®) following parenteral (intravenous or intradermic) injection.

e. Multi-modal detection

In order to combine, in the clinical practice, the advantages of scintigraphy (sensitivity of detection and quantitative concepts) with those of MRI (anatomical location), certain studies (de Vries et al. 2005) have been performed co-injection of cells labeled with $Ox^{111}In$ or IONP.

II. Factors determining the *in vivo* detectability of labeled cells using MRI

In order to generate a contrast in ^1H-MRI, metallic, paramagnetic (gadolinium or manganese) or super-paramagnetic (iron oxide) complexes are used. MRI spectrometry information can be superimposed on the anatomical image information obtained with ^1H-MRI, when using ^{19}F for example.

In order to facilitate their cellular internalization, these agents are integrated within the monomeric and polymeric, lipid or charged structures. The metal is simply chelated to these structures (gadolinium, ^{19}F) or it forms the core particle (iron or gadolinium oxides).

1. The process of intracellular internalization of nanoprobes

The internalization of the macromolecular or particular structures is accomplished via an endocytosis process that corresponds to phagocytosis, pinocytosis, or endocytosis mediated by a receptor.

The importance of the endocytosis process depends on the following factors:

- the nature of the cells: raised in macrophages;
- the size of the particles and especially the charge on their surface: the very large particles, charged negatively or positively, are strongly endocytosed as are certain small, negatively-charged particles (Wilhelm et al. 2003);
- the duration of incubation and the concentration of particles in the incubation medium (Fig. 2).

The direct labeling of non-macrophagic cells requires the transforming and adapting of the particle coating in order to improve cell penetration of the probe. Thus, the use of an organic matrix has been proposed, as it confers to the particles highly-polarized surface charges that form rapid electrostatic bonds with the cellular membrane allowing efficient labeling of human and murine progenitors as well as human carcinoma cells (Bulte et al. 2001). External treatment of the iron core with DMSA provides negative surface charges and facilitates internalization in macrophages (Wilhem et al. 2003) as well as in lymphocytes (Billotey et al. 2005). The coupling of a transfectant agent with the surface of dextran particles such as HIV-TAT peptide permits efficacious labeling of lymphocytes (Moore et al. 2002). But it also leads to an activation of T cells with a risk of inducing apoptosis (Wu and Schlossman 1997), which may interfere with immune processes under evaluation. Other transfectant agents are currently used in clinical studies, such as protamine sulfate (Arbab et al. 2004) or electroporation (Walczak et al. 2005).

In order to promote the intracellular internalization of gadolinium, which is very weak in chelates such as those used as intravascular contrast agents, it is chelated with amino-modified polymers (Modo et al. 2002), lipid monomer (Vuu et al. 2005), cationic liposome (Oliver et al. 2006), or quantum dots (Oostendorp et al. 2010). In a chelated or oxide form, gadolinium can form the core of microemulsion particles (Tseng et al. 2010) or inorganic nanoparticles functionalized on the surface (Fizet et al. 2009).

2. Sufficient cellular content

Corresponds to the concentration of the probe per cell allowing clinical interpretation.

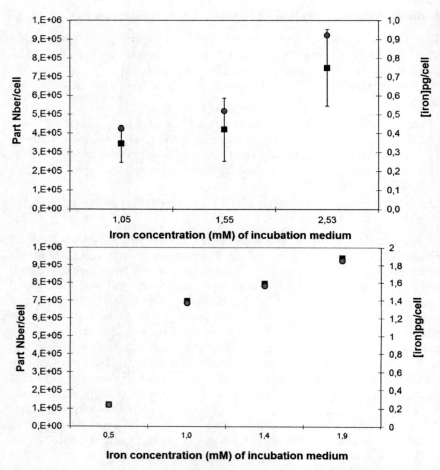

Fig. 2. Measurement of nanoparticles/iron content of CD3+-TC in relation to the iron concentration of the incubation medium (as unpublished material of the author). Probe content are expressed in particle number (black squares) or in iron mass (grey circles) and measured using magnetophorosis after 60 or 100 minutes of incubation. of chase-time (A) or with ESR after 70 minutes of incubation and 14 hours of chase-time (B) On graph A, the standard deviation represents the variation of iron content cell by cell.

The probe cell content can be measured via ICP-MS but also, in the case of metallic-core nanoparticles, by ESR which permits the deduction of the average charge per cell. Magnetophoresis (Whilhelm et al. 2002) measures the distribution of the magnetic content of the cell population providing that the majority of cells are labeled (Fig. 2).

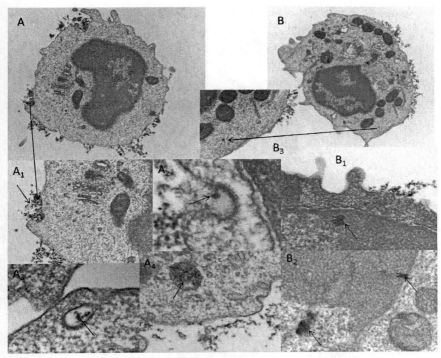

Fig. 3. TEM analysis of the intracellular distribution of anionic IONP within $CD3^+$-TC (as unpublished material of the author). A plot illustrating TEM images of a $CD3^+$-TC labeled following 1 h incubation at a concentration equivalent of 2.5 mM iron and 1 hour "chase-time". Numerous IONP are still visible in extracellular location in form of clusters in contact with the cellular membrane. Zoomed images (different magnification) reveal different stages of IONP internalisation, with invagination of the cellular membrane (A_1), formation of endocytosis vesicles (incorporating IONP) (A_2), intracellular internalization of these vesicles—early endosomes (A_3), and development into lysosomes containing numerous clusters of IONP (A_4). B plot illustrating TEM images of a $CD3^+$-TC labeled following 1 hour incubation at a concentration equivalent of 2 mM iron and after 3 hours "chase-time". There are less extracellular IONP, numerous intracellular clusters (B_1 and B_2), or "free" clusters in the cytoplasm (B_3). This analysis allows us to estimate the *in vivo* evolution of the distribution of these particles within cells following injection. Thus, the *in vitro* observation *after 3 hours chase time* is in accordance with *in vivo* observation 24 days after the transfer of CD3+, revealing the intranuclear predominance of clusters (Fig. 1).

TEM analysis in the case of IONP or confocal microscopy in the case of fluorescent nanoprobes allows researchers to study the intracellular distribution of nanoparticles.

3. Absence of functional alterations

The goal is to achieve optimal intra-cellular concentration of the labeling agent, which permits external detection by MRI without functional alteration.

In addition to cell toxicity attributable directly to the labeling agent, the toxicity associated with the magnetic field has been noted (Shäfer et al. 2008).

In 2011, only the IONPs have obtained approval from regulatory agencies for the *in vivo* labeling of macrophages and are also used in clinical studies with direct cellular labeling (deVries et al. 2005).

4. Obtaining an MRI contrast

The injected cells are locatable using ^{1}H-MRI by modifying locally the relaxation parameters of surrounding protons. The labeling agent can be a paramagnetic (gadolinium) or super-paramagnetic (iron oxide) structure.

Gadolinium provokes a positive signal enhancement on the T1-weighting sequences except at very strong concentrations. As gadolinium ions are very toxic, they must be either chelated or used in the form of oxides, or be encapsulated. The nature of the chelate or the external layer determines the number of water molecules which can interact with the gadolinium and thus, the magnitude of the signal.

Under a strong magnetic field, the presence of the labeled cells appears in the form of a dark spot on the T2*-weighted sequence due to magnetic field inhomogeneities caused by the presence of IONP. The uncertainty in probe relaxivity *in situ* complicated by intrinsic sources of contrast in tissues and saturation phenomena of the T2* effect could explain that the quantification of labeled cells using IONP is prone to error. In clinical studies, under a low magnetic field, the effects of inhomogeneity are small, and the detection is thus based on the T2 effect which is less easily detectable than the T1 effect. This results in a suboptimal sensitivity and the absence of proportionality between the signal and the number of cells.

Thus, T1 agents are theoretically better adapted for cellular imaging, which accounts for the recent developments in the design of gadolinium-based nanoprobes. Most of present studies are based upon *in vitro* results or upon *in vivo* detection of locally injected cells. We have recently proved the concept of monitoring *in vivo* using MRI, the trafficking of human dendritic cells following their *in vitro* labeling with gadolinium oxide core nanoparticles and parenteral injection (Fig. 5).

Another alternative recently proposed (Ahrens et al. 2005) was to associate the anatomical image obtained via 1H-MRI with that obtained by ^{19}F, after having labeled the cells with a 19F-based tracer agent. The objective is to increase the detection sensitivity of the labeled cells outside the confines of the ^{1}H background signal issued from water. This type of imaging, however, requires special equipment and long imaging sequences that are difficult to apply in routine clinical practice.

The intensity of the signal depends on the nature and the density of the metallic agent but also on the structure in which the metallic agent is included The large organic shell around the iron core leads to a lower impact of the iron core on the water proton relaxivity and hence a lower contrast for dextran nanoparticles such as commercialized compounds (Table 2). Much higher relaxivities are observed with dendritic macromolecules (Bulte et al. 2001; Basly et al. 2010) or just a thiolation of the core by DMSA (Billotey et al. 2003).

Table 2. Comparison of relaxivity levels of different IONP (acquired at 1.5T for different commercialized IONP, anionic IONP, and for a recently developed magnetodendrimer) (as unpublished material of the author).

Name of the compound	Coating agent	Hydrodynamic Size (nm)	R_1 $(mM^{-1}.s^1)$	R_2 $(mM^{-1}.s^{-1})$
Endorem®	Dextran T10	120–180	10.1	120
Sinerem®	Dextran T10, T1	15–30	9.9	65
Ferumoxytol®	Carboxymethyl-dextran T10	30	15	89
Resovist®	Carboxydextran	60	9.7	189
Supravist®	Carboxydextran	21	10.7	38
Ferropharm®	Citrate	7	14	33.4
Anionic IONP (Billotey et al. 2005)	DMSA	53	11.6	363.2
Dendronized iron oxide nanoparticles (Basly et al. 2010)		40–50	7.8 ± 10	349 ± 25

The particles actually commercialized and approved for use in man are all dextran particles presenting relatively weak relaxivities which explains the recent developments aimed to improve. The very high number of intra-hepatic macrophage-type cells compensates for the non-optimal contrast properties in hepatic studies.

This underlines that the performance of cellular imaging methods depends on the contrast properties of the labeling agent and the cellular content and also on the "system studied" as well as on the density of the labeled cells in the tissue.

III Actual and in-development Tools for Cellular Imaging Applied to the Exploration of the Pancreas

The purposes of the cellular imaging in the diabetes field are: i) to detect the insulitis level in patients at risk of developing T1D; ii) to quantify and obtain a dynamic assessment of the insulitis in patients at risk of developing T1D in order to monitor an immuno-modulator treatment administered at an early stage; iii) to assess and to quantify the β cells mass.

1. Detection and assessment of the insulitis

Due to the spill-over phenomenon and the very weak density of the islets of Langerhans that represent less than 1% of the pancreatic mass, SPECT cannot be efficiently used for the exploration of insulitis. There is actually no PET tracer available which allows for stable cellular labeling. Only MRI methods can be recommended for use in this indication.

The most direct cellular target is the lymphocyte, directly responsible of the infiltration of the islets of Langerhans (insulitis) and of their specific and irreversible destruction inducing T1D. However, it has been shown (Denis et al. 2004) that macrophage labeling allows for the *in vivo* detection of insulitis.

a. Leukocyte labeling

Only pre-clinical studies using IONP and direct labeling have been published to date. Anionic particles with an high cellular internalization capacity allow for the *in vivo* monitoring via MRI (Billotey et al. 2004) of the different lymphocyte infiltration stages leading to diabetes in NOD mice transferred with autoreactif CD3+ cells (Fig. 4).

Due to the poor endocytosic ability of lymphocytes, a transfection agent is required for the dextran type particles. One approach consisted of specific labeling of a lymphocyte subpopulation (CD8 cells) using dextran particles combined with a specific peptide (Moore et al. 2004).

In order to obtain quantitative data and to increase the sensitivity of detection, the use of T1 agents was suggested. Some preliminary tests (data not published) revealed that nanoparticles with gadolinium oxide core were well suited for the *in vivo* MRI detection of insulitis in NOD mice (cf Fig. 5).

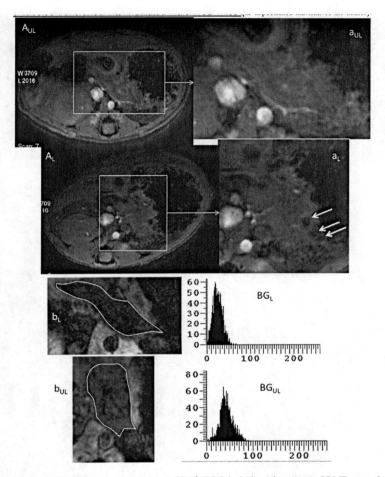

Fig. 4. *In vivo* MRI tracking of activated CD3$^+$-TC labeled with anionic IONP or unlabeled transferred into irradiated male NOD mice (as unpublished material of the author). T2*-weighted (gradient echo sequence) axial views centered on pancreas acquired at 7T with dedicated small animal MR system acquired *in vivo* at Day 11 (A -TR/TE = 13/4 msec) or Day 20 (B-TR/TE = 25/10 msec) after injection of 5x10^6 CD3$^+$-TC *in vitro* labeled with anionic IONP ($-_L$) or unlabeled ($-_{UL}$). a_L & a_{UL} corresponding magnified white boxes centered on pancreas on A_L & A_{UL} images. Evident focal dark spots are visible on pancreas in mouse injected with labeled CD3$^+$-TC (a_L) correspond to the zoomed image within the corresponding white region of interest. In contrast, unlabeled CD3$^+$-TC which had migrated to pancreas node did not generate any change in pancreas MR contrast (b_{UL}). The zoomed images centered on the pancreas, acquired 20 days following the transfer of CD3$^+$-TC, reveal a more negative signal (black), covering in a diffuse manner the entire pancreas, in the mice injected with CD3$^+$-TC labeled (b_L) in comparison to those injected with unlabeled CD3$^+$-TC (b_{UL}), as confirmed in the data value histograms (BG$_L$ and BG$_{UL}$) by signals corresponding to ROI of the whole pancreas (white delineation). This diffusely decreased signal intensity is due the invasion of the pancreatic islets of Langerhans by labeled CD3$^+$-TC.

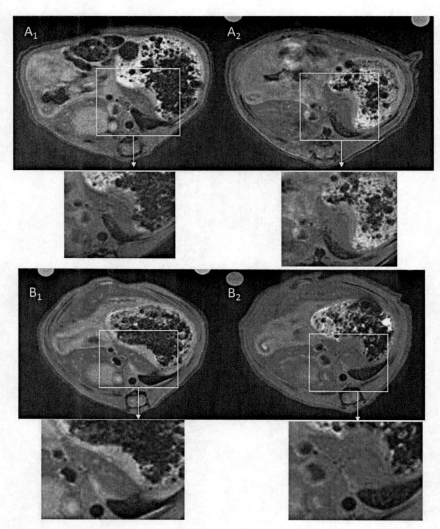

Fig. 5. *In vivo* MRI tracking of labeled with gadolinium based nanoparticles or unlabeled activated $CD3^+$-TC transferred into irradiated male NOD mice (as unpublished material of the author). T1-weighted flash sequence (TR/TE = 157/2.1 msec) axial views centered on pancreas (white boxes with corresponding magnification) acquired at 7T with dedicated small animal MR system acquired *in vivo* at Day 6 (-$_1$) or Day 20 (-$_2$) after injection of 5×10^6 *in vitro* labeled (A) with gadolinium based nanoparticles (Fizet et al. 2009) or unlabeled (B) $CD3^+$-TC into irradiated male NOD mice. It can be seen that the pancreas is enhanced at Day 20 compared to Day 6 in mice injected with labeled $CD3^+$-TC; this difference is not observed in mice injected with unlabeled $CD3^+$-TC. The enhancement in contrast in pancreas ROI between Day 6 and Day 20 was calculated at 49.5 and 3.9%, in mice injected with labeled and unlabeled $CD3^+$-TC, respectively.

1. In vivo or indirect labeling of macrophages

Weissleder's team successfully demonstrated in a murine model of autoimmune diabetes that:

- Microvascular lesions were present during the development of insulitis, allowing for the *in vivo* labeling of activated macrophages having infiltrated the surrounding tissue, and that these lesions could be located *in vivo* via MRI 24 hours after the injection of dextran-IONP (Denis et al. 2004).
- This imaging technique allows for detecting animals that will be good responders to immunomodulator therapy (Turvey et al. 2005).

Very recently (Gaglia et al. 2010), the same team compared the pancreas MRI signal level before and 48 hours following the injection of labeled dextran IONP (Combidex®). In a series of nine recent-onset T1D patients, the authors demonstrated that the decrease in T2 (due to the presence of IONP) was significantly ($p = 0.0005$) more pronounced in the patient group as compared to the control group involving 11 non–diabetic patients.

2. Detection and assessment of the beta mass

The quantification of the β-cell mass by means of imaging implies an indirect cellular labeling, using vectorized probe. When taking into account the poor density of the cells to be located and the proximity of the liver, only PET permits analysis and quantification of pancreatic uptake.

Several radio-labeled ligands have been suggested, but for the time being, none has demonstrated its efficacy to quantify the β-cell mass. Thus, a recent study (Fagerholm et al. 2010) demonstrated that a radio-labeled analog of the vesicular monoamine transporter 2 (VMAT2), the (11)C-dihydrotetrabenazine ((11) C-DTBZ), which is a putative molecular target for the quantitative imaging of pancreatic the β-cell mass by PET, exhibited high pancreatic binding but due to non-specific binding in the exocrine pancreas. Thus, the decrease observed in the pancreatic uptake in case of T1D is not specific for the loss of β cell mass.

Applications to Areas of Health and Disease

Precise assessment of β-cell mass is not possible in patients with either T1D or T2D. In T2D, the role of insulin resistance and β-cell dysfunction is widely recognized, although their respective contribution is highly variable from patient to patient. Monitoring of β-cell mass and lymphocytic islet infiltration in T1D patients and in antibody-positive first-degree relatives is another significant clinical challenge. It is generally admitted that in T1D, β-cells are subjected to specific attacks by immune competent cells, resulting in the progressive reduction of insulin reserve through β-cell apoptosis. However, histological analysis of pancreatic glands from newly-diagnosed diabetic patients has revealed heterogeneous degrees of insulitis within the same pancreatic gland, which may account for the restoration of β-cell function, at least to some extent, in intensively treated patients by recruiting intact βcells.

Key Facts

- Cellular *in vivo* imaging consists in creating/generating a mapping of the distribution of cells tagged by physical agents transmitting a signal that can be detected using medical imaging.
- Due to their high sensitivity, nuclear imaging techniques are the methods of choice of cellular *in vivo* imaging, except for the exploration of the pancreas, because of its proximity to the liver (spill-over phenomenon).
- Limitations account for the development of cellular labeling agents detectable using MRI, which may be used via direct or *in vitro* labeling techniques or *in vivo* via the labeling of macrophages.
- A labeling agent must present properties allowing it to interact with the cellular membrane that may be specific (i.e., endocytosis mediated via a receptor) or non-specific.
- Labeling of cells will be performed *in vitro* (direct labeling) or *in vivo* for macrophages.
- The labeling agent should not alter the functional properties of the cells and must be detectable externally using imaging techniques.
- Although the use of radiopharmaceuticals permits the labeling of numerous human cells, these agents cannot be used for investigating cellular processes over long time periods nor are they useful for exploring the pancreas.
- IONP are employed for the *in vivo* labeling of macrophages. Other nanoprobes, presenting improved cell interaction properties and a larger effect on the MRI signal, are currently under development.

Definitions

Spill-over: phenomenon linked to partial volume effect of. It results in the increase of signal in a region close to an other area with higher activity.

Chase-time: time delay between the end of the *in vitro* labeling process and the analysis of results or the intravenous injection of cells; during this time period, the cells are placed into their normal culture medium, after repeated washing in order to eliminate any labeling agent remained in suspension.

Gammascintigraphy & SPECT: imaging techniques based on the detection of gamma rays emitted by radiopharmaceuticals following intravenous injection to the patient or after ingestion or inhalation by the patient.

PET: imaging technique based on the detection of two 511 keV photons formed by annihilation of a positron emitted by specific radiopharmaceuticals in the tissue. At recent time, PET system provide a higher spatial resolution compared to SPECT.

Insulitis: infiltration of the Langerhans islets by auto-reactive T lymphocytes, resulting in the destruction of cells and the development of T1D.

^{1}H-MRI: Imaging based on the quantification and localization of the resonance signal of protons, providing mapping with enhanced contrast for most human tissues, due to their high water content. Clinical MRI is based on this detection.

Summary Points

- Specific imaging of the pancreas is an important challenge for the early diagnosis, at a reversible stage, of insulitis inducing T1D and for the evaluation of immunomodulator therapy.
- The quantification of β-cell mass in T2D would enable physicians to better adjust treatments.
- SPECT, first choice method for the cellular imaging in general, is limited for the exploration of the pancreas due to the proximity of the liver.
- The use of MRI, which is less sensitive, requires the optimization of a certain characteristics.
- The use of IONP approved by regulatory agencies is proposed to detect, in an indirect manner, insulitis in patients with recent-onset T1D lacking clinical evidence.
- Several developmental and pre-clinical studies assess the use of more sensitive and specific nanoprobes which would allow for the quantification of lymphocyte infiltration.
- The quantitative imaging of β-cell remains a challenge, despite the development of PET tracers which are radio-labeled analogues of carrier ligands specific for these cells.
- The cellular imaging of the pancreas is in its infancy and its future will rely on the design of efficient labeling agents, which are non-toxic even following the lysis of labeled cells, allowing for the quantification and monitoring over time of the cells implied in the development of T1D and T2D.

Abbreviations

DMSA	:	meso-2,3-dimercaptosuccinic acid
ESR	:	electron spin resonance
^{19}F	:	fluor 19
^{18}FDG	:	18F-fluoro-desoxyglucose
ICP-MS	:	inductively plasma-mass spectrometry
IONP	:	iron oxyde nanoparticles
MRI	:	magnetic resonance imaging/1H-MRI = proton MRI
$Ox^{111}In$	:	Indium 111 oxinate
PET	:	positron emission tomography
SPECT	:	single photon emission computed tomography
SBR	:	Signal over background ratio
^{99m}Tc	:	technetium 99m
T1D	:	type 1 auto-immune diabetes
T2D	:	type 2 diabetes
TEM	:	transmission electronic microscopy

References

Ahrens, E.T., R. Flores, H. Xu and P.A. Morel. 2005. *In vivo* imaging platform for tracking immunotherapeutic cells. Nat. Biotechnol. 23: 983–987.

Arbab, A.S., G.T. Yocum, H. Kalish, E.K. Jordan, S.A. Anderson, A.Y. Khakoo, E.J. Read and J.A. Frank. 2004. Efficient magnetic cell labeling with protamine sulfate complexed to ferumoxides for cellular MRI. Blood 104: 1217–1223.

Billotey, C., C. Wilhelm, M. Devaud, J.-C. Bacri, J. Bittoun and F. Gazeau. Cell internalization of anionic maghemite nanoparticles: quantitative effect on magnetic resonance imaging. 2003. Magn. Reson Med. 49: 646–654.

Billotey, C., C. Aspord, O. Beuf, E. Piaggio, F. Gazeau, M.F. Janier and C. Thivolet. 2005. T-Cell Homing to the Pancreas in Autoimmune Mouse Models of Diabetes: *In Vivo* MR Imaging. Radiology. 236: 579–587. 2010 http://www.ncbi.nlm.nih.gov/pubmed/20839261

Bulte, J.W., T. Douglas, B. Witwer, S.C. Zhang, E. Strable, B.K. Lewis, H. Zywicke, I.D. Duncan and J.A. Frank. 2001. Magnetodendrimers allow endosomal magnetic labeling and *in vivo* tracking of stem cells. Nat. Biotechnol. 19: 1141–1147.

Denis, M.C., U. Mahmood, C. Benoist, D. Mathis and R. Weissleder. 2004. Imaging inflammation of the pancreatic islets in type 1 diabetes. PNAS. 101: 12634–12639.

Fagerholm, V., K.K. Mikkola, T. Ishizu, E. Arponen, S. Kauhanen, K. Någren, O. Solin, P. Nuutila and M. Haaparanta. 2010. Assessment of islet specificity of dihydrotetrabenazine radiotracer binding in rat pancreas and human pancreas. J. Nucl. Med. 51: 1439–1446.

Fizet, J., C. Rivière, J.-L. Bridot, N. Charvet, C. Louis, C. Billotey, M. Raccurt, G. Morel, S. Roux, P. Perriat and O. Tillement. 2009. Multi-luminescent hybrid gadolinium oxide nanoparticles as potential cell labeling. J. Nanosci. Nanotechnol. 9: 5717–5725.

Gaglia, J.L., A.R. Guimaraes, M. Harisinghani, S.E Turvey, R. Jackson, C. Benoist, D. Mathis and R? Weissleder. 2011. Noninvasive imaging of pancreatic islet inflammation in type 1A diabetes patients. J. ClinInvest. 121: 442–445.

Modo M., D. Cash, K. Mellodew, S.C.R. Williams, S.E. Fraser, T.J. Meade, J. Price and H. Hodges. 2002. Tracking transplanted stem cell migration using bifunctional, contrast agent-enhanced, magnetic resonance imaging. NeuroImage. 17: 803–811.

Moore A., J. Grimm, B. Hanand and P. Santamaria. 2004. Tracking the recruitment of diabetogenic CD8+ T-cells to the pancreas in real time. Diabetes 53: 1459–1466.

Moore, A., P. Zhe Sun, D. Cory, D. Högemann, R. Weissleder and M.A. Lipes. 2002. MRI of insulitis in autoimmune diabetes. Magn. Reson. Med. 47: 751–758.

Oliver, M., A.Ahmad, N. Kamaly, E. Perouzel, A. Caussin, M. Keller, A. Herlihy, J. Bell, A.D. Miller and M.R. Jorgensen. 2006. MAGfect: a novel liposome formulation for MRI labelling and visualization of cells. Org. Biomol. Che. 4: 3489–3497.

Oostendorp, M. and K. Douma, T.M. Hackeng, M.J Post, M.A.M.J. van Zandvoort and W.H. Backes. 2010. Gadolinium-labeled quantum dots for molecular magnetic resonance imaging: R1 versus R2 mapping. Magn. Reson. Med. 64: 291–298.

Prince, H.M., D.M. Wall, D. Ritchie, D. Honemann, S. Harrrison, H. Quach, M. Thompson, et al. 2008. *In vivo* tracking of dendritic cells in patients with multiple myeloma. J. Immunother. 31: 166–179.

Schäfer, R., R. Bantleon, R. Kehlbach, G. Siegel, J. Wiskirchen, H. Wolburg, T. Kluba, F. Eibofner, H. Northoff, C.D. Claussen and H.P. Schlemmer. 2010. Functional investigations on human mesenchymal stem cells exposed to magnetic fields and labeled with clinically approved iron nanoparticles. BMC Cell Biol. 11: 22.

Signore, A., M. Chianelli, E. Ferretti, A. Toscano, K.E. Britton, D. Andreani, E.A. Gale and P. Pozzilli. 1994. New approach for *in vivo* detection of insulitis in type I diabetes: activated lymphocyte targeting with 123I-labelled interleukin 2. Eur. J. Endoc. 131: 431–437.

Tseng, C.-Li, I-L. Shih, L. Stobinski and F.-H. Lin. 2010. Gadolinium hexanedione nanoparticles for stem cell labeling and tracking via magnetic resonance imaging. Biomaterials. 31: 5427–5435.

Turvey, S.E., E. Swart, M.C. Denis, U. Mahmood, C. Benoist, R. Weissleder and D. Mathis. 2005.Noninvasive imaging of pancreatic inflammation and its reversal in type 1 diabetes. Journal Clin. Invest 115: 2454–2461.

de Vries, I.J.M., W.J. Lesterhuis, J.O. Barentsz, P. Verdijk, J.H. van Krieken, O.C. Boerman, Wim J.G. Oyen, J.J. Bonenkamp, J.B. Boezeman, G.J. Adema, J.W. Bulte, T.W. Scheenen, C.J. Punt, A. Heerschap and C.G. Figdor. 2005. Magnetic resonance tracking of dendritic cells in melanoma patients for monitoring of cellular therapy. Nat. Biotechnol. 23: 1407–1413.

Vuu, K., J. Xie, M.A. McDonald, M. Bernardo, F. Hunter, Y. Zhang, K. Li, M. Bednarski and S. Guccione. 2005. Gadolinium-rhodamine nanoparticles for cell labeling and tracking via magnetic resonance and optical imaging. Bioconj. Chem. 16: 995–999.

Walczak, P., D.A. Kedziorek, A.A. Gilad, S. Lin, et J.W.M. Bulte. Instant MR labeling of stem cells using magnetoelectroporation. 2005. Magn. Reson. Med. 54: 769–774.

Weissleder, R., D.D. Stark, B.L. Engelstad, B.R. Bacon, C.C. Compton, D.L. White, P. Jacobs and J. Lewis. 1989. Superparamagnetic iron oxide: pharmacokinetics and toxicity. AJR Am. J. Roentgenol. 152: 167–173.

Wilhelm, C., C. Billotey, J. Roger, J.N. Pons, J-C. Bacri and F. Gazeau. 2003. Intracellular uptake of anionic superparamagnetic nanoparticles as a function of their surface coating. Biomaterials 24: 1001–1011.

Wilhelm, C., F. Gazeau and J-C.Bacri. 2002. Magnetophoresis and ferromagnetic resonance of magnetically labelled cells. Eur. Biophys. J. 31: 118–125.

Wu, Y.C., S. Parola, O. Marty, M. Villanueva-Ibanez and J. Mugnier. 2005. Structural characterizations and wavegulding properties of YAG thin films obtained by different sol-gel processes. Opt. Mater. 27: 1471–1479.

9

Mucoadhesive Nanoparticles for Oral Delivery of Insulin

Sajeesh S.,[1,#] *Chandra P. Sharma*[1,a] *and Christine Vauthier*[2,*]

ABSTRACT

Insulin was introduced into clinical practice of diabetes treatments soon after it's discovery in 1921. It is still the most effective and safe treatment option available. However, oral insulinotherapy remains a distant reality for diabetic patients, mainly because it is degraded in the gastro-intestinal tract and it is not absorbed under its active form by the gut. Applications of technologies from the nanomedicine represent an opportunity to bypass the bottle neck of the oral administration of insulin and more generally for peptide and protein drugs. Polymeric nanoparticles seem to be

---

[1]Division of Biosurface Technology, Biomedical Technology Wing, Sree Chitra Tirunal Institute for Medical Sciences & Technology (SCTIMST), Thiruvanathapuram, Kerala, India.
[#]*Present address*: Nano-Biomaterials Lab, Department of Biological Sciences, Korea Advanced Institute for Science &, Technology (KAIST), Daejeon 305-701, Republic of Korea; E-mail: sajeeshchem08@gmail.com
[a]E-mails: sharmacp@sctimst.ac.in; drcpsharma@rediffmail.com
[2]Univ. Paris Sud, Physico-chimie, Pharmacotechnie, Biopharmacy, UMR 8612, Chatenay-Malabry, 296, France; E-mail: Christine.vauthier@u-psud.fr
*Corresponding author

List of abbreviations after the text.

the most promising candidate for oral insulin delivery. They are considered as part of the relevant technologies of nanomedicine to achieve oral delivery of insulin. Firstly, they were shown to protect peptides from degradation in the harsh conditions of the gastro-intestinal tract. Secondly, because of their very small size they are believed to have a high potential to enhance drug transport across the barriers found in the gastro-intestinal tract. Aparts form the chemical and biochemical barriers, one of the physical barriers is the mucus that needs to be crossed before absorption sites can be reached on the surface of the epithelium formed by a monolayer of cells. This chapter focuses on the mucoadhesive properties of polymer nanoparticles. These properties are important to consider for the success of nanomedicine in oral delivery of insulin as they will give the formulation suitable properties to reach absorption sites. In a first part, mucoadhesive polymers are presented with their mode of interactions with mucus prior to their use to design mucoadhesive nanoparticles. The second part presents the different mechanisms and modes of action of mucoadhesive nanoparticles that are discussed considering examples from the literature.

1. INTRODUCTION

Insulin is used in therapy of both type I and type II diabetes. It is the only efficient treatment for type I diabetes. As most protein pharmaceuticals, insulin are currently delivered by invasive routes of administration including subcutaneous injections (Ramezan and Sharma 2009). The need to find a non-injectable form (i.e.: non-invasive route of administration) has focused on the oral route (Ramezan and Sharma 2009). Apart from safety and patient compliance, oral administration of insulin is of clinical relevance for the treatment of diabetes (Arbit 2004). Indeed, in a healthy human, physiologically secreted insulin from pancreas enters portal circulation first and inhibits the hepatic glucose production. In this process, insulin undergoes a metabolism in the liver to a significant extent (more than 50%) and thereby the excess glucose is converted and stored in the form of glycogen. Thereafter, insulin not used by the liver finds its way to the peripheral circulation. This mechanism reduces the chance of hypoglycemic effect. This is in contrast with the distribution of exogenous insulin obtained after subcutaneous injection to diabetic patients. In this case, insulin distributes first in the peripheral circulation increasing risks of hypoglycemic effects and being ineffective to deliver the required amount of insulin to the liver to maintain their normal physiological function in glucose metabolism (Arbit 2004). Although oral administration of insulin is indisputably the best route of administration, all efforts put to develop suitable delivery formulation since its discovery remained unsuccessful because peptides and protein drugs are degraded in digestive media and are not absorbed as intact molecules by the intestinal epithelium (Lee and Yamamoto 1989). On the basis of the physiology, challenges to oral delivery of peptide and protein drugs include enzymatic degradation and poor permeation across the gastro intestinal tract (GIT) (Lee and Yamamoto 1989).

Oral administration of insulin required a suitable formulation ensuring protection against degradation and helping absorption of the entire protein. Nanomedicine includes emerging therapeutical approaches taking advantages of the small size of material used as drug carriers to help drugs to bypass biological barriers. It is a very active field of research on the oral delivery of insulin problem (Ramesan and Sharma 2009). Association of insulin with formulations used in nanomedicine improved dramatically the resistance of the hormone against degradation by proteases (Madsen and Peppas 1999). This is in favor to the success of the method as the slow rate of degradation may enhance the amount of proteins available for absorption across the GIT. However, insulin absorption by the epithelium still remains a limitation to the development of oral insulino-therapy (Ramazan and Sharma 2009). To achieve the expected benefit from nanomedicine, it is necessary to develop nano-systems by optimizing their mucoadhesive properties and giving them other functionalities such as permeation enhancing and protease inhibition properties. The present chapter was aimed to summarize the different approaches used to improve mucoadhesive properties of polymer nanoparticles. The first part presents polymers which were used as mucoadhesives in pharmaceutical formulations. The second part summarizes works done on mucoadhesive nanoparticles including those devoted to improve oral delivery of insulin. Transport and translocation of nanoparticles through the epithelium is the subject of another chapter of this book by Woitiski and will not be discussed in the present chapter.

1. MUCOADHESION

The term 'mucoadhesion' refers to adhesion between polymeric material and mucosal surfaces. Polymers, with certain structural features, become adhesive to mucus layer upon hydration. Though exact reason for adhesion may vary with the types of polymer, formulations containing mucoadhesive polymers are interesting because they provide intimate contact with the mucosal layer and thereby introduce drugs directly on the top of the enterocytes responsible for the highly selective permeability of the gut (Fig. 1). Thus, the close contact of the mucoadhesive drug delivery system with the mucus allows to increase the drug concentration gradient at the surface of the epithelium which in turn favors drug absorption.

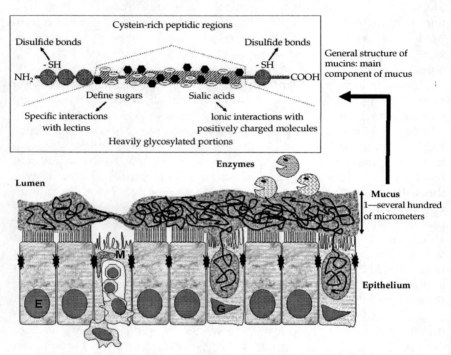

Fig. 1. General structure of the gut mucosa and of mucins, main component of mucus. Cells in the epithelium are enterocytes (E), M cells (M) and goblet cells (G). Inserted frame: The different relevant chemical groups involved in mucoadhesion are indicated as well as the different possible specific interactions that can be promoted with mucoadhesive drug delivery systems. (Unpublished).

The exact mechanism prevailing in mucoadhesion is still unknown but there are four well accepted theories that describe possible interactions involved in the mucoadhesion phenomenon (i.e.: electronic, adsorption, wetting and diffusion theory) (Smart 2005; Ponchel and Irache 1998). All are based on very define types of interactions occurring between the mucoadhesive polymer and mucus glycoproteins which are composed of highly glycosylated proteins with a high content of sialic acid and thiol groups. (Cone 2009; Dodou et al. 2005). In the electronic theory of mucoadhesion, an electron transfer occurring between the mucus and the polymer leads to the formation of a double layer of electrical charges at the mucus/polymer interface. According to the adsorption theory, secondary forces such as hydrogen bonding and Van der Waals interactions are largely responsible for the interaction between the mucus and the adhesive polymers. The wetting theory applies mainly to adhesives in the liquid states in which the ability of the adhesive to swell and spread on the mucus layer depends on the interfacial energy between the two partners. Contact angle measurement can be used to predict interactions of the polymers with the mucus layer. Lower contact angle favors mucoadhesion. In the diffusion theory, interpenetration and physical entanglement of the mucoadhesive polymers with the mucus layer are the key factors for the adhesion process. Phenomena involved in this mucoadhesive mechanism depends on the molecular weight, the degree of cross-linking, the chain length, the flexibility and the spatial conformation of both the polymer and mucins found in the mucus (Smart 2005; Dodou et al. 2005). None of these theories gives by themselves a complete description of the mechanism of mucoadhesion. In general, the global phenomenon of mucoadhesion results from a combination of the four theories with a balance between them which depends on the nature of the mucoadhesive polymer and on the nature and composition of the mucus. In an actual scenario, it can be expected that the polymer gets wet and swells upon contact with the mucosal surface (wetting theory). Thereafter, noncovalent bonds including electrostatic and hydrophobic interactions are created at the mucus–polymer interface (electronic and adsorption theory). The nature and strength of these non-covalent bonds are greatly influenced by the chemical nature of the mucoadhesive polymer. Then, the polymer and mucin chains interpenetrate and entangle together (diffusion theory), to further reinforce the mucus-polymer interactions (Smart 2005).

Polymers should have some general characteristics which will enable them to function as mucoadhesive material (Table 1). The most widely investigated groups of mucoadhesives are hydrophilic macromolecules containing large number of hydrogen bond forming functional groups. The presence of hydroxyl, carboxyl or amine groups on the molecules are favorable to promote adhesion to mucosal surfaces. They are mostly called wet adhesives because they are activated by moistening and adhere to mucosal surfaces through non-specific interactions. Typical example includes poly(acrylic acid), chitosan, sodium alginate and cellulose derivatives (Lowman and Peppas 1999). In general they are used as part of the components of a drug formulation designed to be administered by a mucosal route. In general, mucoadhesive polymers are included in conventional drug delivery formulations dedicated for mucosal routes of administration. At present, they are also widely considered for the development of technologies to be applied in nanomedicine including approaches for oral delivery of insulin.

Table 1. Mucoadhesive polymers: specific functions enhancing mucoadhesive properties and principal characteristics of polymer interactions with mucus.

Polymers and functionalities	Type of interactions	Mode of interaction	Advantage	Disadvantage
Poly (acrylic acid) and derivatives, Carboxymethyl Cellulose, Alginate	Non-specific	Carboxyl groups in these polymers make H-bonding with the hydroxyl groups in the mucus layer	Safe for oral delivery, non-specific binding with the intestinal epithelium, inexpensive	Low *in vivo* efficacy
Chitosan and derivatives, amino dextran, poly (diaminomethyl methacrylates)	Non-specific	Amino groups in the polymer interact with the carboxyl group residues in the mucus layer	Safe for oral delivery, electrostatic interaction with mucus layer	Poor solubility at physiological conditions, strong interactions at mucus layer, poor diffusion property
PEG and their derivatives	Non-specific	Non-adhesive in nature, works as adhesion promoter when used along with a polymeric system such as hydrogel	Non-toxic, no interaction with the intestinal mucosal layer	Excellent *in vitro* results, poor *in vivo* performance
Lectins	Highly specific	Specifically binds to define sugar residues of glycoproteins found on the epithelial surfaces and in mucus	Specific binding, non-toxic and non-immunogenic	Safety issues related to the repeated use, cost factor etc.
Thiols	Specific	Make di-sulphide bonds with mucus glycoproteins	Specific binding mechanism, ability of open epithelial tight junctions	Poor *in vivo* results, stability issue

2. MUCOADHESIVE NANOPARTICLES FOR DRUG DELIVERY APPLICATIONS

Polymeric nanoparticles are particles of less than 1 µm in diameter that are prepared from natural or synthetic polymers (Hans and Lowman 2002). Nanoparticles offer numerous advantages over conventional drug delivery systems as high surface to volume ratio is their salient feature. The high specific surface area promotes a better interaction of a nanoparticle with the mucosal surface compared to conventional drug delivery systems such as gels, patches, tablets or capsules. However, from an experimental point of view, the current methods for mucoadhesion measurements do not provide direct evidence of the nature of polymer–mucin interactions, when applied to a nano-particulate system. The methods currently reported mostly exploit custom-made or modified equipments, and there is no universally accepted test method available yet. Although same hold true for any mucoadhesive material the situation appears more complex in the case of nanoparticles (Peppas and Huang 2004). This is the reason that so few comparisons of inter-group data appeared in the literature and that the definition of "good" mucoadhesives still remains a debatable issue. In general, *in vitro* techniques currently used to evaluate mucoadhesion process involve destruction of the adhesive bond between the polymer and the mucus or tissue predominantly via application of mechanical force. It is obvious that many of these techniques are not applicable with nanoparticles. Therefore, it remains extremely difficult to investigate exact mechanism of nanoparticle mucoadhesion (Lowman and Peppas 1999). Despite these shortcomings, nanoparticles prepared from known mucoadhesive polymers are now widely investigated for developing novel drug delivery systems, especially for proteins and other large molecular weight drug molecules. A classification in four groups can be suggested based on the nature and properties of the material used to prepare them as explained below.

- Nanoparticles from conventional mucoadhesive polymers.
- Mucus penetrating and diffusing nanoparticles.
- Nanoparticles obtained by the functionalization of conventional mucoadhesive polymers.
- Nanoparticles becoming mucoadhesive by surface erosion.

2.1 Nanoparticles from Conventional Mucoadhesive Polymers

Conventional or so-called first generation mucoadhesive nanoparticles are largely based on poly(acrylic acid) (PAA) including their crosslinked derivatives such as Carbopol, and Polycarbophil, chitosan (CS), alginate and cellulose derivatives. Mostly used as pharmaceutical excipients, these mucoadhesive polymers have also gained significance to formulate mucoadhesive nanoparticles.

Polyanionic polymers such as PAA operate mainly via non-covalent interactions with mucus including hydrogen bonding and Van der Waals forces (Table 1, Fig. 2). Hydrogen bonding is expected to take place between carboxylic groups of the polymer chain and the mucin glycoproteins (Fig. 1). However, drug delivery systems formulated with PAA often failed to demonstrate strong mucoadhesive efficacy *in vivo*. This is attributed to the ionization of carboxylic groups of PAA at neutral/alkaline pH of the gut. Indeed, PAA which has a pKa close to 5.0 gets easily ionized at the intestinal pH hence reducing hydrogen bonding interactions with the mucus layer. Another minor effect is attributed to the rigid structure of the polymers. In some examples, PAA was used in combination with poly(ethylene glycol) (PEG) to obtain hydrogel nanoparticles adding several other values to the mucoadhesive formulation (Lowman and Peppas 1999). Indeed, beside mucoadhesive properties, there are several other advantages of using hydrogels. Because of their high water content, they constitute a 'protein-friendly' environment which is an excellent carrier matrix for such fragile molecules. Their preparations are based on organic solvent free methods and their loading with protein can be achieved by the manipulation of their physicochemical properties. One very interesting property in drug delivery is the ability of some hydrogels to respond to environmental stimuli such as variations of temperature and of pH. It is a unique advantage emphasizing developments of pharmaceutical formulations for the oral delivery of proteins. pH sensitive hydrogels contain ionizable groups for which the ionization status depends on the pH of the environment. Change in pH may ionized the acidic or basic

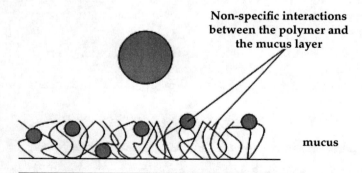

Fig. 2. Interactions of conventional mucoadhesive nanoparticles with the gut mucosa. Nature of interactions is mainly non specific including non-ionic interactions between the nanoparticles and the mucus layer. (Unpublished).

functional groups creating electrostatic repulsion between polymer chains and in turn resulting in the swelling of the hydrogel (Lowman and Peppas 1999). For instance, encapsulated protein can be protected inside a PAA based hydrogel at acidic pH of the stomach (pH 1-2) because the hydrogel network remains collapsed at pH below 5.0. As pH environment changes to alkaline conditions found in the gut, the hydrogel swells allowing the encapsulated protein to diffuse out of the hydrogel (Lowman et al. 1999). Other interesting features of mucoadhesive polymers composed of PAA are noteworthy regarding applications for oral delivery of protein and peptide drugs. PAA can act as calcium chelators resulting in calcium concentration depletion. This has two major benefits for oral delivery of protein and peptide drugs. First, depletion of calcium concentration in the gut inhibits proteolytic activity of the major proteases of the GIT, trypsin and chymotrypsin (Madsen and Peppas1999). Protein and peptide drugs may be protected against degradation and their therapeutic activity may be protected as well. Secondly, it disturbs cell-cell adhesion by loosening the epithelial tight junctions hence drug permeation is improved (Staddon et al. 1995; Lowman and Peppas 1999). Regarding oral delivery of insulin, several types of mucoadhesive formulations were prepared considering polyanionic polymer based nano and microparticles. Results for *in vivo* investigations showed a clear dose dependent effect occurring 2 h after oral administration of the insulin-containing formulation in both healthy and diabetic rats while the effect lasted at least 6 h (Lowman et al. 1999; Lowman and Peppas 2004).

Other conventional mucoadhesive polymers used to formulate mucoadhesive nanoparticles are polysaccharides including chitosan (CS) and alginate.

Chitosan is a polycation prepared from chitin isolated from shrimps. Its strong mucoadhesive properties are explained by hydrogen and ionic bonds occurring between the negatively charged sialic acid residues of mucins and the positive charges of the amino groups of CS (Fig. 1). Apart from the mucoadhesive properties, the polycationic form of CS is also able to open epithelial tight junctions enhancing the paracellular permeation of hydrophilic macromolecular drugs including insulin. It is noteworthy that the effect on the integrity of the epithelium and on the cell membranes of CS is much lower compared to that of known absorption enhancers (Illum 1998). CS then appears as a much safer permeation enhancing agent in addition to its mucoadhesive properties. Mucoadhesive nanoparticles can be formulated with chitosan by ionotropic gelation with tripolyphosphates which are negatively charge oligomers of phosphate ions. The formulations were proposed as drug delivery systems for oral administration of insulin. Efficacy of CS-nanoparticles to enhance oral absorption of pharmacologically active insulin was investigated by monitoring the glycemia of alloxan-induced diabetic rats. Results indicated that the CS-nanoparticles enhanced the intestinal absorption of insulin to a greater extent in comparison with an aqueous solution of CS. CS-nanoparticles increased remarkably the relative bioavailability of insulin compared to a subcutaneous injection of an insulin solution. While the glycemia was reduced over a period of 15 h after administration of 21 IU/kg insulin in the CS-nanoparticles, the average relative pharmacological bioavailability was up to 14.9 % (Pan et al. 2002). By increasing the dose in insulin in the CS-nanoparticles up to 100 IU/kg, the glycemia was markedly reduced at 10 h post-administration. Then, an interesting prolonged pharmacological effect lasted for up to 24 hours (Pan et al. 2002). In another work, CS-nanoparticles were prepared with hydroxypropyl methyl cellulose phthalate (HPMCP) which is a pH sensitive and gastroresistant polymer. The mucoadhesion of the nanoparticles was 2 to 4 times superior to that of the corresponding CS-tripolyphosphate nanoparticles. Penetration of the nanoparticles in the mucosa was improved compared to that of the reference CS-tripolyphosphate nanoparticles. The hypoglycemic effect of insulin obtained after peroral administration of the CS/HPMCP nanoparticles was also markedly increased. It was more than 9.8 and 2.8-folds compared to that produced by oral administration of a solution of insulin and of insulin-loaded CS-tripolyphosphate nanoparticles, respectively (Makhlof et al. 2010). In an earlier work, HPMCP was considered as a coating material for CS capsules to design a delivery system for insulin to the colon (Tozaki et.al. 1997). A significant improvement of the oral bioavailability of insulin in rats was reported after oral administration of a dose of 20 IU. While the oral bioavailability compared to the intravenous route was 5.3%, the glycemia dropped down 6 h after oral administration and the effect lasted for 24 h thereafter. Improvement of the performance of CS-nanoparticles can be obtained by modifying the chemical structure of the polysaccharide on its amino groups. For instance, trimethylation was aimed to improve the solubility of CS in aqueous solutions at pH approaching neutral values. This has dual advantages as the trimethyl chitosan (TMC) is protonated at pH above 6.5. The polymer chains are fully soluble in aqueous media of pH above 6.5 which is in contrast with that of CS which tend to aggregate in these media (Thanou et al. 2001). The TMC mucoadhesion is improved at the intestinal pH compared to that of CS. The second advantage concerns the effect of CS on the opening of the tight junctions which is very important to facilitate the paracellular transport of hydrophilic compounds including insulin. As only the protonated CS occurring at pH below 6.5 can trigger the opening of the tight junctions, TMC providing with a protonated form at pH above 6.5 found in the gut is suitable to improve paracellular permeability of the intestinal epithelium (Thanou et al. 2001). Promising results were obtained from TMC nanoparticles in experiments based on *in vitro* cell cultures of the cell monolayer model of Caco 2 (Thanou et al. 2001). Apart from nanoparticles obtained from TMC, other derivatives of CS were used to develop nano and microparticles for oral delivery of proteins. Typical examples included phthalate, succinate, PEGylated derivatives of CS.

Alginate is another polysaccharide use to formulate mucoadhesive nanoparticles for oral delivery of insulin. Most methods of preparation include an ionotropic pre-gelation of the alginate with divalent calcium followed by complexation with a polyelectrolyte. Calcium crosslinked alginate particles can protect peptides from gastric degradation by virtue of pH dependent release mechanism (Tønnesen and Karlsen 2002; Sarmento et al. 2007). After oral administration to rats, the superiority of alginate nanoparticles to improve bioactivity and absorption of insulin was attributed to the mucoadhesive properties of the formulation and to an improvement of the internalization of the peptide within the intestinal mucosa (Sarmento et al. 2007).

2.2 Mucus Penetrating and Diffusing Nanoparticles

Efficacy of conventional mucoadhesive nanoparticles can be compromised by the interaction of the drug delivery device with soluble mucins found in the lumen of the GIT (Peppas and Huang 2004). As a consequence, they are removed quickly from the intestine before they can reach the epithelium where drug absorption takes place. In order to counterbalance this effect, it was suggested to develop mucoadhesive formulations that bind specifically to the mucus layer. The rational behind this approach was to improve the inter-diffusion and inter-penetration of polymeric system in the mucosal barrier by reducing the interfacial energy and glass transition temperature of the parent polymer composing the nanoparticles. To this purpose, adhesion promoters were introduced in conventional mucoadhesive polymers. A typical example of adhesion promoter is composed of a PEG containing macromer which can be easily incorporated in the mucoadhesive PAA/poly(methacrylic acid) (PMAA) copolymer by copolymerization. Resultant material possesses good chain mobility/flexibility in order to diffuse across the mucosal layer and avoid soluble mucin binding through a steric effect. PEG was also used to coat poly(lactide-co-glycolide) (PLGA) nanoparticles and poly(sebacic acid) nanoparticles (Lai et al. 2009) (Fig. 3). The diffusivity of the PEG-PLGA nanoparticle across the mucus was only reduced by 4 and 12 times respectively compared to the diffusivity evaluated for the same nanoparticles in water. In comparison, the corresponding non-coated nanoparticles were 3300 folds lower in mucus than in water. Results of diffusivity measurements suggest that surface modification of the

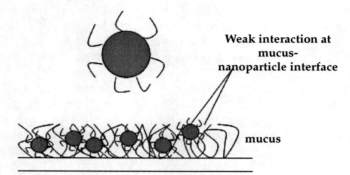

Fig. 3. Interactions of mucus penetrating and diffusing nanoparticles with gut mucosa. The nanoparticles are coated with polymer chains enabling the nanoparticles to diffuse through the mucus layer avoiding strong interactions with mucins. (Unpublished).

nanoparticles can greatly improve diffusivity of the nanoparticles in the mucus layer overcoming the soluble mucus barrier. A new generation of mucoadhesive nanoparticles is about to emerge while it is now necessary to combine these findings with developments of nanoparticles showing tailored drug release profile (Lai et al. 2009).

Alternatively to PEG, polysaccharides are other interesting material to be used to decorate polymeric nanoparticles aiming to improve their mucoadhesion properties. This approach was mainly developed in a series of core-corona nanoparticles composed of a poly(isobutyl cyanoacrylate) hydrophobic core surrounded by an hydrophilic corona composed of polysaccharides such as CS and dextran (Bertholon et al. 2006; Bravo-Osuna et al. 2007). Large amounts of the CS-coated nanoparticles were entrapped in the mucus layer even at low nanoparticle concentration. With dextran-coated nanoparticles, the mucoadhesive phenomena was non saturable in contrast to what was observed with CS decorated nanoparticles. This finding agreed with the fact that CS may adhere to mucus components through electrostatic interactions where the number of sites is finite giving saturable phenomenon. It can be concluded that bioadhesive properties of nanoparticles can be finely tuned by modifying surface properties of the nanoparticles.

2.3 Functionalization of Nanoparticles Made of Conventional Mucoadhesive Polymers

Highly specific mucoadhesive properties can be given to the nanoparticles by decorating them with defined motives (Fig. 4). The rational behind this is to promote adhesion of nanoparticle drug delivery system on a define portion of the intestine where absorption sites of the drug are located. Improvements of mucoadhesive properties are obtained by modifying

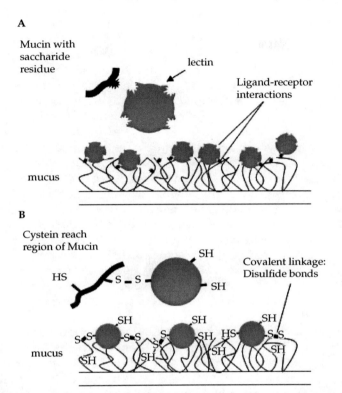

Fig. 4. Interactions of functionalized nanoparticles with the gut mucosa. Nanoparticles can be functionalized with lectins (A) or thiol (B) groups promoting specific interactions with define sugar residues of mucins or thiol groups of the cystein-rich region of mucin respectively. (Unpublished).

surface properties of nanoparticles. This can be achieved by chemical or physical coupling methods of ligands interacting with mucus on already prepared nanoparticles or on the polymer prior to the synthesis of the nanoparticles (Lehr 2000; Wood et al. 2008).

Lectins are typical examples of relevant ligands. They are non-immunologic glycoproteins recognizing very define sugar molecules (Lehr 2000). They can bind to glycosylated membrane components overlaying the GIT mucosa surface. For instance, the wheat germ agglutinin (WGA) extracted from plants can bind to N-acetylglucosamine and sialic acid residues found in the GIT. *In vitro*, it shows the highest binding rate to intestinal cells of human origin compared to lectins from other plant sources. Interesting studies were reported in the literature considering the use of lectins to improve specificity of mucoadhesion of nanoparticles in the GIT (Ponchel and Irache 1998; Gao 2006). Only a limited number of works have considered application of the approach on biodegradable nanoparticles and were aimed to improve oral delivery of peptides. A lectin modified PLGA nanoparticles loaded with a synthetic pentapeptide used in immunomodulation therapy, the thymopentin, showed an enhanced interactions with pig mucus by 1.8 to 4.2 folds compared with interactions measured in same conditions with the unmodified PLGA nanoparticles (Yin et al. 2006). In another example considering oral insulin delivery, WGA was conjugated to alginate microparticles (Kim et al. 2005). The hypoglycemic effect of the WGA-decorated particles was the most pronounced. However, it was not demonstrated that the effect was actually due to an effective prolonged residence time in the GIT tract thanks to an improvement of the mucoadhesiveness of the particles. In another work, the mucoadhesive nature of PMAA-grafted-ethylene glycol hydrogel particles was enhanced with WGA thanks to specific interactions with carbohydrates found on mucins (Wood et al. 2008) (Fig. 1 and 4A). Improved mucoadhesive properties was accompanied by enhanced absorption of insulin as evaluated *in vitro* suggesting that the WGA decorated PMAA grafted ethylene glycol holds great promise as an oral formulation for the delivery of insulin improving diabetes treatments (Wood et al. 2008).

Another approach improving mucoadhesion with mucus is based on the use of thiol (sulphide) groups of the cystein rich region of mucins to promote strong and specific interactions with nanoparticles (Fig. 1 and 4B). In this aim, the nanoparticle surface is decorated with thiol groups. The rational is to enhance adhesion of nanoparticles on the mucus by formation of covalent disulfide bonds between thiol groups available on the surfaces of the two partners. The strategy was initially suggested by the group of Bernkop-Schnurch. (Bernkop-Schnürch et al. 2003). It was widely used for the development of drug delivery platforms for oral administration of hydrophilic macromolecules such as insulin, heparin and salmon calcitonin (Bernkop-Schnürch et al. 2003). Addition of thiol groups can induce transient and reversible increase of the permeability of the intestinal epithelium thanks to specific interactions with proteins of the tight junctions. This interesting effect was recognized to increase transport of drugs by the paracellular pathway and could be applied to improve oral bioavailability of protein and peptide drugs.

Thiol groups can be introduced on the surface of nanoparticles by coupling molecules such as cysteine, thioglycolic acid, cysteamine on mucoadhesive polymers like PAA, CS, alginate, carboxymethyl cellulose. Standard conjugation methods are suitable and the approach was recently applied on nanoparticles designed for oral delivery of insulin. In general, results showed improvement of both the efficacy of the thiol modified drug delivery system and of the mucoadhesion. For instance, nanoparticles containing PAA-cysteine increased 2.3 folds the area under the curve (AUC) of insulin administered orally to diabetic rats compared with the efficacy given by the corresponding non modified PAA formulation (Deutel et al. 2008). Although the incorporation of cystein in one of the nanoparticle formulations improved the efficacy of insulin delivery, the relative bioavailability of the peptide remained extremely low (0.2 and 0.1 respectively for PAA-Cysteine and PAA systems) compared to the subcutaneous injection of insulin. In another example, thiol modification achieved by grafting of cystein on surface of poly(methacrylic acid)-chitosan-poly(ethylene glycol) (PCP) particles (diameter around 1 µm) significantly improved transport of insulin across Caco 2 cells and mucoadhesion on excised rat intestinal tissue compared with the unmodified PCP particles (Sajeesh et al. 2010). *In vivo*, the thiol modified PCP particles were more effective in reducing blood glucose level in diabetic rats. The benefit brought by the introduction of thiol groups on the drug delivery systems to promote mucoadhesion and efficacy of oral administration of insulin was further confirmed in a recent work considering TMC nanoparticles. The introduction of cysteine residues in the structure of the TMC nanoparticles improved mucoadhesive properties by 4.7 folds compared with the non-modified TMC nanoparticles. This improvement correlated well with the increase of insulin transport through the rat intestine which ranged from 1.7 to 2.6 for the TMC nanoparticles and from 3.3 to 11.7 folds with the TMC-cysteine nanoparticles (Yin et al. 2009). Further work is required to elucidate whether the improvement of the mucoadhesion was the only mechanism responsible for the increase of the oral efficacy of insulin delivered by the thiol modified nanoparticles or whether the permeation of the epithelium was increased because of the presence of the thiol groups. It is probable that both mechanisms have contributed to the enhanced delivery of insulin observed with the thiol modified particles but the part of each mechanism in this contribution remained to be determined.

2.4 Nanoparticles Becoming Mucoadhesive by Surface Erosion

In general, hard bioerodible thermoplastic polymers such as polyesters and polyanhydrides are not mucoadhesives. However, they were found interesting potential mucoadhesive polymers to formulate nanoparticles for oral drug delivery applications. Actually, polymers like poly(fumaric-co-sebacic) (PFASA) anhydride can rapidly degrade in contact with aqueous media revealing mucoadhesive properties thanks to the appearance of carboxylic acid groups resulted from the hydrolytic cleavage of the anhydride bond (Fig. 5). The increase of the number of carboxylic acid groups appearing at the nanoparticle surface during

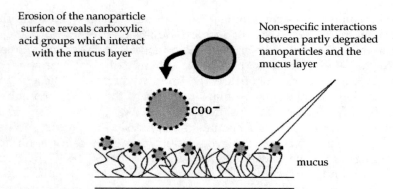

Fig. 5. Interactions of surface eroding nanoparticles with the gut mucosa. Typical example are poly(anhydride)-coated nanoparticles that reveal carboxylic group during hydrolysis. (Unpublished).

degradation enhances the ability of the polymer to form hydrogen bonds with mucus and epithelial cell glycoproteins promoting the adherence of the nanoparticles on the mucosa. Mucoadhesive properties shown by such polyanhydride nanoparticles on small intestinal tissue of rats appeared stronger than those of all other nanoparticles made of other polymers (Chickering et al. 1996). *In vivo*, the transit of the nanoparticles is delayed while the relative bioavailability of a drug can be improved. It was proposed that formulations including such hydrophobic polymers can be retained in the GIT and possibly increase intestinal absorption of drugs (Chickering et al. 1996). The prolonged residence time of particles in the GIT may increase the chance of particles to be taken up by enterocytes which were observed to cross the mucosal epithelium and the follicle-associated epithelium covering the lymphatic elements of Peyer's patches (Mathiowitz et al. 1997). Particles of p(FASA) with a diameter around 5 μm were applied to the oral delivery of insulin on type 1 diabetic rats and dogs. Different bioavailabilities of insulin were observed depending on the type of animals and on the administration schedule. In diabetic rats, the fed double dose experiment produced a relative bioavailability of 23.3% while the fed single dose experiment produced a bioavailability of only 5.5 ± 1.7% (Furtado 2008). In fasted diabetic dogs, the relative bioavailability was 5.5 ± 3.4%. Oral delivery of active zinc insulin was also achieved with another type of particles formulated with a combination of PLGA and oligomers of fumaric acid as polyanhydrides. The formulation was able to control plasma glucose levels when faced with a simultaneously administered glucose challenge. The efficacy was 11.4% of that of an intraperitoneal delivery of zinc insulin. The rational behind the use of these polymers revealing mucoadhesive properties upon degradation is a tricky approach allowing to immobilize the drug delivery device in the mucus releasing high local concentration of drugs close to absorption sites of the intestinal epithelium.

Conclusion

Bioadhesive polymers are promising materials in designing drug delivery systems for mucosal barriers. However, oral delivery of proteins and peptides still remains a bottle-neck. Attempts to improve oral delivery of proteins and peptides have been quite unsuccessful so far in obtaining satisfactory and consistent results. For instance, a therapeutic protein like insulin used in the treatment of diabetes needs to be administered on a daily basis with a very precise and robust delivery method able to regulate the pharmacological response at an optimum level. In this review, it was shown that technologies from nanomedicines can improve oral insulin delivery just by improving mucoadhesion of the nanoparticles. Although still insufficient, further improvements can be expected from a better optimization of the mucoadhesive properties and adding to the delivery systems other functionalities including permeation enhancing properties. A lot of work remained to be done but technologies from nanomedicine bring new hopes to succeed in obtaining an oral treatment of insulin available to diabetic patients.

Summary Points

- Penetration of drugs in mucus is required to reach absorption sites stranded on the gut epithelium.
- Mucoadhesion is a complex phenomenon that results from the combination of 4 main mechanisms.
- Certain polymers exhibit mucoadhesion phenomenon.
- Mucoadhesive nanoparticles promote oral delivery efficacy of active insulin.
- Mucoadhesion should take place in the mucus layer and not with soluble mucins of the lumen.
- Mucus penetrating/diffusing nanoparticles showed better mucoadhesive performance.
- Mucoadhesion of polymeric nanoparticles can be improved by specific modification approach.
- Thiol modification improves mucoadhesion of nanomedicine by formation of disulfide bonds with cystein rich region of mucins and confers enhancing permeation properties promoting drug absorption through the paracellular pathway.

- Mucoadhesive technologies from nanomedicine bring true hope for the development of a treatment of diabetes based on oral insulinotherapy.

Definitions

Copolymer: polymer made of at least two monomers

Copolymerization: Method of synthesis of copolymer based on the polymerization of at least two types of monomer together.

Homopolymer: Macromolecule formed from one type of repeated unit

Hydrogel: polymeric material formed by a tridimensional network of polymer chains swelled with water.

Macromer: Monomer of high molecular weight.

Monomer: Generally small chemical molecule which can polymerize meaning that they can be react together to form long polymer chains.

Mucin: Glycoproteins found in mucus.

Mucoadhesion: adhesion to mucosa.

Mucus: hydrogel material composed of mucins found on the top of the epithelium in mucosa.

Nanoparticles: small polymer nanoparticles characterized by a diameter ranging from 1 to 1000 nm. Most nanoparticles designed as drug delivery systems have size ranging from 70 to 300 nm or around 1 µm for the largest.

Polymer: General term to design macromolecule resulted from the self addition of the same monomer unit or by the addition of several types of monomer units.

Polymerization: method of synthesis of polymer.

Key Facts

- Key facts on oral delivery of insulin: Oral delivery of insulin is facing many challenges due to the physiology of the gastrointestinal tract.
- Key facts on nanomedicine technologies: Technologies from nanomedicines provide opportunities to improve drug delivery efficacy and targeting. It is belieaved that it can make oral insulinotherapy a reality to patients in near future.
- Key facts on mucoadhesion: Mucoadhesion promotes interactions of exogenous particles with mucus overlaying the epithelial cells in the gastrointestinal tract. This can help mucoadhesive nanoparticles to reach absorption sites found on the gut epithelium.
- Key facts on designing mucoadhesive nanoparticles: Mucoadhesiveness of nanoparticles depends on nanoparticle surface

properties which in turn govern the different types and strength of interactions promoted between particles and mucus components. Major difficulty is to find a proper balance between strength of interactions allowing adequate adherence of nanoparticles onto the mucus and at the same time allowing particles to penetrate and diffuse through the mucus layer to access enterocytes. Modulating surface properties of nanoparticle delivery system seems to be the key in improving their efficacy towards oral insulin delivery.

Abbreviations

AUC	:	area under the curve
CS	:	chitosan
GIT	:	gastro intestinal tract
HPMCP	:	hydroxypropyl methyl cellulose phtalate
PAA	:	poly(acrylic acid)
PCP	:	poly(methacrylic acid)-chitosan-poly(ethylene glycol)
PEG	:	poly(ethylene glycol)
PFASA	:	poly(fumaric-co-sebacic acid)
PLGA	:	poly(lactic-co-glycolic acid)
PMAA	:	poly(methacrylic acid)
TMC	:	trimethylchitosan
WGA	:	wheat germ agglutinin

References

Arbit, E. 2004. The physiological rationale for oral insulin administration. Diabetes Technol. Ther. 6: 510–517.

Bernkop-Schnurch, A., C.E. Kast and D. Guggi. 2003. Permeation enhancing polymers in oral delivery of hydrophilic macromolecules: thiomer/GSH systems. J. Control Release 93: 95–103.

Bertholon, I., G. Ponchel, D. Labarre, P. Couvreur and C. Vauthier 2006. Bioadhesive properties of poly(alkylcyanoacrylate) nanoparticles coated with polysaccharide. J Nanosci. Nanotech. 6: 1–8.

Bravo-Osuna, I., C. Vauthier, A. Farabollini, G.F. Palmieri and G. Ponchel. 2007. Mucoadhesion mechanism of chitosan and thiolated chitosan-poly(isobutyl cyanoacrylate) core-shell nanoparticles. Biomaterials 28: 2233–2243.

Chickering, D., J. Jacob and E. Mathiowitz. 1996. Poly(fumaric-co-sebacic) microspheres as oral drug delivery systems. Biotechnol. Bioeng. 52: 96–101.

Cone, R.A. 2009. Barrier properties of mucus. Adv. Drug Deliver. Rev. 61: 75–85.

Deutel, B., M. Greindl, M. Thaurerm and A. Bernkop-Schnürch. 2008. Novel insulin thiomer nanoparticles: *in vivo* evaluation of an oral drug delivery system. Biomacromolecules. 9: 278–285.

Dodou, D., P. Breedveld and P.A. Wieringa. 2005. Mucoadhesives in the gastrointestinal tract: revisiting the literature for novel applications. Eur. J. Pharm. Biopharm. 60: 1–16.

Furtado, S., D. Abramson, R. Burrill, G. Olivier, C. Gourd, E. Bubbers and E. Mathiowitz. 2008. Oral delivery of insulin loaded poly(fumaric-co-sebacic) anhydride microspheres. Int. J. Pharm. 347: 149–155.

Gao, X., W. Tao, W. Lu, Q. Zhang, Y. Zhang, X. Jiang and S. Fu. 2006. Lectin-conjugated PEG–PLA nanoparticles: Preparation and brain delivery after intranasal administration. Biomaterials 27: 3482–3490.

Hans, M.L., and A.M. Lowman. 2002. Biodegradable nanoparticles for drug delivery and targeting. Curr. Opin. Solid St. M. 6: 319–327.

Illum, L. 1998. Chitosan and its use as a pharmaceutical excipient. Pharm. Res.15: 1326–1231.

Kim, B-Y., J.H. Jeong, K. Park and J.D. Kim. 2005. Bioadhesive interaction and hypoglycemic effect of insulin-loaded lectin–microparticle conjugates in oral insulin delivery system. J. Control Release 102: 525–538.

Lai, S.K., Y.Y. Wang and J. Hanes. 2009. Mucus-penetrating nanoparticles for drug and gene delivery to mucosal tissues Adv. Drug Deliver. Rev. 27: 158–171.

Lee, V.H.L., and A. Yamamoto. 1989. A Penetration and enzymatic barriers to peptide and protein absorption. Adv. Drug Deliver. Rev 4: 171–207.

Lehr, C.M. 2000. Lectin-mediated drug delivery: The second generation of bioadhesives. J. Control Release. 65: 19–29.

Lowman, A.M., M. Morishita, M. Kajita, T. Nagai and N.A. Peppas. 1999. Oral delivery of insulin using pH-responsive complexation gels. J. Pharm. Sci. 88: 933–937.

Lowman, A.M., N.A. Peppas. Hydrogels. pp. 397–418. *In:* E. Mathiowitz [Ed.] 1999 Enclyopedia of Controlled Drug Delivery- Vol 1, Wiley, New York, USA.

Madsen, F., and N.A. Peppas. 1999. Complexation graft copolymer networks: swelling properties, calcium binding and proteolytic enzyme inhibition. Biomaterials. 20: 1701–1708.

Makhlof, A., Y. Tozuka and H. Takeuchi. 2010. Design and evaluation of novel pH-sensitive chitosan nanoparticles for oral insulin delivery. Eur. J. Pharm. Sci. (In Press).

Mathiowitz, E., J.S. Jacob, Y.S. Jong, C.P. Carino, D.E. Chickering, P. Chaturvedi, C.A. Santos, K. Vijayaraghavan, S. Montgomery, M. Bassett and C. Morell. 1997. Biologically erodable microspheres as potential oral drug delivery systems Nature 386: 410–414.

Pan, Y., Y.J. Li, H.Y. Zhao, J.M. Zheng, H. Xu, G. Wei, J.S. Hao and F.D. Cui. 2002. Bioadhesive polysaccharide in protein delivery system: chitosan nanoparticles improve the intestinal absorption of insulin *in vivo*. Int. J. Pharm. 249: 139–47.

Peppas, N.A., and Y. Huang. 2004. Nanoscale technology of mucoadhesive interactions. Adv. Drug Deliver. Rev. 56: 1675–1687.

Ponchel, G., and J. Irache. 1998. Specific and non-specific bioadhesive particulate systems for oral delivery to the gastrointestinal tract. Adv. Drug Deliver. Rev. 34: 191–219.

Ramesan, M.R., and C.P. Sharma. 2009. Challenges and advances in nanoparticle-based oral insulin delivery Expert Rev. Med. Devices 6: 665–676.

Sajeesh, S., C. Vauthier, G. Gueutin, G. Ponchel and C.P. Sharma. 2010. Thiol functionalized polymethacrylic acid based hydrogel microparticles for oral insulin delivery. Acta. Biomater. 06: 3072–3080.

Sarmento, B., A. Ribeiro, F. Veiga, P. Sampaio, R. Neufeld and D. Ferreira. 2007. Alginate/chitosan nanoparticles are effective for oral insulin delivery. Pharmceu. Res. 24: 2198–206.

Smart, J.D. 2005. The basics and underlying mechanisms of mucoadhesion Adv. Drug Deliver. Rev. 57: 1556–1568.

Staddon, J.M., K. Herrenknecht, C. Smales and L.L. Rubin. Evidence that tyrosine phosphorylation may increase tight junction permeability. J. Cell Sci. 108: 609–619.

Thanou, M., J.C. Verhoef and H.E. Junginger. 2001. Chitosan and its derivatives as intestinal absorption enhancers. Adv. Drug Deliv. Rev. 50: S91–S101.

Tozaki, H., J. Komoike, C. Tada, T. Maruyama, A. Terabe, T. Suzuki, A. Yamamoto and S. Muranishi. 1997. Chitosan capsules for colon drug delivery: improvement of insulin absorption from rat colon. J. Pharm. Sci. 86: 1016–1021.

Tønnesen, H.H., and J. Karlsen. 2002. Alginate in drug delivery systems. Drug Dev. Ind. Pharm. 28: 621–630.

Wood, K.M., G.M. Stone and N.A. Peppas. 2008. Wheat germ agglutinin functionalized complexation hydrogels for oral insulin delivery. Biomacromolecules. 9: 1293–1298.

Yin, L., J. Ding, C. He, L. Cui, C. Tang and C. Yin. 2009. Drug permeability and mucoadhesion properties of thiolated trimethyl chitosan nanoparticles in oral insulin delivery. Biomaterials. 30: 5691–5700.

Yin, Y., D. Chen, M. Qiao, Z. Lu and H. Hu. 2006. Preparation and evaluation of lectin-conjugated PLGA nanoparticles for oral delivery of thymopentin. J. Control Release 116: 337–345.

10

Insulin-nanoparticles for Transdermal Absorption

Jiangling Wan,[1,a,]* Huibi Xu[1,b] and Xiangliang Yang[1,c]

ABSTRACT

Insulin is the first clinical choice for diabetes with rapid and precise clinical effect. The subcutaneous route has been the preferred method of insulin delivery until now. However, the burden of daily injections, physiological stress, pain and inconvenience as well as high cost remains problems. The quest to replace needles with non- or less-invasive insulin delivery system has driven a great number of pharmaceutical research in this area. Among all these newly developed insulin delivery system, the transdermal route attracts more and more attention due to its advantages in delivery insulin. On the other hand, the nanotechnology has been applied in pharmaceutics industry for decades, and it also shows advantages in this area. The combination of these two technologies provides a potential strategy for insulin transdermal delivery. This review exams some recent attempts of transdermal delivery of insulin using nanotechnology, especially nanoparticles. In all

---

[1]National Engineering Research Center for Nanomedicine, College of Life Science and Technology, Huazhong University of Science and Technology, Wuhan 430074, P. R. China.
[a]E-mail: wanjl@mail.hust.edu.cn
[b]E-mail: hbxu@mail.hust.edu.cn
[c]E-mail: yangxl@mail.hust.edu.cn
*Corresponding author

List of abbreviations after the text.

these attempts, some new chemical and physical technologies are applied to help the insulin-loaded nanoparticles to overcome the skin barrier, which is the main problem with the transdermal delivery system, including chemical enhancers, iontophoresis, sonophoresis and microneedles. In addition, this article also concentrates on the effects and disadvantages of all these methods in insulin transdermal delivery.

INTRODUCTION

Insulin is the first clinical choice for diabetes with rapid and precise clinical effect. The subcutaneous route has been the preferred method of insulin delivery until now (Khafagy et al. 2007). Although this parenteral route has been satisfactory in terms of efficacy in majority, it results in stimulation of smooth muscle cell proliferation, peripheral hyperinsulinemia, and incorporation of glucose into the lipid of arterial walls which might be the causative factor in diabetic micro- and macroangiopathy (Gwinup et al. 1990). Moreover, the burden of daily injections, physiological stress, pain, and inconvenience as well as high cost remains problems (Kennedy 1991).

Over the past few decades, the skin has been a great deal of interest as a portal for the systemic delivery of drugs (Ting et al. 2004). Since skin is the largest organ of the human body, transdermal drug delivery (TDD) is an appealing alternative to subcutaneous delivery. TDD is gaining prominence over other forms of drug delivery due to its potential advantages, including reduced systemic side effects, noninvasiveness, increased patient compliance, large area of interface, potential for continuous and controlled delivery (Gallo et al. 1997; Karande et al. 2006). All these advantages make TDD an ideal way for insulin administration. However, the main obstacle limited the development of TDD is stratum corneum (SC). SC often referred to as a "brick and mortar" structure (Kalia et al. 2004), is the outermost few microns of skin tissue. The barrier function of the SC is essential to maintaining internal homeostasis, but it is also a major impediment to drug penetration. The SC exhibits low permeability to foreign molecules, especially large hydrophilic molecules which leads to the low bioavailability of the drug which TDD delivers.

In order to overcome the skin barrier, there are two approaches available: first is to make it more "leaky" towards hydrophilic molecules; and another way is to modify the therapeutic molecule to render it more hydrophobic and therefore "acceptable" to the membrane. For both

188 Nanotechnology and Nanomedicine in Diabetes

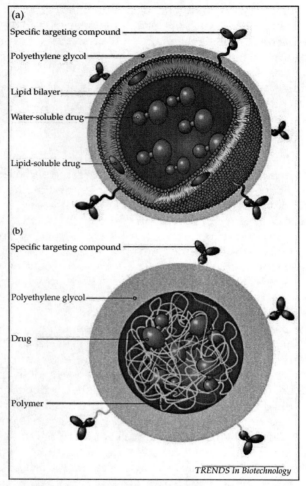

Fig. 1. Nanoparticles for non-parenteral drug delivery: (a) Liposome nanoparticle. (b) Polymer nanoparticle. This figure shows the basic structure of nanoparticles used for drug delivery. Drugs are encapsulated in the core of the nanoparticles or mixed with the polymer. The whole particles were guided by the specific targeting compounds to the targets and thus release the drug specifically. (Antosova et al. 2009).

Color image of this figure appears in the color plate section at the end of the book.

approaches above, nanotechnology is playing a more and more important role to overcome the skin barrier function as one of the hottest areas in pharmaceutical industry.

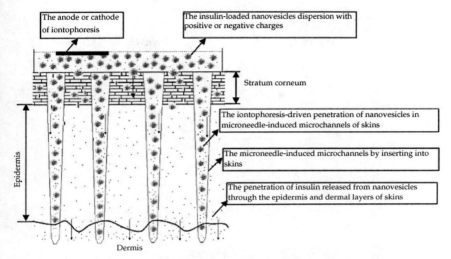

Fig. 2. The sketch of the transdermal delivery of insulin by utilizing iontophoresis-driven penetration of nanovesicles through microneedle-induced microchannels of skins. This figure shows the whole process of the transdermal delivery of insulin by utilizing iontophoresis-driven penetration of nanovesicles. After the microneedle injection, some microchannels were introduced into the skin. The insulin-loaded nanovesicles dispersion with charges penetrate in the microchannels driven by the iontophoresis and eventually release the insulin through the epidermis and dermal layers of skins. (Chen et al. 2009).

NANOTECHNOLOGY IN STRATEGIES TO OVERCOME THE SKIN BARRIER

In the last decade, scientists found many ways to overcome the skin barrier for transdermal drug delivery system (TDDS), including: insulin-loaded carrier, iontophoresis, sonophoresis, microneedles and so on. Nanotechnology is applied in all these methods.

Insulin-loaded Nanoparticle

Nanoparticle has already been used in drug delivery system for a long period of time and proved to be aneffective way for drug delivery. Because of the properties of the SC, hydrophilic molecules or drugs can't penetrate the SC easily. Encapsulation of these molecules or drugs with lipophilic material helps to overcome the skin barrier. Encapsulation consists of the entrapment of drugs within delivery systems such as liposomes, nanoparticles, transfersomes, ethosomes, solid lipid nanoparticles, noisomes and nanodisperisons (Dhamecha et al. 2009).

Table 1. Transdermal formulation technologies of insulin delivery system. The table shows different formulations and technology for transdermal insulin delivery.

Formulation/technology	Application	Outcome
Synthetic peptides	STZ-induced diabetic rats	Sustained supperssion of blood glucose levels for 11 h
Flexible lecithin vesicles	Mice	More than 50% reduction in blood glucose levels in 18 h
Nanoinsulin	Mice	Significant reduction in blood glucose levels
Iontophoresis	STZ-induced diabetic rats	Significant reduction in blood glucose levels
Iontophoresis/penetration enhancer	Rabbits	Significant reduction in stratum corneum barrier
Iontophoresis/LA, OA, LOA or LLA	Excised rat skin	Significant insulin flux enhancement
Iontophoresis/Poloxamer P407 gel	STZ-induced diabetic rats	36%–40% reduction in blood glucose levels
Sonophoresis	Hairless rats	Significant reduction in blood glucose levels for 60 min
Sonophoresis	Rabbits	Significant hypoglycemic response
Microneedles	STZ-induced diabetic hairless rats	0.5–7.4 ng/mL plasma insulin levels and 80% reduction in blood glucose levels

The results of these animal study show that all these methods have significant effect for reducing the blood glucose level. (Khafagy et al. 2007). Abbreviations: STZ, streptozotocin; LA, lauric acid; OA, oleic acid; LOA, linoleic acid; LLA, linolenic acid.

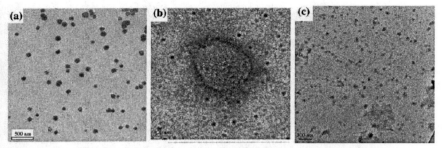

Fig. 3. The TEM and HR-TEM images of insulin-loaded nanovesicles. (a) ILV1 consisting of 1.5% soybean lecithin, 3.0% propylene glycol and 40 IU/ml insulin with the average diameter of 91.0 nm and zeta potential of −16.2mV, (b) ILV4 at the magnification times of 150,000, (c) the control solution containing free insulin. This figure shows the TEM and HR-TEM images of insulin-loaded nanovesicles.(a) is the image of formulation ILV1 which consists of 1.5% soybean lecithin, 3.0% propylene glycol and 40 IU/ml insulin. The particles are with the average diameter of 91.0 nm and zeta potential of −16.2 mV, (b) is the image of formulation ILV4 at the magnification times of 150,000, (c) is the control solution containing free insulin (Chen et al. 2009).

Xinhua Zhao's group prepared uniform spherical insulin nanoparticles by supercritical antisolvent (SAS) micronazation process and optimized the formulation and the SAS process (Zhao et al. 2010). The insulin nanoparticles were with a mean particle size of 68.2 ± 10.8 nm. The SAS process has not been induced degradation of insulin. The *in vitro* study showed that insulin nanoparticles were accorded with the Fick's first diffusion law and had a high permeation rate.

A recently study used $CaCO_3$-nanoparticles as the TDDS for delivering insulin to mice (Higaki et al. 2006). The study evaluated the pharmacokinetic and pharmacodynamic effects of the insulin-nanoparticles in normal mice and those with diabetes. The prepared insulin loaded $CaCO_3$-nanoparticles were transdermally applied to the back skin of normal MY mice and dB/dB and kkAy mice with diabetes after fasting for 1h. The serum insulin got to the maximum point of 67.1 ± 25.9μ IU/mL at 4h after transdermally administrated 200μg insulin loaded nanoparticles, While after subcutaneous injection of 3μg of monomer insulin was 462± 20.9μIU/mL at 20min. Transdermal nanoinsulin reduced glucose levels in a dose-dependent manner. In ddY mice, insulin bioavailability was 0.9% with transdermal nanoinsulin for 6h based on serum insulin levels, and 2.0% based on pharmacodynamic blood glucose-lowering effects. The significant sustained decrease of blood glucose in normal mice and those with diabetes proved that this $CaCO_3$-nanoparticle system successfully delivered insulin transdermally.

Another study developed lecithin vesicles as a carrier for transdermal delivery of insulin (Guo et al. 2000). Guo and his group characterized the preparation of flexible lecithin vesicles containing insulin and to assess the enhancing effect of these flexible vesicles on the transdermal delivery. The flexible lecithin vesicles were prepared by reverse-phase evaporation and treated further by sonication. The particle size of the flexible vesicles was 87.1 nm with a polydispersity index of 15.6%. The entrapment efficiency of the vesicles was 81%. When vesicles were nonocclusively applied onto the abdominal mice skin at a dose of 0.90 IU/cm2, *in vivo* hypoglycemic study showed the drop percentage of blood glucose by flexible vesicles was 21.42 ± 10.19% at 1 h, reached 61.48±8.97% at 5h, and was larger than 50% within 18h. The controlled group of conventional vesicles and saline had no hypoglycemic effect.

Yoshiro Tahara's group demonstrated that a solid-in-oil (s/o) nanodispersion, an oil-based nanodispersion of insulin, effectively enhanced the permeation of proteins into the skin (Tahara et al. 2008). The mean particle size of s/o nanodispersions with insulin was 257 ± 14 nm with polydispersion index (PDI) of 0.16–0.37. The prepared nanodispersions are of high stability. There were no precipitates after the storage of the s/o nanodispersions at room temperature for 1 month. The result of skin penetration study showed that 1.02 ± 0.24μg/cm2 of insulin penetrated into the YMP skin including the SC, epidermis and dermis, after treated with the s/o nanodispersion 48h. As a comparison, only 0.14±0.08 μg/cm2 of insulin penetrated into the skin when treated with the aqueous solution.

Nanotechnology in Iontophoresis and Electroporation

Iontophoresis is a new technology developed in the 20th century. It is a technology used to enhance the transdermal delivery of charged and neutral compounds through the skin by the application of a small electric current. The small electric current leads to the processes of electromigration and electro-osmosis that facilitates the transport of permeates across the skin. This new technology has been used to facilitate and regulate the transdermal delivery of insulin in order to control blood glucose levels in diabetic rats. A considerable reduction of blood glucose levels has already been achieved by using a pulse current with a lower current intensity and shorter application time (Liu et al. 1988). At present, scientists focus on resolving skin toxicity and other problems, to make the technology a commercial reality (Panchagnula et al. 2000).

Some other studies' result showed that the combinations of iontophoresis with absorption enhancer, electroporation and sonophoresis proved to be more effective and lower-toxic.

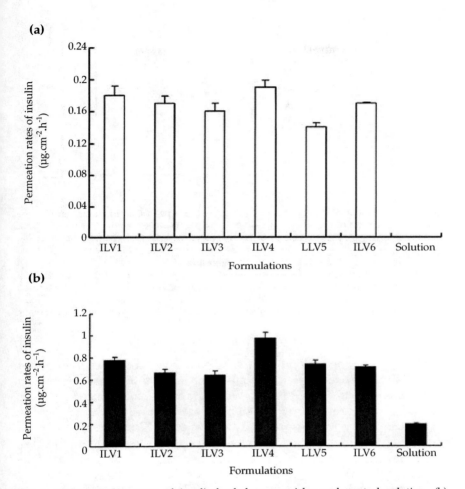

Fig. 4. (a) Permeation rates of insulin-loaded nanovesicles and control solution, (b) Iontophoretic permeation rates of insulin from nanovesicles and control solution through integrity skins. This figure implies the enhanced permeation effect by using insulin-loaded nanoveiscles and inotophoresis methods. Both insulin-loaded nanovesicles and naovesicles with iontophoresis show much higher insulin permeation rate than using insulin solution alone. Besides, using iontophoresis can increase the permeation rate of insulin significantly. Of all the formulations, ILV4 has the highest permeation rate of insulin by using both methods. (Chen et al. 2009).

H.B. Chen and his teammates combined inotophoresis technology with microneedle and insulin-loaded nanovesicles (Chen et al. 2009). They reported a novel active strategy for enhancing transdermal delivery of insulin using-loaded nanovesicles with various charges

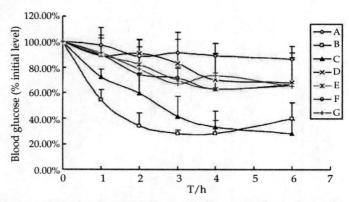

Fig. 5. The blood glucose levels of various treatment groups (n=5). This figure implies the hypoglycemic effects of six different treatment groups.(A) is used as the Negative control. While under the treatment of (B) Free insulin administrated by subcutaneous injection (1.0 IU/kg), the blood glucose decreased largely, which may lead to glucopenia. so is the result of group (C) ILV4 combined with iontophoresis and microneedles. The groups of (D) Free Insulin solution combined with iontophoresis and microneedles, (E) ILV4 combined only with microneedles, (F) ILV4 combined only with Iontophoresis, and (G) ILV4 based onpassive penetration all show similar effects in decreasing the blood glucose. The Glucose levels of streptozotocin-induced diabetic rats were normalized against the initial (0 h) value (Chen et al. 2009).

Color image of this figure appears in the color plate section at the end of the book.

driven by iontophoresis through skin containing microneedle-induced microchannels. The prepared insulin-loaded nanoparticles with diameters ranged from 90 nm to 175 nm and membrane thickness of 3–5 nm. The nanovesicles had a high entrapment efficiency of 89.05 ± 0.91%. The permeation study was separated into three parts: Permeation studies of nanovesicles; Permeation studies of nanovesicles driven by iontophoresis; and Permeation studies of nanovesicles combined with iontophoresis and microneedles. Their results showed that the positive zeta potential and small diameters of nanovesicles were significantly advantageous to the penetration of the insulin-loaded nanovesicles. When this kind of nanovesicles was combined with iontophoresis and absorption enhancers, the penetration rate was much higher. The *in vivo* study showed that the glucose levels of the rats were 33.3% and 28.3% of the initial levels at 4 and 6h after treated with the combination method, which were comparable to those treated with subcutaneous injection.

Electroporation involves exposure of the skin to relatively high voltages for a short time, which create intense electric fields across the thin SC. The enhancement of molecules transport through skin is due to a variety of mechanisms, including enhance diffusion through the aqueous pathways produced in the lipid bilayers, electrophoretic movement and electroosmosis.

Recently, Rastogi's group applied nanosized vesicles combined with electroporation technology to enhance the permeation of insulin through the skin barrier (Rastogi et al. 2010). The nanosized vesicles were made with block copolymers of poly (ε-caprolactone) (PCL) and polyethylene glycol (PEG). The insulin-loaded vesicles were delivered to rat abdominal skin through application of high voltage and current. The blood glucose level showed that electroporaion method of delivering insulin mimics the blood glucose profile of the IV injection. However, the former showed a delayed but prolonged hyperglycemic effect. Their work also revealed the relationship between the insulin permeation rate and parameters of electroporation. Insulin flux was measured under variable voltages, number of pulses and pulse length. The results showed that the rate of permeation increases linearly with increase in voltage, pulse number and pulse length.

Sonophoresis

Application of sonophoresis has been showed to enhance the permeability of various drugs including macromolecules. Therapeutic ultrasound at various frequencies ranged 20 KHz to 16 MHz has been used to enhance the delivery and activity of various drugs. Addition of absorption enhancers can achieve higher permeability. Although ultrasound over a wide frequency range has been used for delivery of drugs, transdermal transport enhancement mediated by low-frequency ultrasound was more significant than the tansdermal transport enhancement mediated by high-frequency ultrasound. Evidences have showed ultrasound in the delivery of drugs have three effects, including thermal, cavitational and acoustic streaming (Lavon et al. 2004).

Microneedle

The microneedle concept had been raised in the 1970s (Gerstel et al. 1976). But the microneedle was not demonstrated experimentally until 1990s when the microelectronics industry provided the microfabrication tools needed to make such small structures. Since the transdermal drug delivery was firstly studied in 1998 (Henry et al. 1998), attention has been drawn to the possibility of using microneedle as a novel drug delivery tools.

The microneedles are usually made of silicon, metal, polymer, and glass that can be divided in two groups: solid and hollow microneedles. To make the microneelde suitable for various drugs, it was further modified to many other variations: "poke with patch", "coat and poke", and "dip and scrape" (Prausnitz 2004). In contrast to the solid microneedles, microneeldes containing a hollow bore offer the possibility of transporting drugs through the interior of well-defined needles by diffusion or, for more rapid rates of delivery by pressure-driven flow.

In the early studies of the usage of microneedles for insulin delivery, the size and the shape are investigated to increase the penetrating dose by changing. McAllister et al. (McAllister et al. 2003) reported that using solid microneedles could increase the permeability of the skin. The microchannels are large enough for the compounds like insulin as large as 100 nm to go through. Changing the shape of the microneedles, Martanto et al. (Martanto et al. 2003) delivered insulin to diabetic hairless rats *in vivo*. The insulin solution was put on the top of the microneedles, the microneedle arrays were inserted into the skin with a high-velocity injector. The arrays were left on the skin for 4h and blood glucose was found to steadily decreaseas much as 80%. The blood glucose of the negative control whose skin was placed by insulin did not have significant change. To control the insulin release, Nordquist et al. (Nordquist et al. 2007) reported another kind of microneedle dispenser. Yan Wu et al. (Yan Wu et al. 2010) studied the effect of duration of microneedle treatment, insulin concentrantion and area of microneedle treatment. They found that the area of microneedle treatment have significant effect on the glucose level. Conventional microneedle patches are limited in the areas that contact with the skin. To solve this problem, microneedle roller was developed. Park et al. (Park et al. 2010) introduced a carboxy-methyl-cellulose (CMC) microneedle roller and then studied its fabrication and mechanics. They also detected its ability to pierce skin and the lateral diffusion in skin after microneedle treatment. Their results showed that the roller can increase skin permeability for medical and other applications. Using microneedle roller for the transdermal delivery of insulin, Cui-Ping Zhou et al. (Zhou et al. 2010) investigated a series of experiment *in vivo*. They studied the effect of the length of microneedle rollers and the insulin patch concentration *in vivo*. They found that the 250 and 500μm microneedle rollers can pierce the skin during manual application and are the most promising tools for *in vivo* delivery of biologically active proteins such as insulin.

Despite of all these improvement in the microneedles, people begin to pay attention to the vertical of the insulin recently. Huabing Chen et al. (Chen et al. 2009) connected iontophoresis with microneedle to increase the insulin penetration. They prepared insulin-loaded nanovesicles with various average diameters and zeta potentials. They compared the permeation rate of these nanovesicles with iontophoresis or microneedle or both of these two methods. The results showed that the positive zeta potential and small diameters of nanovesicles are significantly advantageous to the penetration of the insulin-loaded nanovesicles when combined with iontophoresis and microneedles.

CHALLENGES AND PROBLEMS IN TRANSDERMAL INSULIN DELIVERY

Nano-carriers

Although nanotechnology has been applied to drug delivery for a long time, there are still many problems we have to face the process of preparation of insulin-loaded nano-carriers may have huge effects on the structure of insulin as a protein. The processes are often with high energy or surfactants and co-surfactants. These factors may destroy the original chemical structure of insulin which leads to low bioactivity. What's more, some materials and the surfactants are harmful to human skin or body though they are suitable for transdermal drug carriers. All these problems need to be solved with more studies.

Iontophoresis and Electroporation

As iotophoresis is always used combined with absorption enhancers, the skin irritation is a great concern in the application. A suitable semi-solid formulation which is compatible with both the devices and the skin is needed as the insulin carrier. What's more, the devices used with inotophoresis are expansive for the patients. Thus, the development of a portable, cost-effective devices as well as a suitable insulin-loaded carrier are the major challenges in this area.

The same problems are also involved in the method of electroporation. First of all, the devices used in this method are very expansive for most of people. What's more, according to Rastogi's study (Rastogi et al. 2010), a "needle", although blunt, was still used. To some people, this method didn't lower patients' fear of needles. Another concern is the degradation of nanoparticles *in vivo*. A study reported that the PCL-PEG-PCL polymer can last 45 days *in vivo* (Ma et al. 2010). The side effects caused by electroporation process are also concerned and the formulation of vesicles needs to be optimized. As the results of the Rachna Rastogi's study, a high voltage is used to increase the rate of insulin permeation. However the high voltage may cause many side effects, including sustained pain, skin irritation and contamination problems.

Sonophoresis

The major concern with this method is the effect of ultrasound on protein conformation and/or activity.

PROSPECTS

Diabetes mellitus (DM) is a common metabolism disease with the morbidity increased continually. It is realized that insulin is not only indispensable to the patients with typeI DM, but applied to 10~15% of those with typeII DM. To date, injection is still the common method for insulin therapy, yet with great pain, discomfort and inconvenience. Therefore, developing a new effective, safe, non-injection route for insulin delivery to wholly or partly supplant injection method has attracted more attention of pharmaceutical researchers around the world.

As one of the most prospective delivery systems of insulin for new administration trials, the study of TDDS combined with nanotechnology still has a long way to go to increase the permeation and bioavailability of insulin, as well as the safety, convenience and financially-available to patients. What is more, the crossover between different new physical or chemical technologies to incorporate different advantages may be an important orientation for the future research of the insulin transdermal delivery system.

Key Facts for Insulin-nanoparticle of Transdermal Absorption

- The transdermal delivery of Insulin is limited by high molecular weight, aggregation to hexamer, and electric charges, which could reduce the skin penetrability of Insulin.
- Chemical and physical penetration enhancers and delivery systems prepared by biotechnology would improve the transdermal absorption of Insulin.
- With nanotechnology, the novel Insulin delivery systems are prepared using monomers or small micelles coated with nano-sized delivery systems, which could improve the dispersibility of Insulin micelles in human medium, change electric charges, and enhance the skin penetrability of Insulin.
- The transdermal absorption of Insulin could be improved greatly by the combination of nanotechnology and physical technology, which makes the transdermal delivery of Insulin become the ideal one.

Summary Points

- The transdermal delivery of Insulin shows great property in place of the defectable intravenous injection way, which is also limited in the skin penetration due to the high molecular weight, aggregation to hexamer, and electric charges.
- Nanotechnology such as various kinds of nanomaterials and absorption enhancers is largely implied into transdermal drug delivery system (TDDS), including insulin-loaded carrier, iontophoresis, sonophoresis, microneedles.
- Most of the nanomaterials and absorption enhancers would show some degree of cytotoxicity to human cells and may also influence the structure or function of the insulin.
- There are still lots of space to develop the newly method of iontophoresis, sonophoresis, microneedles such as optimization of the conditions and addition of voltage stimulation.
- In general, the requirement for a portable, cost-effective devices as well as a suitable insulin-loaded carrier is the major challenges in the application of the transdermal drug delivery system (TDDS).
- The future research of the insulin transdermal delivery system would focus on the increase of the permeation and bioavailability of insulin, as well as the safety, convenience and financially-available to patients; or hunting for new physical or chemical technologies and their crossover with the existing technology.

Definitions

Transdermal drug delivery (TDD): a method of delivering drugs through the skin and into the bloodstream.

Supercritical antisolvent (SAS): Supercritical antisolvent precipitation is based on the fast dissolution of a liquid solution in a supercritical fluid.

Iontophoresis: It is a technique using a small electric charge to deliver a medicine or other chemical through the skin.

Sonophoresis: It is a process that exponentially increases the absorption of semisolid topical compounds into the epidermis, dermis and skin appendages.

Chemical enhancers: The chemical agents that had been shown permeation enhancement in the transdermal delivery of various drugs.

Surfactant: Surfactants are compounds that lower the surface tension of a liquid, the interfacial tension between two liquids, or that between a liquid and a solid. Surfactants may act as detergents, wetting agents, emulsifiers, foaming agents, and dispersants.

Zeta potential: Zeta potential is a scientific term for electrokinetic potential in colloidal systems.

Abbreviations

CMC	:	carboxy-methyl-cellulose
DM	:	Diabetes mellitus
PCL	:	poly (ε-caprolactone)
PEG	:	polyethylene glycol
SAS	:	supercritical antisolvent
SC	:	stratum corneum
TDD	:	transdermal drug delivery
TDDS	:	transdermal drug delivery system

Literature

Antosova, Z., M. Mackova, V. Kral and T. Macek. 2009. Therapeutic application of peptides and proteins: parenteral forever? Trends Biotechnol. 27: 628–635.

Chen, H.B., H.D. Zhu, J.N. Zheng, D.S. Mou, J.L. Wang, J.Y. Zhang, T.L. Shi, Y.J. Zhao, H.B. Xu and X.L. Yang. 2009. Iontophoresis-driven penetration of nanovesicles through microneedle-induced skin microchannels for enhancing transdermal delivery of insulin. J. Control. Release 139: 63–72.

Dhamecha, D.L., A.R. Rathi, M. Saifee, S.R. Lahoti and M.H.G. Dehghan. 2009. Int. J. Pharm. Sci. 1: 24–46.

Gallo S.A., A.R. Oseroff, P.G. Johnson and S.W. Hui. 1997. Characterization of electric pulse induced permeabilization of porcine skin using surface electrodes. Biophys. J. 72: 2805–2811.

Gerstel, M.S., and V.A. Place. 1976. Drug delivery device. US Patent No. 3,964,482.

Guo, J., Q. Ping and L. Zhang. 2000. Transdermal delivery of insulin in mice by using lecithin vesicles as a carrier. Drug Deliv. 7: 113–116.

Gwinup, G., A.N. Elias and N.D. Vaziri. 1990. A case for oral insulin therapy in the prevention of diabetic micro- and macroangiopathy. Int. J. Artif. Organs. 13: 393–395.

Henry, S., D. McAllister, M.G. Allen and M.R. Prausnitz. 1998. Microfabricated microneedles: a novel method to increase transdermal drug delivery. J. Pharm. Sci. 87: 922–925.

Higaki, M., M. Kameyama, M. Udagawa, Y. Ueno, Y. Yamaguchi, R. Igarashi, T. Ishihara and Y. Mizushima. 2006. Transdermal delivery of $CaCO_3$-nanoparticles containing insulin. Diabetes Technol. Ther. 8: 369–374.

Kalia, Y.N., A. Naik, J. Garrison, R.H. Guy. 2004. Iontophoretic drug delivery. Adv. Drug Deliv. Rev. 56: 619–658.

Karande, P., A. Jain and S. Mitragotri. 2006. Relationships between skin's electrical impedance and permeability in the presence of chemical enhancers. J. Control Release 110: 307–313.

Kennedy, F.P. 1991. Recent developments in insulin delivery techniques: current status and future potential. Drugs 42: 213–227.

Khafagy, E.S., M. Morishita, Y. Onuki and K. Takayama. 2007. Current challenges in non-invasive insulin delivery systems: a comparative review. Adv. Drug Deliv. Rev. 59: 1521–1546.

Lavon, I., and J. Kost. 2004. Ultrasound and transdermal drug delivery. Drug Discov. Today 9: 670–676.

Liu, J.C., Y. Sun, O. Siddiqui and Y.W. Chien, W.M. Shi and J. Li. 1988. Blood glucose control in diabetic rats by transdermal iontophoretic delivery of insulin. Int. J. Pharm. 44: 197–204.

Ma, G.L., B.L. Miao and C.X. Song. 2010. Thermosensitive PCL-PEG-PCL Hydrogels: Synthesis, Characterization, and Delivery of Proteins. J. Appl. Polym. Sci. 116: 1985–1993.

Martanto, W., S. Davis, N. Holiday, J. Wang, H. Gill and M. Prausnitz. 2003. Transdermal delivery of insulin using microneedles *in vivo*. Proceedings of International Symposium on Controlled Release Bioactive Material, NO 666.

McAllister, D.V., P.M. Wang, S.P. Davis, J.H. Park, P.J. Canatella, M.G. Allen and M.R. Prausnitz. 2003. Microfabricated needles for transdermal delivery of macromolecules and nanoparticles: novel fabrication methods and transport studies. Proc. Natl. Acad. Sci. USA 100: 13755–13760.

Nordquist, L., N. Roxhed, P. Griss and G. Stemme. 2007. Novel microneedle patches for active insulin delivery are efficient in maintaining glycaemic control: an initial comparison with subcutaneous administration. Pharmaceut. Res. 24: 1381–1388.

Panchagnula, R., O. Pillai, V. Nair and R. Poduri. 2000. Transdermal iontophoresis revisited. Curr. Opin. Chem. Biol. 4: 468–473.

Park, J.H., S.O. Choi, S. Seo, Y.B. Choy and M.R. Prausnitz. 2010. A microneedle roller for transdermal drug delivery. Eur. J. Pharm. Biopharm. 76: 282–289.

Prausnitz, M.R. 2004. Microneedles for transdermal drug delivery. Adv. Drug Deliv. Rev. 56: 581–587.

Rastogi, R., S. Anand and V. Koul. 2010. Electroporation of polymeric nanoparticles: an alternative technique for transdermal delivery of insulin. Drug Dev. Indl. Pharm. 36: 1303–1311.

Tahara, Y., S. Honda, N. Kamiya, H.Y. Piao, A. Hirata, E. Hayakawa, T. Fujii and M. Goto. 2008. A solid-in-oil nanodispersion for transcutaneous protein delivery. J. Control. Release 131: 14–18.

Ting, W.W., C.D. Vest and R.D. Sontheimer. 2004. Review of traditional and novel modalities that enhance the permeability of local therapeutics across the stratum corneum. Int. J. Dermatol. 43: 538–547.

Wu, Y., Y.H. Gao, G.J. Qin, S.H. Zhang, Y.Q. Qiu, F. Li and B. Xu. 2010. Sustained release of insulin through skin by intradermal microdelivery system. Biomed. Microdevices 12: 665–671.

Zhao, X.H., Y.G. Zu, S.C. Zu, D. Wang, Y. Zhang and B.S. Zu. 2010. Insulin nanoparticles for transdermal delivery: preparation and physicochemical characterization and *in vitro* evaluation. Drug Dev. Ind. Pharm. 36: 1177–1185.

Zhou, C.P., Y.L. Liu, H.L. Wang, P.X. Zhang and J.L. Zhang. 2010. Transdermal delivery of insulin using microneedle rollers *in vivo*. Int. J. Pharm. 392: 127–133.

11

Applications of Carbon Nanomaterials for MALDI-TOF-MS and Electrochemical Analysis of Insulin

Stefan A. Schönbichler,[1,a] Lukas K.H. Bittner,[1,b] Johannes D. Pallua,[1,c] Verena A. Huck-Pezzei,[1,d] Christine Pezzei,[1,e] Günther K. Bonn[1,f] and Christian W. Huck[1,g,*]

ABSTRACT

At present, different carbon nanomaterials like carbon nanotubes, graphitic nanofibers or fullerenes have been explored to achieve highly sensitive methods for analyzing insulin. Two main fields

---

[1]Institute of Analytical Chemistry and Radiochemistry, Leopold Franzens University, Innrain 52a, 6020 Innsbruck, Austria.
[a]E-mail: stefan.schoenbichler@uibk.ac.at
[b]E-mail: lukas.bittner@uibk.ac.at
[c]E-mail: johannes.pallua@uibk.ac.at
[d]E-mail: verena.huck-pezzei@uibk.ac.at
[e]E-mail: christine.pezzei@uibk.ac.at
[f]E-mail: guenther.bonn@uibk.ac.at
[g]E-mail: christian.huck@uibk.ac.at
*Corresponding author

List of abbreviations after the text.

of applications for carbon nanomaterials for this purpose are discussed in this chapter: applications for qualitative matrix-assisted laser desorption/ionization time-of-flight mass-spectrometry analysis of insulin down to trace level, and the usage of electrochemical methods for quantitative investigations. Carbon nanotubes were used for material-enhanced laser desorption/ionization time-of-flight mass-spectrometry, as a matrix for matrix-assisted laser desorption/ionization time-of-flight mass-spectrometry and as catalyst for electrochemical devices. Several modifications like oxidation, the attachment of immobilized metal ion affinity chromatography groups or adsorbtion of citric acid were performed. Silica was derivatized with fullerenes to obtain a material for solid phase extraction (Vallant et al. 2007). A water-soluble fullerene was synthesized to precipitate insulin for further matrix-assisted laser desorption/ionization time-of-flight mass-spectrometry analysis (Shiea et al. 2003). Derivatized graphitic nanofibers were utilized as material-enhanced laser desorption/ionization time-of-flight mass-spectrometry material and modified with immobilized metal ion chromatography-functionalities to ensure high affinity to insulin (Greiderer et al. 2009).

Because of their electrical conductivity and catalytic properties, carbon nanotubes are used for biosensors and electrochemical sensors. To achieve low detection limits for insulin, different approaches were adopted: A carbon nanotube/dimethylformamide slurry was applied to a cleaned glassy-carbon-electrode (Wang and Musameh 2004). A ruthenium oxide glassy-carbon-electrode was produced, before adding a carbon-nanotube-layer (Wang et al. 2007). Using the same method, a multi-walled carbon nanotube/3,4-Dihydro-2H-pyran-, a multi-walled carbon nanotube/dihexadecyl phosphate- and a multi-walled carbon nanotube-chitosan-layer coated glassy-carbon-electrode were prepared (Snider et al. 2008). The lowest detection limit (1nM) was achieved with the ruthenium oxide glassy-carbon-electrode (Wang et al. 2007).

INTRODUCTION

Nanoscience denotes an interdisciplinary field, dealing with phenomena, or creation of devices in a range smaller than 1 µm (typically 1 to 100 nm). Two main approaches were made to produce nano-particles and devices: the bottom-up, and the top-down approach. The bottom-up approach works at molecular level by chemical composition. The top-down approach reaches its aims by bringing larger entities to nanolevel without affecting the molecular level (Merkoci 2009).

Nanomaterials were implemented in bioanalytical applications to achieve highly sensitive and robust methods. By decreasing particle size to nanolevel, a change of the physicochemical properties take place. The increased surface-to-volume ratio and the unusual binding affinities depending on size, shape and molecular structure are interesting aspects for analytical applications (Merkoci 2009).

CARBON NANOMATERIALS

Because of their unique chemical, physical and biological properties, carbon nanomaterials were explored in different scientific fields like analytical chemistry, drug delivery, electronics, optics and medicine. Carbon appears in different allotropes like diamond, graphite, amorphous carbon or fullerene (Merkoci 2009).

Diamond

In a perfect diamond crystal every carbon atom is sp^3 hybridized. The atom in the middle is covalently bound to four carbon atoms, which are placed in the angles of a tetrahedron. Due to the 3-dimensional, very stable geometric structure, diamond is the hardest known natural mineral. A consequence of the single bond structure is the optical transparency over a broad range and the insulator property of diamond (Merkoci 2009).

Graphite

Carbon atoms in graphite are sp^2 hybridized and form layers consisting of hexagons. The delocalized π-electrons cause the electrical conductivity of graphite (Merkoci 2009).

Fullerenes and carbon nanotubes were previously described in chapter: Carbon nanotubes coupled to siRNA generate efficient transfection and are a tool for examining glucose uptake in skeletal muscle.

Amorphous Carbon

Amorphous carbon consists of sp^2 and sp^3 hybridized atoms which are arranged without any crystal structure. No long-range order and no delocalized π-electrons appear in this system (IUPAC 1997).

Carbon nanomaterials are used in a broad range of applications. The high surface/volume ratio and strong binding ability for different types of molecules makes them suitable for enrichment, purification and desalting of biomolecules (Vallant et al. 2007). Due to their optical and energy-transferring properties they can be used as MALDI-matrix or assisting material (Shiea et al. 2003; Chen et al. 2004; Greiderer et al. 2009). The catalytic nature of some carbon nanomaterials and the conductivity are utilized in electrodes for electrochemical detection of biomolecules (Wang 2005).

MALDI-TOF-MS

Besides electrospray ionization (ESI) matrix-assisted laser desorption/ionization (MALDI) represents a soft ionization method used in mass spectrometry (MS). "Soft" in this case means the possibility to ionize a molecule without heavy fragmentation. On the one hand the fragmentation-process is important for structure-determination; on the other, it is a disturbing factor. In the 1960´s a laser desorption/ionization- (LDI) MS was introduced. An organic sample, placed as thin film on a metal surface was irradiated with a laser beam. This method was restricted to masses below 1000 Da, a strong fragmentation took place and a poor sensitivity could be achieved (Hillenkamp and Jasna 2007).

In 1985 Karas and co-workers introduced the term "matrix-assisted" laser desorption. It was found that a better desorption and ionization of alanine took place by mixing it with tryptophan. The "matrix" tryptophan absorbs the laser energy and increases the desorption/ionization-process of alanine (Karas et al. 1985).

One year later, in 1986, the spectrum of the bee venom mellitin, a 2845 Da oligogopeptide was published using tryptophan as matrix (Karas and Bahr 1986).

Now even larger molecules like proteins, RNA, and DNA were accessible by this method.

To this day many matrices like 2,5-dihydroxybenzoic acid, sinapinic acid, α-Cyano-4-hydroxycinnamic acid or caffeic acid have been investigated. The ability to absorb light in the wavelength of the laser is the most important property of the matrix. Regarding the instrument-configuration the components can differ, and even ion trap MS (QIT-MS) or Fourier transform MS (FT-MS) can be used to detect the MALDI ions, but the most common detector is the time-of-flight MS (TOF-MS). Usually the lasers used are generating a pulsed and not a continuous ion beam. Therefore the MS has to be adequate to these requirements. Due to the fact that magnetic sector- and quadrupole instruments show poor performance concerning sensitivity with pulsed ion sources, and the TOF-MS is working properly under these conditions, the MALDI-TOF-MS has become the dominant technology. Including the capability to ionize masses larger than 1 MDa with predominant single charged ions, the TOF-MS with its theoretically unlimited mass ranges perfectly suited for MALDI. In general the MALDI-TOF-MS consists of a laser, an ion-accelerator, a flight tube and a detector. Common laser-types used are: Nitrogen lasers with a wavelength of 337 nm and ND:YAG lasers with typically 355 and 266 nm wavelength. The aforementioned lasers are operating in the ultraviolet-range, but also infrared lasers like the ER:YAG laser (2.94 µm) and tunable optical parametric oscillators (OPO, 1.7 to 2.5 µm) are applicable.

The ion accelerator consists of the sample target and a nearby grid with applied potential.

The flight tube is a field free drift region with a typical length of one meter which leads to the detector (see Fig. 1). The detector, a microchannel plate (MCP) is connected with an oscilloscope to plot the flight time versus the ion current signal. For precise time-measurement the time-metering starts triggered by a small fraction of the laser light hitting a photo-diode.

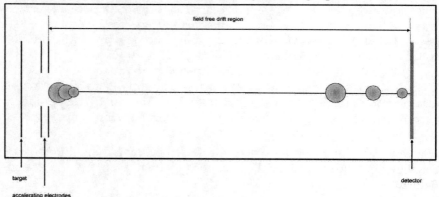

Fig. 1. Principle of a linear time-of-flight mass-analyser. Ions with different masses have different arrival-times at the reflector. (drawn by the author).

The kinetic energy E_k of the ions accelerated by a potential difference U can be calculated by the following equation including the charge state z and the elementary charge e.

$$E_k = zeU = \frac{mv^2}{2}$$

Two ions with the same kinetic energy E_k, but different masses m, discriminate by their velocity v.

The arrival time t after passing the flight tube length L will be:

$$t = L\sqrt{\frac{m}{2E_k}} = L\sqrt{\frac{m}{2zeU}}$$

This equation can be inverted to:

$$\frac{m}{z} = 2eU \left(\frac{t}{L}\right)^2$$

The potential difference U typically is 20kV. It is shown, that $\frac{m}{z}$ is proportional to t^2. With a known flight time through the field free drift region the mass of the ion can be calculated. Irregular surface of the desorption plume and velocity distribution causes differences in the flight time of ions of the same mass resulting in bad mass resolution. There are three common ways to enhance mass resolution by correcting the velocity distribution: pulsed ion extraction, the use of a reflectron and increasing the acceleration potential. The pulsed ion extraction is also called delayed extraction or time-lag focusing. The applied potential is first kept low or zero and then, typically after tens of nanoseconds, raised to the end level. That causes a higher acceleration for slower ions which passed a shorter way from the desorption plume to the nearby grid. Ions with higher acceleration caused by the desorption-process covers a larger distance, and less effect of the potential difference take place. Due to the higher acceleration of the initially slower ions, both, initially slow and fast meet in one point in the flight tube. At this "space focus" the detector should be placed. The disadvantage of this method is based on the fact that the focus-point is mass-dependent. Only a small mass-range can be corrected by pulsed ion extraction and it is limited to masses below 30 kDa. The use of a reflectron or ion mirror is the second option to reduce the loss of mass resolution caused by initial velocity distribution, and was introduced by Mamyrin and coworkers. The reflectron acts like a u-turn for ions and consists of a series of ring electrodes. The electrodes produce an electric field which slows down the incoming ions and repel them back near their origin. Faster ions of the same mass possess higher kinetic energy, penetrates the reflectron deeper before being reflected, and have to pass a greater distance. This reduces the distribution of kinetic energy. The mass detector is placed at the end of the loop-like flight path, in the focus point (see Fig. 2). The ion-mirror increases the mass resolution, but decreases the sensitivity. Modern MALDI-TOF-MS includes two detectors, and the user can choose between higher sensitivity or higher mass resolution. Another method to minimize the initial velocity distribution is to increase the acceleration potential. This will decrease the part of the initial energy which is summed up with the acceleration energy. However, the use of very high voltages is limited in laboratories, therefore the reflectron and pulsed-ion extraction are common technologies in modern MALDI-TOF-MS (Hillenkamp and Jasna 2007).

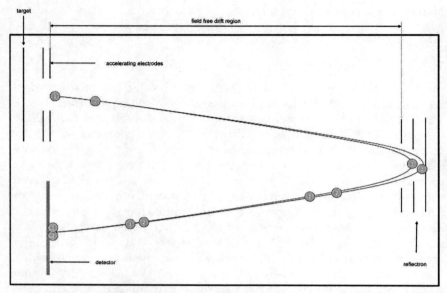

Fig. 2. Principle of a reflectron time-of-flight mass analyser. Ions 1 und 2 have the same mass, but different E_{Kin}. The reflectron focuses both ions on the detector to achieve the same flight time. (drawn by the author).

MATERIAL ENHANCED LASER DESORPTION/IONIZATION (MELDI)

Because of the complexity of biological fluids, pretreatments are required before applying them to MALDI-MS. In the MELDI approach additional to the matrix a third material is used for purification, enrichment and desalting of the sample. Materials derivatized for this purpose include cellulose, silica, poly (glycidyl-methacrylate/divinylbenzene (Feuerstein et al. 2006)) and carbon nanomaterials (diamond powder (Feuerstein et al. 2006), nanotubes (Najam-ul-Haq et al. 2006), nanofibers (Greiderer et al. 2009) and C_{60} fullerenes (Vallant et al. 2007). To achieve different affinities, immobilized metal affinity chromatography (IMAC), reversed phase (RP) and ion exchange functionalities are introduced onto the surface to capture compounds of interest. Beside the chemical properties even morphological and physical characteristics determine the binding ability of the MELDI-Material (Greiderer et al. 2009).

The MELDI-process starts with incubation of the biofluid with the MELDI-Material. After a washing step the MELDI-material-slurry and a matrix are spotted on a target and analyzed by MALDI-TOF-MS. In addition the enriched compounds can be eluted, separated by μ-LC and identified by tandem-MS (see Fig. 3). The MELDI approach was developed for protein-pattern analysis for early diagnosis of diseases (Feuerstein et al. 2006). Because of the high affinity of insulin to most of the MELDI materials, many of them are suited for insulin-enrichment and desalting of complex samples.

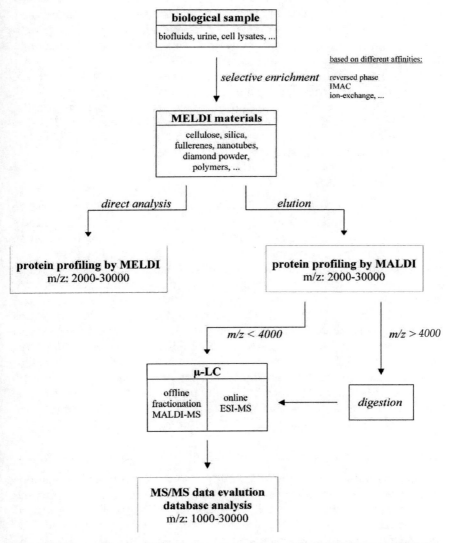

Fig. 3. Protein profiling workflow. Proteins are enriched using a MELDI-material and then analysed direct or after elusion. (reproduced with permission from (Vallant et al. 2007)).

QUALITATIVE INSULIN ANALYSIS BY MALDI-TOF-MS

Graphitic Nanofibers

Graphitic nanofibers (GNFs) are stacked graphene layers (Melechko et al. 2005) occurring in different structures depending on growth conditions and metals which are used as catalysts. The most reported structures are: the ribbon like GNFs, which displays graphene layers parallel to growth axis, the platelet GNFs, which shows graphene layers perpendicular to the growth axis, in case of the herringbone GNFs the stacked layers are angled to the growth axis; and finally the cup-stacked GNFs, which are assembled in cylindrical and conical structures (see Fig. 4) (Serp et al. 2003). GNFs are obtained by catalytic chemical vapor deposition (c-CVD) and occur in diameters generally higher than nanotubes around 100–200 nm (Melechko et al. 2005), but can also reach 500 nm (Serp et al. 2003). The length varies typically between 5 and 20 µm (Melechko et al. 2005).

Applications of GNFs

By derivatization different affinities can be achieved for selective enrichment of proteins and peptides. Immobilized metal ion affinity chromatography (IMAC) loaded with Cu^{2+}, cation and anion exchange materials are obtained by modifying GNFs. The lipophilic nature of the GNFs, in combination with the hydrophilic functional groups, constitutes the binding-characteristic of this substance. Four derivatization-steps are required to obtain the final product. In the first step the GNFs are oxidized with H_2SO_4/HNO_3. In the second step the resulting carboxylic acid groups are reacted to acid chloride by using thionylchloride. As a next step the acid chloride is turned into an amide by reacting with iminodiaceticacid (IDA). Finally, the GNF-IDA is loaded with Cu^{2+} ions by stirring it with $CuSO_4$ to form a bidentate complex (see Fig. 5). The insulin loading capacity for the synthesized material was found to be 30 µg per mg GNF-IDA-Cu^{2+} and the recovery rate was determined around 55%. The detection limit found by MALDI analysis was 50 fmol μl^{-1} (Greiderer et al. 2009).

Key facts of Immobilized Metal Ion Affinity Chromatography (IMAC)

The complex-building-ability of histidine and cysteine with metalions (in decreasing order: Cu > Zn > Ni >Mn) was utilized in an approach to catch proteins with immobilized metal ions. Porath et al. introduced the material *bis*-carboxymethyl amino agarose. Oxiran-activated-cellulose was reacted with iminodiacetic acid and flushed with $CuSO_4$ to load the Cu^{2+} ions. The affinity to human serum proteins was tested by using the

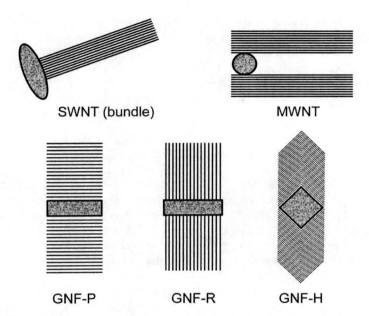

Fig. 4. Types of carbon nanotubes and graphitic nanofibers. Multiwalled carbon nanotubes are concentric arranged single walled carbon nanotubes. Graphitic nanofibers are differing in growth axis (reproduced with permission from (Serp et al. 2003)).

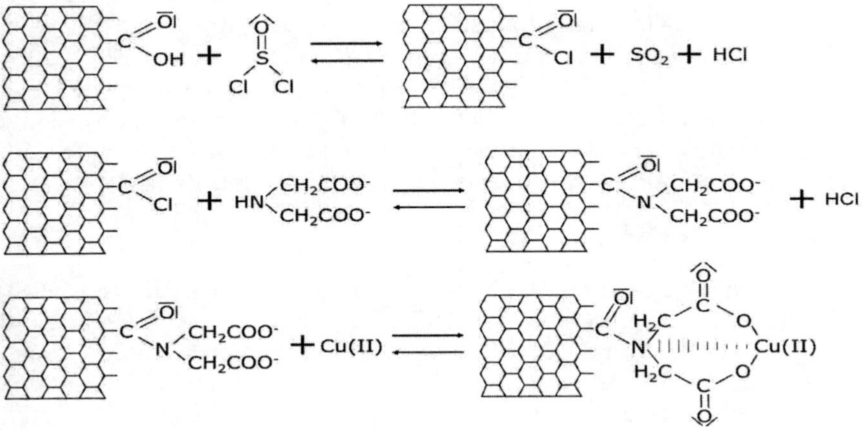

Fig. 5. Derivatization pattern for oxidized fullerenes, carbon nanotubes and graphitic nanofibers to obtain IMAC-functionalities. After an oxidation step the carbonic acid is converted in an acid chloride and reacted with iminodiacetic acid. The product is loaded with Cu^{2+} (reproduced with permission from (Najam-ul-Haq et al. 2006)).

new IMAC-material in column chromatography. Retained proteins like albumin, γ-globulin, prealbumin were identified after elution by gel-electrophoresis (Porath et al. 1975).

The high affinity of albumin to the IMAC-material can affect the binding capacity regarding other proteins. Even when this method cannot be called "specific" like immuno-affinity-chromatography, it is a useful tool for the purification and enrichment of histidine-containing or his-tagged proteins.

Fullerenes

The idea that a graphite-layer can be formed to a hollow ball came up in 1966. To achieve this unique structure, the addition of some pentagons between the hexagons is required. Eiji Osawa was the first who predicted the existence of the C_{60}-molecule (Osawa 1970). In 1985 the first fullerene was produced by laser evaporation of graphite. Buckminster Fuller, who invented the geodesic domes, was the eponym of this new class of molecules. The term fullerene denotes all closed carbon cages including 12 pentagons and more than one hexagon which form a sp^2 network (Kroto et al. 1991). Commonly available fullerene-molecules are C_{60}, C_{70}, C_{76}, C_{78} and C_{84}. Macroscopic amounts of C_{60} were produced for the first time in 1990. The product of the graphite evaporation was either sublimated gently or solved in benzene to separate it from the soot. The solid form of the substance was named fullerite (Krätschmer et al. 1990). Till today different methods to obtain fullerenes have been developed. The fullerene yield of a toluene soxhlet extract is about 20–40%. Thereof 65% is C_{60}, 30% C_{70} and the rest are higher fullerenes. Separation and purification of C_{60} and C_{70} can be done by aluminium oxide column chromatography using hexan/toluene as solvent. Higher fullerenes can be purified by C18-RP-HPLC and acetonitrile/toluene (Diederich and Whetten 1992).

Applications of Fullerenes

Two C_{60}-fullerene bound silica materials for solid phase extraction (SPE) were synthesized (Vallant et al. 2007). For this purpose two different types of silica (Prontosil 300-5-Si (Bischoff Analysentechnik und Geräte GmbH, Leonberg, Germany); Kovasil 100-5-Si (Zeochem AG, Uetikon, Switzerland)) was derivatized with an aminopropyllinker (Buszewski et al. 2001). C_{60}-epoxyfullerene was reacted with the amino-group of the linker to yield the SPE-material. An affinity-test was performed with a mixture of insulin, ubiquitin and myoglobin. All three proteins were retained by the C_{60}-aminosilica and could be eluted with 80% ACN (see Fig. 6). Recovery tests were carried out against Sep-Pak C18 and Oasis-

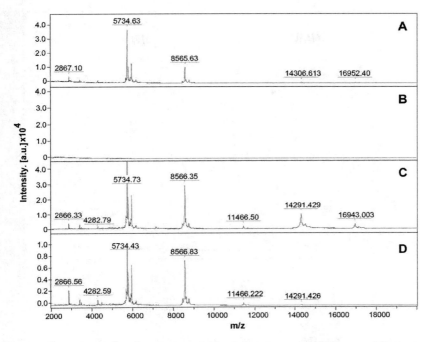

Fig. 6. MALDI-MS analysis of protein mixture including insulin (5734 Da). (A) MALDI-MS of untreated protein mixture; (B) flow through of loading-step on C_{60} amino-silica; (C) MALDI-MS of protein-elute from Prontosil 300-Å C_{60}-Aminosilica; (D) MALDI-MS of protein-elute from Kovasil 100-Å C_{60}-Aminosilica. Matrix solution: HCCA in 50% ACN/0,1% TFA (reproduced with permission from (Vallant et al. 2007)).

material (Waters, Milford, United States). The recovery-rate of insulin (50 µg/ml) for C_{60}-aminosilica was 97.2%, for the SEP-PAK C18 91.8% and for the Oasis-material 91.7%. Also for flavonoids good recovery-rates up to 99.6% were achieved (Vallant et al. 2007).

A star like water-soluble fullerene derivative was synthesized to precipitate positively charged surfactants, amino acids, peptides and proteins (Shiea et al. 2003). When fullerenes react with sodium naphtalide, followed by derivatization with 1,4-butane sultone a hexa(sulfonbutyl)fullerene (HSBF) can be obtained (Chi et al. 1998). A mixture of insulin, ubiquitin, cytochrome c, lysozyme, myoglobin, β-lactoglobulin, β-casein, carbonic anhydrase was added to a HSBF-solution. The precipitate was measured with MALDI using sinapinic acid as matrix. It was shown that insulin could be purified from the other proteins (see Fig. 7) (Shiea et al. 2003).

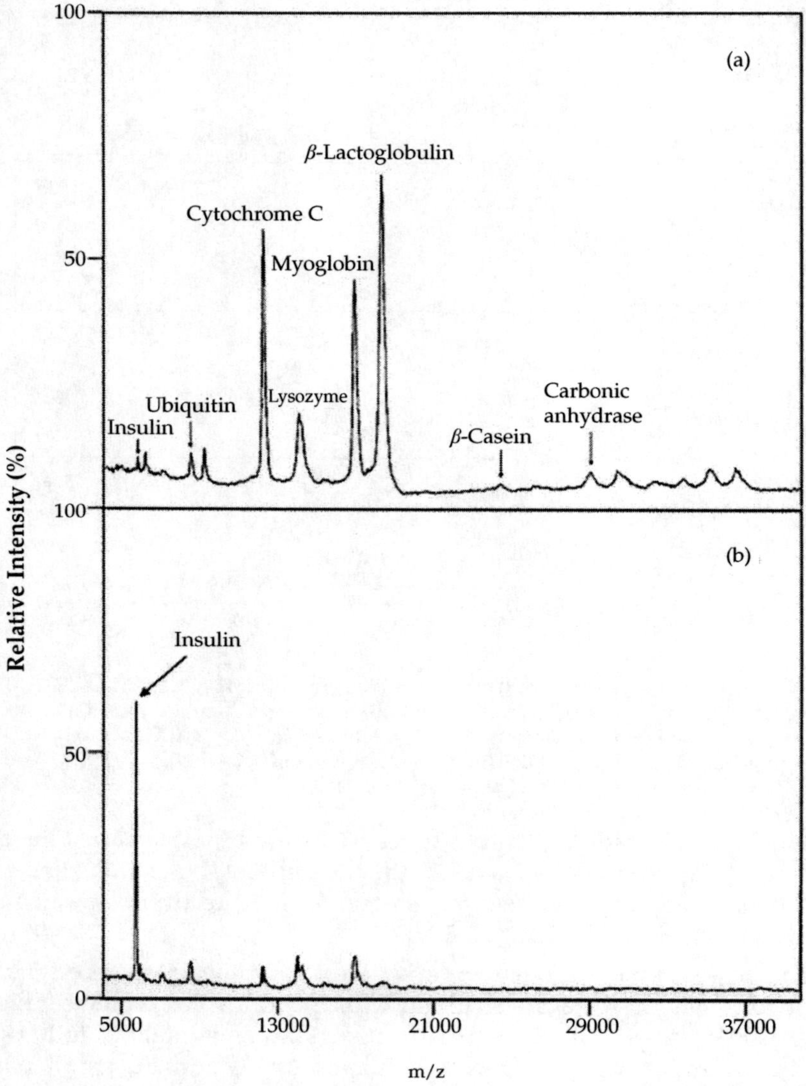

Fig. 7. Separation of insulin from seven other proteins. (a) MALDI-MS analysis of a solution of eight proteins including insulin. (b) MALDI-MS analysis of the ion pair precipitate of obtained by mixing a HSBF solution with the solution of the eight proteins; matrix in both cases: sinapinic acid (reproduced with permission from (Shiea et al. 2003)).

Vallant et al. used different fullerene-derivatives as MELDI-material. Proteins of human serum were enriched and desalted by fullerenes before measuring with LDI-TOF-MS. Especially fullerenoacetic acid shows good affinities in the low molecular weight range from 2300–6300 Da. Though insulin (5733 Da) is not tested explicitly in this study, this material can be a promising tool for insulin-enrichment (Vallant et al. 2007).

Carbon Nanotubes

The structure of carbon nanotubes (CNTs) were previously described in chapter: Carbon nanotubes coupled to siRNA generates efficient transfection and is a tool for examining glucose uptake in skeletal muscle. Different methods for qualitative insulin-analysis using CNTs as matrix or MELDI-material have been reported.

Applications of CNTs

Chen et al. introduced an approach for measuring insulin by utilizing citric acid-treated CNTs as a matrix. For this purpose CNTs were produced by performing the following steps: a piece anodic aluminum oxide (AAO) was covered with NaH and heated to achieve a silvery plate (Na@AAO). This was exposed to hexachlorobenzene-vapor. After removing the by-product NaCl by heat, the CNT@AAO was washed in a HF-solution to get rid of the AAO. For using the CNTs as a matrix they were suspended in an ethanol/citric acid buffer. The adsorbtion of the citric acid was confirmed by using infrared spectroscopy. First the CNT-suspension and then the sample mixed with citric acid buffer was spotted on an AnchorChip target. The suitabitlity to serve as affinity probes was tested by incubating a trace amount of insulin with CNTs. It was shown that insulin can be detected in femtomol-range (87 fmol). The method can also enrich proteins and peptides in the presence of a surfactant, like Triton X-100 and urea. The citrate buffer serves as proton source and reduces the appearance of sodium adducts (Chen and Chen 2004; Chen et al. 2004).

Oxidized CNTs were used as matrix to detect Insulin and neutral carbohydrates like xylose and sucrose. The nanotubes were oxidized by refluxing them with nitric acid. As previously mentioned the sample and the CNT-Matrix were spotted sandwich-like, first the CNTs and then the sample as a second layer. The detection limit for insulin was assumed to be around 300 fmol and for xylose 10 fmol. It was shown that a better shot-to-shot reproducibility can be achieved by oxidized CNTs compared to untreated CNTs (Ren and Guo 2005).

Najam-ul-Haq introduced IMAC-CNTs as a MELDI-support material. The CNTs were derivatized like the GNFs in the above mentioned work of Greiderer et al. The CNT-IDA-Cu^{2+}-material was not explicitly tested for insulin-affinity, but the structural similarity to GNF-IDA-Cu^{2+} indicates viable properties to fulfill this task. GNF-IDA and GNF-IDA-Cu^{2+} were compared and the higher protein binding ability of the second material was clearly demonstrated. Also the untreated and IMAC-CNTs were checked against each other, and due to the enormous higher binding capacity of CNT-IDA-Cu^{2+} it can be concluded that the chemosorption takes the main-part, and physisorbtion plays only a little role in the adsorbtion-process. This is in contrary to diamond, which shows high physisorption-abilities (Najam-ul-Haq et al. 2006).

QUANTITATIVE ELECTROCHEMICAL ANALYSIS OF INSULIN USING CNTs

Introduction

Since their discovery, CNTs have been investigated intensively. CNTs possess a variety of unique properties being able to display metallic, semiconducting and also superconducting electron transport, having a hollow core where guest molecules can be stored and showing the largest elastic modulus of any known material (Wang 2005).

There are reports that CNTs promote electron transfer reactions of proteins (even those where the redox center is embedded deep within the glycoprotein shell) and enhance the electrochemical reactivity of important biomolecules (Wang 2005).

CNTs are carbon based materials used in electrochemistry and are superior to established materials such as glassy carbon (GC), graphite and diamond showing enhanced electronic properties, large edge plane/basal plane ratios and fast electrode kinetics. Catalytic and physical properties (high electrical conductivity, chemical stability and mechanical strength) make CNTs optimal for deployment in sensors (Jacobs et al. 2010).

All these unique properties make CNTs useful in electrocatalysis, direct electrochemistry of proteins or other electroanalytical applications like electrochemical sensors and biosensors (Gong et al. 2005).

Generally CNT technology based sensors show higher sensitivities, lower limits of detection and faster electron transfer kinetics compared to classical carbon electrodes (Jacobs et al. 2010).

The compatibility between CNTs and biomolecules has led to numerous new hybrid systems for the development of nanoscale devices with applications in biology, medicine and electronics (Merkoci 2009).

Preparation of Electrodes

Due to van der Waals forces between the tubes, CNTs aggregate in most organic and aqueous solvents, which leads to difficulties in making CNT composite. Most reported CNT based electrodes were hitherto mainly prepared dispersing the CNTs randomly onto an electrode or by confining the CNTs onto a polymer substrate such as Teflon or Nafion. Only recent new methods, e.g., aligned CNTs, have been developed. Conventional methods for the preparation of CNT-electrodes include cast coating, chemical deposition and chemical linkage. As newer ways to modify electrodes the following methods are reported: confining the CNTs with polymers, paste and sol-gel matrix or assembling the CNTs with layer by layer techniques (Gong et al. 2005).

We are focusing on the preparation of CNT electrodes utilized for the analysis of Insulin.

CNT-modified GCE

Wang et al. prepared a CNT modified glassy carbon electrode (GCE) casting sonicated CNT in DMF on the cleaned surface of a GCE (Wang and Musameh 2004).

RuOx/CNT modified GCE

For the preparation of a Ruthenium oxide (RuOx)-modified GCE a RuOx film was applied by electrochemical deposition of $RuCl_3$ using a 12.5 min potential cycling. The CNTs were sonicated with DMF and the resulting slurry was applied to the surface of the electrode (Wang et al. 2007).

MWCNT/dihexadecyl phosphate

An aqueous suspension of MWCNTs was mixed with dihexadecyl phosphate and the films were formed by drop casting the suspension onto the electrode surface (Snider et al. 2008).

MWCNT/DHP

An aqueous suspension of MWCNT was mixed with 3,4-Dihydro-2H-pyran (DHP) and the films were formed by drop casting the suspension onto the electrode surface (Snider et al. 2008).

CHIT CNT modified GCE

For the preparation a dispersion of CNTs in a Chitosan (CHIT) solution was casted onto the surface of the GCE and dried (Zhang et al. 2005).

Electrochemistry of CNTs

Like GCE, graphite and diamond CNTs are electrochemical inert and do not show a voltammetric response by applying usual potentials. To prevent impurities like nanocrystal metal catalysts, amorphous carbon and carbon nanoparticles, the raw CNTs have to be pretreated for purification with methods such as chemical oxidation or thermal treatment. This shortens the CNTs and produces functional oxygenated groups especially along the sidewalls which modifies the electronic and structural properties of the CNTs. It was reported that CNTs possess good electrocatalytic activities which is ascribed to the edge-plain-like component of the CNTs. Several amino acids, glucose, NADH, dopamine, insulin and other molecules can be detected through electrocatalysis of CNTs (Gong et al. 2005).

Insulin Analysis

For the electrochemical analysis of proteins a number of sensors have been developed based on techniques including direct electrochemistry and indirect electrochemistry following a selective reaction. CNTs have been introduced to utilize its faster electron transfer kinetics and to provide a wire to the redox site of a protein. Applications cover a very broad range of proteins (Jacobs et al. 2010).

Wang and coworkers developed a system for direct amperometric and voltammetric determination of insulin. The use of CNTs increased the electron transfer kinetics dramatically. The limit of detection (LOD) was as low as 14 nM. Figure 8 compares the hydrodynamic voltammograms using a N,N-dimethylformamide (DMF)/GCE and a CNT/GCE for insulin (Wang and Musameh 2004).

Following up Wang's group added a layer of Ruthenium oxide onto the CNTs which increased the catalytic activity further and improved the stability and sensitivity dramatically reaching a LOD of 1 nM (Wang et al. 2007).

Zhang et al. developed a CHIT CNT modified GCE generating stable insulin currents and reaching a LOD of 30 nM (Ren et al. 2005).

Snider et al. integrated a DHP/CNT sensor into a microfluidic chip to continuously sample insulin secretion from islet cells (Snider et al. 2008).

Applications to Areas of Health and Disease

Sensitive Methods like CNT-based electrochemical detection can be used for *in vivo* measurement of insulin-release or hormone monitoring in chromatographic effluents (Wang and Musameh 2004). These tools are also promising devices for fast quantitative determination of pharmaceutical insulin formulations and quality control of pancreatic islets prior to their transplantation (Zhang et al. 2005).

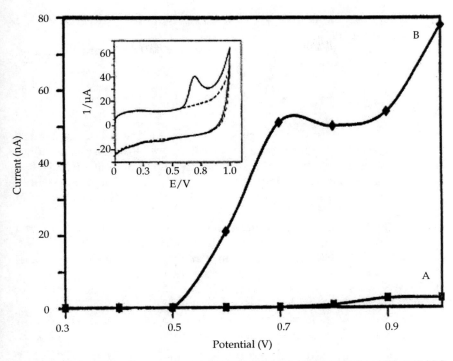

Fig. 8. Hydrodynamic voltammograms for insulin (800 nM). (A) DMF/GC and (B) CNT/GC electrodes. Supporting electrolyte: phosphate buffer (0.05 M, pH 7.4, 0.02% (v/v) Tween-80); CNT loading: 20 µg; flow rate: 1 ml/min. The inset shows a cyclic voltammogram for 40 µM insulin recorded in the same electrolyte solution using a scan rate of 50 mV/s. The dotted line reflects the voltammogram of the blank phosphate buffer solution (reproduced with permission from (Wang and Musameh 2004)).

Summary Points

- Nanomaterials have a particle size below 1 µm (usually 1 to 100 nm).
- Carbon nanomaterials can be used for enrichment, purification and desalting of proteins including insulin.
- Carbon nanotubes, due to their electrical conductivity and catalytic properties, are suitable for sensitive electrochemical insulin detection.
- With carbon nanotubes as a matrix for matrix-assisted laser desorption/ionization time-of-flight mass-spectrometrya detection limit of 87 fM for insulin could be reached.
- Derivatized graphitic nanofibers can be used as material-enhanced laser desorption/ionization time-of-flight mass-spectrometry material to achieve a detection limit of 50 fM for insulin.

- Solid phase extraction by fullerene-derivatized silica showed a recovery rate of 97.2% for insulin.
- Utilizing carbon nanotubes, low detection limits of 1 nM for insulin quantification can be achieved.

Key Terms

- **MALDI-TOF-MS:** The matrix-assisted laser desorption/ionization mass-spectrometry is a method to detect a molecule mass by ionization and desorption of the analyte by a laser with the help of a matrix.
- **TOF-MS:** The time-of-flight mass-analyser determines the mass of a molecule by measuring the flight-time of the accelerated molecule in a flight tube of certain length.
- **MELDI-TOF-MS:** The material-enhanced laser desorption/ionization mass-spectrometry is a type of MALDI-TOF-MS with an enrichment, desalting or purification step prior to analysis by a material what acts like a solid phase extraction and/or a matrix.
- **Matrix:** is a small molecule with light-absorbing properties which can act as a proton donor to desorb and ionize the analyte.
- **SPE:** Solid phase extraction is a method to enrich or purify one or more substances by distribution between a solid and a liquid phase.

Abbreviations

AAO	:	anodic aluminium oxide
ACN	:	acetonitrile
c-CVD	:	catalytic chemical vapor deposition
CHIT	:	chitosan
CNT	:	carbon nanotube
DHP	:	3,4-Dihydro-2H-pyran
DHP/CNT	:	3,4-Dihydro-2H-pyran/carbon nanotube
DMF	:	dimethylformamide
ESI	:	electrospray ionization
FT-MS	:	Fourier transform mass spectrometer
GC	:	glassy carbon
GCE	:	glassy carbon electrode
GNF	:	graphitic nanofiber
GNF-IDA	:	graphitic nanofiber-iminodiacetic acid
GNF-IDA-Cu^{2+}	:	graphitic nanofiber-iminodiacetic acid-copper
HSBF	:	hexa(sulfonbutyl)fullerene
IDA	:	iminodiacetic acid
IMAC	:	immobilized metal ion affinity chromatography

IPR	:	isolated pentagon rule
LDI	:	laser desorption/ionization
LOD	:	limit of detection
MALDI	:	matrix-assisted laser desorption/ionization
MCP	:	microchannel plate
MELDI	:	material enhanced laser desorption/ionization
MS	:	mass spectrometer
MWCNT	:	multi-walled carbon nanotube
QIT-MS	:	quadrupole ion trap mass spectrometer
RP	:	reversed phase
$RuCl_3$	:	ruthenium trichloride
RuOx	:	ruthenium oxide
SPE	:	solid phase extraction
SWCNT	:	single-walled carbon nanotube
TOF	:	time-of-flight
TOF-MS	:	time-of-flight mass spectrometer

References

Buszewski, B., M. Jezierska-Switala, R. Kaliszan, A. Wojtczak, K. Albert, S. Bachmann, M. Matyska and J. Pesek. 2001. Selectivity tuning and molecular modeling of new generation packings for RP HPLC. Chromatographia 53: 204–212.

Chen, C., and Y. Chen. 2004. Desorption/ionization mass spectrometry on nanocrystalline titania sol–gel deposited films. Rapid Commun. Mass Spectrom. 18: 1956–1964.

Chen, W., L. Wang, H. Chiu, Y. Chen and C. Lee. 2004. Carbon nanotubes as affinity probes for peptides and proteins in MALDI MS analysis. J. Am. Soc. Mass Spectrom. 15: 1629–1635.

Chi, Y., J. Bhonsle, T. Canteenwala, J. Huang, J. Shiea, B. Chen and L. Chiang. 1998. Novel water-soluble hexa (sulfobutyl) fullerenes as potent free radical scavengers. Chem. Lett. 27: 465–466.

Diederich, F., and R. Whetten. 1992. Beyond C60: The higher fullerenes. Acc. Chem. Res. 25: 119–126.

Feuerstein, I., M. Najam-ul-Haq, M. Rainer, L. Trojer, R. Bakry, N. Aprilita, G. Stecher, C. Huck, G. Bonn and H. Klocker. 2006. Material-Enhanced Laser Desorption/Ionization (MELDI)—A New Protein Profiling Tool Utilizing Specific Carrier Materials for Time of Flight Mass Spectrometric Analysis. J. Am. Soc. Mass Spectrom. 17: 1203–1208.

Gong, K., Y. Yan, M. Zhang, L. Su, S. Xiong and L. Mao. 2005. Electrochemistry and electroanalytical applications of carbon nanotubes: a review. Anal. Sci. 21: 1383–1393.

Greiderer, A., M. Rainer, M. Najam-ul-Haq, R. Vallant, C. Huck and G. Bonn. 2009. Derivatized graphitic nanofibres (GNF) as a new support material for mass spectrometric analysis of peptides and proteins. Amino acids 37: 341–348.

Hillenkamp, F., and P.-K. Jasna, Eds. 2007. MALDI MS a practical Guide to Instrumentation, Methods and Applications. Wiley-VCH, Weinheim. Germany
IUPAC (1997). Compendium Of Chemical Terminology.
Jacobs, C.B., M.J. Peairs and B.J. Venton. 2010. Review: Carbon nanotube based electrochemical sensors for biomolecules. Anal. Chim. Acta. 662: 105–127.
Karas, M., D. Bachmann and F. Hillenkamp. 1985. Influence of the wavelength in high-irradiance ultraviolet laser desorption mass spectrometry of organic molecules. Anal. Chem. 57: 2935–2939.
Karas, M. and U. Bahr. 1986. Laser desorption mass spectrometry. TrAC Trends in Analytical Chemistry 5: 90–93.
Krätschmer, W., L. Lamb, K. Fostiropoulos and D. Huffman. 1990. Solid C60: a new form of carbon. Nature 347: 354–358.
Kroto, H., A. Allaf and S. Balm. 1991. C60: Buckminsterfullerene. Chem. Rev. 91: 1213–1235.
Melechko, A., V. Merkulov, T. McKnight, M. Guillorn, K. Klein, D. Lowndes and M. Simpson. 2005. Vertically aligned carbon nanofibers and related structures: Controlled synthesis and directed assembly. J. Appl. Phys. 97: 041301.
Merkoci, A., Ed. 2009. Biosensing Using Nanomaterials. Wiley Series Nanoscience and Nanotechnology. John Wiley & Sons, Inc, Hoboken, New Jersey
Najam-ul-Haq, M., M. Rainer, T. Schwarzenauer, C. Huck and G. Bonn. 2006. Chemically modified carbon nanotubes as material enhanced laser desorption ionisation (MELDI) material in protein profiling. Anal. Chim. Acta. 561: 32–39.
Osawa, E. 1970. Superaromaticity. Kagaku 25: 854–863.
Porath, J., J. Carlsson, I. Olsson and G. Belfrage. 1975. Metal chelate affinity chromatography, a new approach to protein fractionation. Nature 258: 598–599.
Ren, S., and Y. Guo. 2005. Oxidized carbon nanotubes as matrix for matrix assisted laser desorption/ionization time of flight mass spectrometric analysis of biomolecules. Rapid Commun. Mass Spectrom. 19: 255–260.
Ren, S., L. Zhang, Z. Cheng and Y. Guo. 2005. Immobilized carbon nanotubes as matrix for MALDI-TOF-MS analysis: Applications to neutral small carbohydrates. J. Am. Soc. Mass Spectrom. 16: 333–339.
Serp, P., M. Corrias and P. Kalck. 2003. Carbon nanotubes and nanofibers in catalysis. Appl. Catal., A 253: 337–358.
Shiea, J., J. Huang, C. Teng, J. Jeng, L. Wang and L. Chiang. 2003. Use of a water-soluble fullerene derivative as precipitating reagent and matrix-assisted laser desorption/ionization matrix to selectively detect charged species in aqueous solutions. Anal. Chem. 75: 3587–3595.
Snider, R.M., M. Ciobanu, A.E. Rue and D.E. Cliffel. 2008. A multiwalled carbon nanotube/dihydropyran composite film electrode for insulin detection in a microphysiometer chamber. Anal. Chim. Acta 609: 44–52.
Vallant, R., Z. Szabo, S. Bachmann, R. Bakry, M. Najam-ul-Haq, M. Rainer, N. Heigl, C. Petter, C. Huck and G. Bonn. 2007. Development and application of C60-fullerene bound silica for solid-phase extraction of biomolecules. Anal. Chem. 79: 8144–8153.

Vallant, R., Z. Szabo, L. Trojer, M. Najam-ul-Haq, M. Rainer, C. Huck, R. Bakry and G. Bonn. 2007. A new analytical material-enhanced laser desorption ionization (MELDI) based approach for the determination of low-mass serum constituents using fullerene derivatives for selective enrichment. J. Proteome Res. 6: 44–53.

Wang, J. 2005. Carbon-nanotube based electrochemical biosensors: a review. Electroanalysis 17: 7–14.

Wang, J. and M. Musameh. 2004. Electrochemical detection of trace insulin at carbon-nanotube-modified electrodes. Anal. Chim. Acta 511: 33–36.

Wang, J., T. Tangkuaram, S. Loyprasert, T. Vazquez-Alvarez, W. Veerasai, P. Kanatharana and P. Thavarungkul. 2007. Electrocatalytic detection of insulin at RuOx/carbon nanotube-modified carbon electrodes. Anal. Chim. Acta 581: 1–6.

Zhang, M., C. Mullens and W. Gorski. 2005. Insulin oxidation and determination at carbon electrodes. Anal. Chem. 77: 6396–6401.

Section 4: Drugs and Treatments

12

Second Generation Sulfonylurea Glipizide Loaded Biodegradable Nanoparticles in Diabetes

Swarnlata Saraf,[1,a,*] *Shailendra Saraf*[1,b] *and Lan-Anh Le*[2]

ABSTRACT

Glipizide is an oral sulfonylurea drug used in the management of type II diabetes mellitus. Due to its poor aqueous solubility and variable bioavailability as an oral preparation, there is growing research and development into efficient and safe alternative drug delivery systems offering sustained activity. Nanoparticle systems allow for rapid preparation and biodegradation and therefore are helpful options for Glipizide transportation. By adapting the drug loading systems through formulation, processing or optimization techniques a great deal of flexibility in the overall Glipizide effect can be seen. For example, slow release preparations avoid the

---

[1]University Institute of Pharmacy, Pt. Ravishankar Shukla University, Raipur, Chhattisgarh 492010, India.
[a]E-mail: swarnlata_saraf@rediffmail.com
[b]E-mail: shailendrasaraf@rediffmail.com
[2]Rosemead Surgery, 8A Ray Park Avenue, Maidenhead, East Berkshire, SL6 8DS, UK; E-mail: organisedlan@gmail.com
*Corresponding author

List of abbreviations after the text.

sudden peaks seen in current oral preparations and can extend the short plasma half-life (2–4 h), stabilizing the bioavailability fluctuations, reducing dosing frequency and improving patient compliance. Such carriers are designed with PLGA and PEGylated (surface engineering by PEG) using oil-in-water (o/w) emulsion solvent extraction/evaporation techniques.

As demonstrated by electron microscopy, Glipizide is entrapped in spherical and porous nanoparticle system with high drug loading and encapsulation efficiencies. Such NPs could improve patient compliance and reduce side effects in drug treatment for chronic illnesses.

1. INTRODUCTION

Although opportunities to develop nanotechnology-based efficient drug delivery systems extend into all therapeutic classes of pharmaceuticals, many therapeutic agents have not been successful because of their limited ability to reach the target tissue. In addition, faster growth opportunities are expected in developing delivery systems for diabetes agents, hormones and vaccines because of safety and efficacy shortcomings with current administration modalities.

Nanoparticulate drug delivery systems offer plenty of advantages over conventional systems, such as improved efficacy, reduced toxicity, enhanced biodistribution and improved patient compliance (Govender et al. 2000; Yoo et al. 1999). Pharmaceutical nanoparticles (NPs) are subnanosize structures, which contain drug or bioactive substances within them and are made of several tens or hundreds of atoms or molecules, varying in size from 5 nm to 300 nm and in morphologies, e.g., amorphous, crystalline, spherical and needles (Brigger et al. 2002; Panyam and Labhasetwar 2003). It is necessary to use additives like surfactants, dispersants and metals to obtain uniform and stable particles and with further processing, nanostructured powders and dispersions can even be used to coat components or device (Hou et al.2003).

2. DIABETES THERAPEUTIC STRATEGIES

Type 2 diabetes mellitus is a chronic metabolic disorder characterized by high blood glucose concentrations (hyperglycemia) caused by insulin deficiency that is often combined with insulin resistance. Type 2 diabetes is the commonest form of diabetes (Nwobu and Johnson 2007), and sadly the prevalence, of this disorder of insulin action and secretion, is increasing worldwide. The International Diabetes Federation (IDF) 2006 report stated an approximate 246 million people worldwide were affected and estimated an increase to 380 million within 20 years. Physical activity, low intake of dietary fats (especially saturated fats) and adequate intake of complex carbohydrate and fibre can improve insulin action and secretion, enhancing glucose disposal, and controlling the disease. Unfortunately, patients often struggle to make these necessary lifestyle changes in order to manage their diabetes and prevent disease progression without additional pharmacological interventions.

Traditional treatment strategies, including various oral hypoglycemic agents, are unable to slow the progressive decline in pancreatic function, and despite combination therapies in the later stages of diabetes, exogenous insulin therapy is often required. Numerous novel new treatments are the focus of much research in the area and include receptor targeting of acetyl-CoA carboxylase 2, I kappa kinase (IKK), beta (IKKB), glucagon-like peptide-1 (GLP-1), dipeptidyl peptidase IV inhibitors, CSII (insulin pump therapy), nuclear receptors as drug targets, modulators of peroxisome proliferator-activated receptors (PPAR), glucagon receptor antagonists, insulin receptor activators and protein tyrosine phosphatase inhibitors (Jain and Swarnlata 2010). Furthermore islet cell transplantation, gene expression profiling and combination therapies research also offers potential for the next generation of safe efficacious therapeutics in diabetes care.

One goal of diabetes management is to establish and maintain normoglycaemia and although newer drugs, such as thiazolidinediones, newer sulfonylureas and meglitinides increase choice from the old sulfonylureas and metformin, society demands more cost effective treatments (Abegunde et al. 2007). The long half life available from future nanoparticle drugs may well offer further choice and in particular a delay in disease progression in type 2 diabetes.

3. SECOND GENERATION SULFONYLUREA—GLIPIZIDE

Available since 1984 and marketed by Pfizer under the brand name of Glucotrol in the USA, GPZ is now available from a number of companies since coming off patent. Glucotrol and the slow release, Glucotrol XL, are available in 5 and 10 mg doses and the latter is also available in the lower dose of 2.5mg. Due to the larger cyclo or aromatic group, it undergoes enterohepatic circulation, one hundred fold less than with first generation sulfonylureas and so only requires once daily dosing (Melero et al. 2008).

GPZ 1-cyclohexyl-3-(p-(2-(5-methyl pyrazine-carboxamido) ethyl) phenyl) is a medium-to-long acting sulfonylurea and appears to lower blood glucose by stimulating the release of insulin from the pancreas, an effect dependent upon functioning beta cells in the pancreatic islets. Extra pancreatic effects include an increase in insulin sensitivity and a decrease in hepatic glucose production (Inzucchi 2002). However the mechanism by which GPZ lowers blood glucose during long-term administration has not been clearly established. Stimulation of insulin secretion by GPZ in response to a meal is undoubtedly of major importance however, in long term GPZ use, fasting insulin levels are not elevated but the postprandial insulin response continues to be enhanced after at least 6 months of treatment (Jamzad and Fassihi 2007). Some patients fail to respond initially, or gradually lose their responsiveness to sulfonylurea drugs despite gastrointestinal absorption of GPZ being uniform, rapid, and essentially complete. Protein binding is 98–99% and is primarily to albumin and major metabolites are products of aromatic hydroxylation and on hepatic biotransformation show no hypoglycemic activity. A minor metabolite which accounts for less than 2% of a dose, an acetylaminoethyl benzene derivative is reported to have 1/10 to 1/3 of the hypoglycemic effect of the parent compound and its half life is 2–5 hours (http://en.wikipedia.org).

Research reveals that GPZ has a good general tolerability, comparatively low incidence of hypoglycemia and a low rate of secondary failure (Negre-Salvayre 2009). In addition it has potential for slowing the progression of diabetic retinopathy. In general, rapid gastrointestinal (GI) absorption is required for oral hypoglycemic drugs, to prevent a sudden increase in the blood glucose level after food intake (http://www.medclik.com and Youn et al. 2007). Several studies using healthy volunteers or patients revealed that the time to reach peak serum GPZ concentration ranged from 0.5 to 1 hour following oral administration. Slow absorption of drug usually originates from poor permeability of the drug across the GI membrane. Despite the extended release version of GPZ there remains a need for the development of the even longer sustained release formulation of GPZ and nanoparticles may well offer the key to such a multiparticulate delivery systems (Kitchell and Wise 1985).

4. BIODEGRADABLE NANOPARTICLES

Biodegradable nanoparticle design usually begins with a conceptual ideal. Biodegradable polymers used in the delivery system need to be compatible with the tissues they target, and designed to break down after a given time period. Biodegradability of GPZ nanoparticulate drug delivery technology, with the capability to provide a biomaterial with a broad range of amphiphilic characteristics offers a long half life, targeted action at key sites, limited fluctuations within a therapeutic range, reduced side effects, decreased dosing frequency, and improved patient compliance (Yadav et al. 2007). The biodegradable polymers are the key in designing such biodegradable nanoparticulate delivery systems, as they do not require design consideration regarding the drug release (Couvreur and Karine 2005). The most extensively used polymers in preparation of nanoparticles are the polyesters, which include poly (lactic acid) (PLA) and poly (glycolic acid) (PGA), as well as their copolymers like Polylactic co-glycolic acid (PLGA) (Fessi and Doelker 1998). Such polymers are of interest because of the human body's ability to absorb them. Copolymers such as poly (lactic acid)-co-caprolactone are starting to be used successfully for preparation of drug delivery systems. The degradation rate depends on the composition of the polymer, its molecular weight, the size of particles, its solubility and environmental conditions including pH (Swarnlata 2009). Biodegradability can be engineered into polymers by the judicious addition of chemical linkages such as anhydride, ester, or amide bonds, among others as per application of nanoparticles. The usual mechanism for Biodegradable Nanoparticles degradation is by hydrolysis or enzymatic cleavage of the labile heteroatom bonds, resulting in a scission of the polymer backbone. Macrophages can degrade polymers, and also initiate a mechanical, chemical, or enzymatic aging.

5. DESIGNING OF GPZ NANOPARTICLES

The development of biodegradable nanoparticles is crucial for effectively delivering many drugs and is reliant on pharmaceutical technologies. It is important to use an effective methodology during formulation development to counter these complex technical challenges for biodegradable nanoparticles application. For example Gu et al. 2008 developed polymer-based nanotechnologies and suggested a better fundamental knowledge of the *in vivo* interaction of nanoparticles with biological fluids. Long-circulating polymeric colloidal carriers, or 'stealth' systems, are able to avoid the opsonization process and recognition by macrophages. The design of such carriers is based on the physico-chemical concept of 'steric repulsion'. For example, by grafting polyethylene glycol chains at the surface of nanoparticles, the adsorption of steric proteins is markedly reduced due to steric hindrance and the drug has a longer half life. Eto et al. 2008 also prepared biodegradable nanoparticles (nanospheres and nanocapsules) from preformed polymers.

His group focused on the thorough analysis of preparative procedures like emulsification evaporation, solvent displacement, salting-out and emulsification diffusion. They described the mechanism of nanoparticle formation for each technique from a physicochemical perspective, by observing the effects of preparative variables on nanoparticle size and drug-entrapment efficiency.

There are several techniques for preparing solid polymer nanoparticles. Additionally, numerous methods exist for incorporating drugs into the particles (Swarnlata S. 2009). For example, drugs can be entrapped in the polymer matrix, encapsulated in a nanoparticle core, surrounded by a shell-like polymer membrane, chemically conjugated to the polymer, or bound to the particle's surface by adsorption. The most common method used for the preparation of solid, polymeric nanoparticles is the emulsification-solvent evaporation technique. This technique has been successful for encapsulating hydrophobic drugs, like GPZ, but has had poor results incorporating hydrophilic compounds. Solvent evaporation is carried out by dissolving the polymer and the compound in an organic solvent. The emulsion is prepared by adding water and a surfactant to the polymer solution. In order to minimize the particle size, the interface between the oil and water phase must be increased. To overcome this energy barrier and form the emulsion, energy must be added to the system, usually in the form of sonication or homogenization. The surfactant stabilizes the droplets formed during the energy input. The nanoparticles then "harden" as the solvent is removed by evaporation, extraction or diffusion. The nanoparticles are usually collected by centrifugation and lyophilization. This modification favours encapsulating hydrophobic drugs such as GPZ. First, GPZ and the polymer are dissolved in an organic solvent, termed the oily phase. The primary emulsion is prepared by dispersing the organic phase in to an aqueous phase containing a dissolved surfactant. This is then re-emulsified in an outer aqueous phase also containing stabilizer. Maintenance of drug activity during nanoparticle formation is a design challenge for emulsification procedures. Drug activity can be affected by solvent selection and different emulsification methods. The method of producing polymeric nanoparticles has several independent variables. Consequently, total drug loading, nanoparticle stability and release characteristics may vary with slight changes in processing parameters as illustrated in Table 1. First, one must consider the selection of the components used in the nanoparticle production, including the polymer, the polymer molecular weight, the surfactant, the drug, and the solvent. As energy input into an emulsion increases the resulting particle size decreases. In addition there are four separate concentrations that can be

Table 1. Processing variables for Glipizide loaded Poly (lactic-co-glycolic acid) nanoparticles.

Optimization Parameter	Variables Used
Drug-Polymer ratios (GPZ:PLGA)	1:4
	1:2
	1:1
	2:1
Organic Solvents	Methanol
	Dichloromethane
Stirring Speed (RPM)	300
	1000
	3000
Surfactants	PVA
	Polysorbate 80

GPZ- Glipizide; PLGA - Poly (lactic-co-glycolic acid).

altered: the polymer, drug, surfactant, and solvent (Jain and Swarnlata 2009). Finally, the recovery of the particles can be changed depending on the method of lyophilization or centrifugation. Different surfactants may produce particles of different sizes. The amount of stabilizer used will also have an effect on the properties of the nanoparticles. Often a low concentration of surfactants results in a high degree of polydispersity and aggregation. Alternatively, if too much of the stabilizer is used, the drug incorporation could be reduced due to interaction between the drug and stabilizer. However, when the stabilizer concentration is between the "limits", adjusting the concentration can be a means of controlling nanoparticle size. For example, increasing the PVA concentration has been shown to decrease particle size. In the emulsification solvent evaporation method, PVA concentrations of 0.5–1.0% are ideal in creating smaller nanoparticles of 300 nm in diameter (Jain and Swarnlata 2009). Biodegradable nanoparticles production needs to select optimal processing parameters like time, stirring speed, stirring rate, time of adding excipients and active pharmaceutical ingredients (APIs) and formulation additives such as drug polymer ratios, solvent types and drug concentrations, polymer types and their ratios. Optimizing design methodologies such as factorial design, response–surface design, composite design, randomized design, resolution design, screening design, Planckett–Burman design, quadrate design, Box– Behnkm design and mirror image fold-over design, can provide and produce nanoparticles on an industrial scale. Among all design methodology, experimental designing is among the best methodology because it provides a practical point of view (individual and simultaneous factors) to clarify the problems in formulation for up scaling nanoparticle production (Swarnlata 2009).

6. PREPARATION TECHNIQUES

The preparation techniques largely determine the inner structure predicting the *in vitro* release profile and the biological fate of these polymeric delivery systems. The two types of systems differ by their inner structures, with matrix type systems consisting of an enlargement of oligomer or polymer units (nanoparticle/ nanospheres) whilst reservoir systems comprise of an oily core surrounded by an embryonic polymeric shell (nanocapsule). The drug can either be entrapped within the reservoir or the matrix or otherwise be adsorbed on the surface of these particulate systems. The polymers are strictly linked to a nanometer size range using appropriate methodologies (Swarnlata 2010b). For preparation of GPZ nanoparticles, amphiphillic macromolecules are cross linked by heat or chemically then undergo polymerization. These methods include polymerization of monomers in situ, emulsion, micellar, polymerization, dispersion polymerization, interfacial condensation polymerization, interfacial complexation and polymer precipitation methods like Solvent Extraction or Evaporation, Solvent displacement, nanoprecipitation and Salting out. The preparation technique efficiency of nanoparticles is generally determined by their size, surface morphology entrapment efficiency and release kinetics in the body. The efficiency of the particle can be achieved by suitable designs of processing parameters and formulation compositions.

6.1 Polymer precipitation Technique

Polymer precipitation techniques for GPZ nanoparticle preparation starts with hydrophobic polymer and drug dissolution in a particular organic solvent followed by its dispersion in a continuous aqueous phase, in which the polymer is insoluble. The external phase also contains the stabilizer PVA/Polysorbate. Depending upon solvent miscibility techniques this is called solvent extraction or evaporation. The polymer precipitation occurs as a consequence of the solvent extraction or evaporation, which can be brought about by:

- Increasing the solubility of the organic solvent in the external medium by adding an alcohol.
- By incorporating additional amount of water in to the ultra-emulation (to extract or diffuse the solvent).
- By evaporation of the organic solvent at room temperature or at accelerated temperatures or by using vacuum.
- Using an organic solvent that is completely soluble in the continuous aqueous phase.

6.2 Solvent Extraction/Evaporation Technique

This technique involves the formation of a conventional O/W emulsion between a water immiscible organic solvent (containing the drug and polymer), and water (containing the stabilizer). The subsequent removal of solvent (solvent evaporation method) or the additions of water to the system so as to affect diffusion of the solvent to the external phase (emulsification diffusion method) are two variance of the solvent extraction method (Fig. 1).

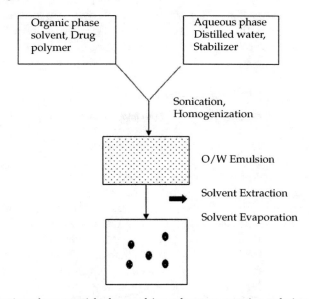

Fig. 1. Preparation of nanoparticles by emulsion solvent evaporation technique.

For instance, in the classic preparation of PLGA nanoparticles, the polymer is dissolved in a solvent (chloroform) and dispersed in a surfactant solution by sonication to yield O/W emulsion, and then the solvent is evaporated usually by high speed pressure homogenization.

The homogenizer breaks the initial coarse emulsion into nanodroplets (nanofluidization), yielding nanospheres with a narrow size distribution. Recently, emulsification diffusion methods have been commonly used for solvent extraction (Jacobs and Helmut 2002; Arulmozhi et al. 2004). The solvents used are often poorly miscible in the dispersion phase and thus diffuse and evaporate out slowly on continual stirring of the system. On the contrary, dispersion medium miscible polymer solvents, such as acetone and alcohol, instantaneously diffuse in to the aqueous phase and the polymer consolidates and precipitates as tiny nanosphere. For example,

GPZ loaded PLGA nanoparticles can be created by dissolution in methanol/ dichloromethane using a vortex shaker to form a homogeneous solution of drug and polymer emulsification then use of the solvent evaporation method. The solution is then added slowly to aqueous surfactant solution (PVA/Polysorbate 80) using high pressure homogenizers to prepare the emulsion. The emulsion formed is then stirred on laboratory magnetic stirrer for 6 h at 25°C. The contents undergo complete solvent evaporation via a vacuum oven, and then the remaining aqueous phase is centrifuged for a few minutes at high shear. After centrifugation the supernatant is discharged and the pallets obtained are washed by using the same volume of distilled water as that of the supernatant and again centrifuged. The washing process is repeated and the washed nanoparticles can be collected after freeze dried.

6.3. Stealth Nanoparticles

Stealth nanoparticles are relatively successful at prolonging the half life of colloidal particles in the blood by creating a steric surface barrier of sufficient density. Because of the possible immunological consequences associated with some bacterial polysaccharides and the high cost of recombinant complement regulators, tremendous efforts have been directed to design synthetic polymers that can fulfill these criteria. The concept of surface modification of particulate carriers to control them from opsonization processes and the specific and non specific interactions with serum components have been appreciated with a number of materials. These materials are coated or grafted with covalently coupling hydrophilic moieties such as poloxamer, poloxamine (poly-oxyethylene and polyoxypropylene block copolymer), and high-molecular-weight polyethylene glycols to the GPZ nanoparticles. A large numbers of nonionic polymers, which are hydrophilic and flexible, e.g., PEG, poloxomer, poloxamine, polyethylene oxide, and more recently some biodegradable copolymers such as polylactide-polyethylene glycol copolymers also impart stealth properties. PEGylation is a recent and effective technique for steric stabilization of GPZ nanoparticles and can be achieved by adsorption, incorporation, or by covalent attachment of PEG to surfaces of particles.

7. CHARACTERISTICS

7.1 Surface Morphology

The surface morphology of the PLGA and pegylated GPZ nanoparticles is spherical (Fig. 2) and porous in nature. The degree of porosity is the fundamental stage offering GPZ its controlled release properties and is adaptable with processing parameters of nanoparticles preparation like drug polymer ratio, surfactant, stirring speed etc.

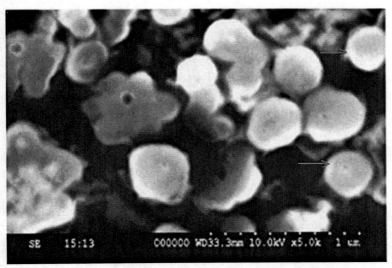

Fig. 2. Scanning electron microscopy photograph of Poly (lactic-co-glycolic acid)—Polyvinyl alcohol Nanoparticles.

7.2 Size

Particle size is one of the most important parameters of nanoparticles. The goal of the nanoparticle delivery system determines the size desired. For example, if the goal is delivery to extra cellular targets then the ideal size is between 30 and 100 nm. For systemic circulation, particles should be in the range of 50 to 500 nm. Cells can internalize particles ranging from 30 to 300 nm. It also appears that there is a lower size limit for ideal delivery, with 75% of particles with a diameter of 30nm cleared by the liver 3 hours post injection. However, 70nm particles had less than 50% taken by the liver. Therefore, it appears that particles below a certain size limit can penetrate the fenestrations of the liver endothelial lining and accumulate. There are tradeoffs when optimizing particle size by altering the molecular weight of the polymer. Smaller diameter solid nanoparticles can be prepared with lower molecular weight polymers, at the expense of reduced drug encapsulation efficiency.

Particle and sub optical particulate sizing is a different procedure, as it involves not only procedural variability, but some of the surface properties may change. The techniques used to determine the particle size distribution of nanoparticles are photon correlation spectroscopy (PCS) and electron microscopy (EM) whilst particle size can be determined by scanning electron microscopy (SEM), transmission electron microscopy (TEM) and freeze-fracture techniques. The size evaluation of nanoparticle dispersion demonstrates better results with PCS as a quantitative method. The particle size distribution at nanosize level is impacted by nanoparticle production processing variables. The mean particle size of the GPZ nanoparticles analyzed at a scattering angle of 900 and temperature of 250 C by SEM is shown in Table 2. The size of the PLGA GPZ nanoparticles is variable when using organic solvent, the drug polymer ratio, stirring speed and stabilizer.

Table 2. Glipizide entrapment, Glipizide content and particle size of Poly (lactic-co-glycolic acid) nanoparticles.

Parameters	Variables Used	Glipizide entrapment (%)	Glipizide content (%)	Particle size (nm)	Zeta potential (mV)
Drug-Polymer ratios (GPZ:PLGA)	1:4	40.36 ± 2.16	10.15 ± 2.45	505± 12.13	
	1:2	57.23 ± 3.16	24.45 ± 3.13	24.45± 3.13	
	1:1	62.89 ±1.97	25.78 ± 2.21	918 ±14.65	
	2:1	90.31 ± 4.56	39.94 ± 3.12	39.94± 3.12	
Surfactant	PVA	57.23± 3.16	24.45 ± 3.13	522 ± 5.24	-18.2 ± 0.5
	Polysorbate	41.09 ± 1.39	21.21± 3.23	781 ± 11.49	-6.57 ±2.1
Stirring Speed (RPM)	300	36.06±3.14	11.28± 1.11	1191± 23.29	
	1000	45.32 ± 2.87	16.03 ± 2.13	977 ± 14.35	
	2000	51.02 ± 2.18	23.22 ± 3.29	781 ± 11.02	
	3000	57.23 ± 3.16	24.45 ± 3.13	522 ± 5.24	

PVA - Polyvinyl alcohol; GPZ- Glipizide; PLGA - Poly (lactic-co-glycolic acid). Adapted from Jain and Swarnlata 2009. All readings are expressed as ± standard deviations.

7.3 Surface Charge

The nature and intensity of the surface charge of nanoparticles is very important as it determines their interaction with the biological environment as well as their electrostatic interaction with bioactive compounds. The surface charge of nanoparticles can be determined by measuring the particle velocity in an electric field. Laser light scattering technique including laser Doppler anemometry or velocimetry has become available as a fast and high resolution technique for the determination of nanoparticle velocity. GPZ nanoparticles surface charge can be characterized with respect to zeta (ζ) potential by using Laser Doppler Anemometry (Malvern Zeta seizer IV, Malvern Instruments Ltd, Malvern, UK) 1 mM HEPES buffer (adjusted to pH 7.4 with 1 M HCl) in order to maintain a constant ionic strength. The zeta potential is an index of stability of nanoparticles in the body.

The higher the magnitude, irrespective of the charge type, the higher the stability expected. GPZ nanoparticles exhibit low zeta potentials ranging from 6.57+ 2.1 mV to 31.03 + 1.7 mV (Table 2). Generally highly negative zeta potential values are expected for pure PLGA nanoparticles due to presence of carboxyl groups on the polymeric chain extremities, but in the case of PLGA - GPZ nanoparticles the zeta potential value is close to zero. The factor which might be responsible for such an effect is the presence of residual stabilizers on the nanoparticles surface. The residual stabilizer on the nanoparticles surface can create a shield between the surface and surrounding medium. The stabilizers would mask the possible charged groups existing on the particle surface leading to zeta potential values close to zero.

7.4 Drug Entrapment

Ideally, a successful nanoparticulate system should have a high drug-loading capacity reducing the quantity of matrix materials needed for administration. Drug loading can be done by two methods:

- Incorporation method—Incorporating at the time of nanoparticles production.
- Adsorption or absorption method—Absorbing the drug after formation of nanoparticles by incubating the carrier with a concentrated drug solution.

Drug loading and entrapment efficiency very much depend on the solid-state drug solubility in matrix material or polymer (solid dissolution or dispersion), which is related to the polymer composition, the molecular weight, the drug polymer interaction and the presence of end functional groups (ester or carboxyl). The PEG moiety has no or little effect on drug loading. For small molecules like GPZ, studies shows the ionic interaction and increase in polymer concentration between the drug and matrix materials can be a very effective way to increase the drug loading.

Structurally GPZ is a complex chemical structure, having polycyclic rings imparting the lipophilic characters. GPZ shows solubility in various organic solvents such as methanol and dichloromethane and very less solubility in water. Our research group found that when methanol is used in preparation of nanoparticles, a very low drug entrapment (23.25 + 1.36%) and drug content (8.36 + 2.15%) are obtained with drug polymer ratio 1:2; this is mainly due to the highly hydrophilic nature of the methanol. With dichloromethane sufficient GPZ entrapment and content are enough to maintain blood level. This occurs due to its hydrophobic nature which enabled GPZ to be retained in the hydrophobic nanoparticle matrix by modified emulsification solvent evaporation techniques.

7.5 Release Kinetics

A sustained release system should release the required quantity of drug with predetermined kinetics in order to maintain an effective drug plasma concentration. To achieve this, the delivery system should be formulated so that it releases the drug in a predetermined and reproducible manner. Release kinetics from biodegradable nanoparticles is governed by diffusion, erosion or a combination thereof, and are dependent on the polymer (Mw, copolymer ratio and crystallinity), drug properties, as well as the device characteristics (preparation conditions, particle size, morphology, porosity and drug loading) and the dissolution conditions. PLGA nanoparticles in phosphate buffer pH 7.4 releases GPZ from nanoparticles prepared by using PVA and polysorbate-80 stabilizer in two components (Fig. 3, Table 3). First an initial exponential phase releases 8.7 perent and 9.78 percent of the drug in 1st hour which is almost identical with the conventional dose of a 5 mg tablet, followed by a slow phase release up to 19 percent of the drug in 12 h, respectively. The release slowed but continued until the 72 h (91 percent drug release), because the diffusion controlled release pattern is completed and the erosion controlled release is probably started. The total drug release is supplied by two uncoupled pools with one pool of fast drug diffusion (burst release) and another pool of slow diffusion by polymer degradation. The relative dominance between diffusion and erosion plays a major role in the release kinetics. The initial burst release is sometimes attributed to the rapid release by diffusion of dissolved drug initially deposited inside the pores. The explanation of the burst is that some drug particles could have migrated to the surface during drying of nanoparticles. The release of GPZ from PLGA -NP continues and releases more than 95% drug at the end of seven days while Stealth nanoparticles

Table 3. Physicochemical Stability of Poly (lactic-co-glycolic acid) Nanoparticles.

Test duration (month)	Test conditions 8C/RH	Mean particle size	Kinetics		
			Zero-order (r^2)	First-order (r^2)	Higuchi (r^2)
1	$40\pm2^0/75\pm5\%$ $30\pm2^0/65\pm5\%$ $25\pm2^0/60\pm5\%$	523 ± 6.24 523 ± 7.34 522 ± 8.94	0.9821	0.9581.	0.981
3	$40\pm2^0/75\pm5\%$ $30\pm2^0/65\pm5\%$ $25\pm2^0/60\pm5\%$	524 ± 4.54 523 ± 6.13 523 ± 3.23	0.9829	0.9554	0.9805
6	$40\pm2^0/75\pm5\%$ $30\pm2^0/65\pm5\%$ $25\pm2^0/60\pm5\%$	524 ± 4.21 523 ± 4.24 523 ± 6.12	0.9827	0.9577	0.9803

Adapted from Jain and Swarnlata 2009. All readings are expressed as $\pm$ standard deviations

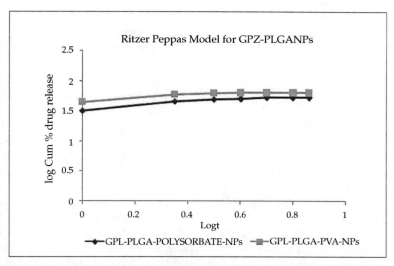

Fig. 3. Ritger Peppas model for Glipizide Poly (lactic-co-glycolic acid) Nanoparticles (Adapted from Jain & Swarnlata 2009).GPL-PLGA-POLYSORBATE-NPs: Glipizide Poly (lactic-co-glycolic acid) polysorbate Nanoparticles; GPL-PLGA-PVA-NPs Glipizide Poly (lactic-co-glycolic acid) polyvinyl alcohol Nanoparticles.

i.e., pegylated PLGA-NP, PLGA-mPEG NP releases only about 80% drug because of the presence of Polysorbate 80 as a stabilizer.

Thus drug release from the nanoparticulate system is dependent on the initial drug load, matrix porosity and leaching solvent. The most probable mechanism of GPZ release from the PLGA matrix appeared to be by diffusion of GPZ from the insoluble porous matrices into the dissolution medium due to swelling of the PLGA matrix.

7.6 Physicochemical Stability

Stability is an important concern since any pharmaceutical product is required to provide sufficient bioactive or medicament to elicit optimal pharmacodynamic response in order to cure or prevent any pathological condition. Storage stability is very important especially in cases of biodegradable system. GPZ loaded NPs consisting of PLGA and PLGA-mPEG are stable. Stability studies on GPZ loaded PLGA NPs containing 0.5% PVA (formulation A) as surfactant, reveal that the NPs were stable at the end of 6 months in all the test conditions (Table 3) according to ICH guidelines. No significant changes in particle size and release profile were observed after the end of 1, 3 and 6 months.

8. ANTIDIABETIC EFFECTS: ANIMAL MODEL

Antidiabetic effect of Biodegradable nanoparticles of GPZ allows the delivery system to release the drug in a pre determined manner then the carrier system degrades once this goal is achieved. The antidiabetic effect of GPZ NP in animal models has shown appreciable result with PLGA NPs and PLGA-m PEG NPs of GPZ (Jain S. 2008). In this study, GPZ nanoparticles prepared with PVA 0.5 percent and polysorbate with PLGA (50:50) and m-PEG-PLGA polymers showed antidiabetic effect. The linear regression coefficient from this animal model study into albino rats found a value of $r2<0.5$, indicating the pharmacological response was not linear and so was not dependent on drug release. One-way ANOVA analysis (multiple comparisons) was followed by the Turkey test for *in vivo* anti diabetic activity of GPZ nanoparticles. Group 3 (standard GPZ treatment, 500 µg/kg t.i.d.) showed lower glucose levels compared to group 2 ($p < 0.05$) showing a significant difference in mean blood glucose levels. Group 4 (treated with GPL-PLGA-NPs) showing a remarkable decrease in blood glucose levels up to 24 h and a further rise in glucose levels. Group 5 (treated with GPZ-PLGA-mPEGNPs) clearly shows a long circulatory effect as the blood glucose levels were significantly lower (<0.05) when compared to group 2. Comparison between group 3 and group 4 showed that PLGA nanoparticles were effective therapeutically but the duration of action was short ($p > 0.05$). Comparison between groups 3 and 5 showed that PEGylated nanoparticles are effective therapeutically and have a longer duration of action in comparison to convention GPZ ($p > 0.05$). Comparison between groups 4 and 5 showed PEGylated nanoparticles had a longer circulating effect while PLGA nanoparticles were effective for 24 hours only (Fig. 4) probably due to recognition of the latter by liver macrophages.

9. CONCLUSION

The nanoparticulate system ensures the sustained delivery of glipizide for an extended duration compared to conventional GPZ. Reductions in dosing frequency, improvements in patient compliance, cost and the side-effect profile are all benefits of using these advanced nanoparticulate systems. Long circulating nanoparticles of Glipizide can be prepared successfully by utilizing the emulsification, solvent and evaporation techniques and optimization of processing variables to create different effects. Such nanoparticles can be stabilized by using polyvinyl alcohol as

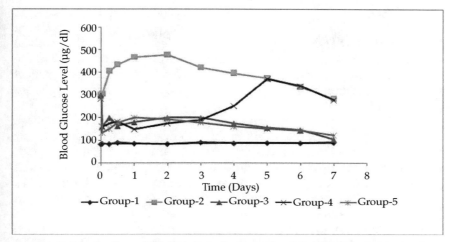

Fig. 4. Antidiabetic effect of Glipizide Poly (lactic-co-glycolic acid) Nanoparticles in albino rats: Tukey test, (Unpublished work). Group I-normal control,Group II-diabetic control, Group 3—standard Glipizide treatment, 500 µg/kg t.i.d., Group 4-treated with Glipizide Poly (lactic-co-glycolic acid) Nanoparticles,Group 5- treated with Glipizide methoxy Polyethylene Glycol- Poly (lactic-co-glycolic acid) Nanoparticles

surfactant and diabetic animal model studies show mPEG-PLGA based formulations are more successful in managing Type 2 diabetes mellitus than those containing PLGA.

10. Key Facts

- Type 2 diabetes mellitus (T2DM) is a common chronic disease and is characterized by fasting and postprandial hyperglycemia and relative insulin insufficiency. It causes significant morbidity and mortality with a considerable health burden to patients, their families' and society. It has become a worldwide epidemic with an incidence increasing particularly in Africa, Asia and South America compared to in Europe or the U.S.
- Although previously only sulfonylureas and metformin were available many newer diabetic drugs, enhancing either insulin action or secretion, are now widely used. Glipizide is a second generation sulfonylurea, with high permeability and low solubility and is classified as a biopharmaceutical class II drug.
- The GI absorption rate of conventionally engineered GPZ is slow, whilst using nanoparticles, with amphiphilic properties, and further surfactant and dispersant additives, uniform and stable particles offering sustained blood plasma levels and targeted delivery can be produced.

- Nanomedicine can manufacture multiparticulate delivery systems which prolong or control drug delivery, as well as improve bioavailability and drug stability (Swarnlata et al. 2006). The biomedical benefits of NPs therefore include minimal drug fluctuations and hence maintenance of therapeutic dosing, an improvement in the side effect profile, decreased dosing frequency, and improved patient compliance (Jain and Swarnlata 2008).
- Conventional carriers are rapidly cleared from the systemic circulation, ending up almost exclusively in the mononuclear phagocyte system (MPS), mainly in macrophages in the liver and spleen. This extensive clearance is disadvantageous in terms of dosing frequency for insulin therapy and second generation sulfonylurea drugs including GPZ. For this reason, there has been increasing interest in the development of stealth NPs as drug carrier systems.
- One important approach has been to use protein-repellent materials, such as poloxamers, poloxamines, and polyethylene glycol (PEG) (Kim et al.2005). Modification of the surface of NPs using these polymers imparts "stealth" characteristics to polymeric NPs. The biodegradable PEG-coated NPs have significant therapeutic potential for use as colloidal delivery system for the controlled release of drugs (Swarnlata 2010a).
- The ongoing development of Glipizide loaded nanoparticle design incorporates factors such as increased surface area, enhanced solubility, increased rate of dissolution, increased oral bioavailability, rapid onset of therapeutic action, reduction in drug dosage required, decreased fed and fasted plasma level variability and decreased patient-to-patient variability.

11. Summary Points

- The solubility of the Glipizide loaded nanoparticles improves Glipizide poor bioavailability and therefore can be of benefit in the management of Type II diabetes mellitus, the advantages of using such nanoparticles for drug delivery are due to the following two properties. Nanoparticles, because of their small size, can penetrate through smaller capillaries and are taken up by cells, which allow efficient drug accumulation at the target site. Secondly, the use of biodegradable materials for nanoparticle preparation allows sustained drug release within the target site over a period of days or even weeks.
- The excipients and processing variables, affect the drug loading and size of the particles.

- The Glipizide nanoparticles are stabilized by using polyvinyl alcohol and polysorbate as surfactant.
- The nanoparticulate system of Glipizide offer sustained release of the drug for extended periods of time.
- Pegylated biodegradable nanoparticles of Glipizide offer long circulation in the body.

Abbreviations

NPs	:	Nanoparticles
GPZ	:	Glipizide
MPS	:	Mononuclear Phagocyte System
PEG	:	Polyethylene Glycol
PLGA	:	Poly (lactic-co-glycolic acid)
T2DM	:	Type 2 diabetes mellitus
PEGylation:		Polyethylene Glycolation
mPEG-PLGA:		methoxy- Polyethylene Glycol- Poly (lactic-co-glycolic acid)
PVA	:	Polyvinyl alcohol

References

Abegunde, D.O., C.D. Mathers, T. Adam, M. Ortegon and K. Strong. 2007. The burden and costs of chronic diseases in low-income and middle-income countries. Lancet 8: 1929–38.

Arulmozhi, D.K., A. Veeranjaneyulu and S.L. Bodhankar. 2004. Neonatal streptozotocin-induced rat model of Type 2 diabetes mellitus: A glance, Indian J. Pharmacol. 36: 217–221.

Brigger, I., C. Dubernet and P. Couvreur. 2002. Nanoparticles in cancer therapy and diagnosis. Adv. Drug Deliv. Rev. 54: 631–51.

Couvreur, P., and A. Karine. 2005. Nanotechnologies for brain and ocular delivery, 15th international symposium on Microencapsulation, September 18–21, Parma, Italy.

Eto, Y., Y. Yoshioka, Y. Mukai, et al. 2008. Development of PEGylated adenovirus vector with targeting ligand. Int. J. Pharm. 354 : 3–8.

Fessi, H., and K. Doelker. 1998. Preparation techniques and mechanisms of formation of biodegradable nanoparticles from preformed polymers. Drug Dev. Ind. Pharm. 24 : 1113–28.

Govender, T., T. Riley, S. Stolnik, M.C. Garnett, L. Illum and S.S. Davis. 2000. PLA-PEG nanoparticles for site specific delivery: Drug incorporation studies. J. Control Rel. 64: 269–347.

Gu, F., L. Zhang, B.A. Teply, et al. 2008. Precise engineering of targeted nanoparticles by using self-assembled biointegrated block copolymers. Proc. Natl. Acad. Sci. USA 105: 2586–91.

Hou, D., C. Xie, K. Huang and C. Zhu. 2003. The production and characteristics of solid lipid nanoparticles (SLNs). Biomaterials 24: 1781–5.

Inzucchi, S.E. 2002. Oral antihyperglycemic therapy for type 2 diabetes. Scientific Review 287: 360–372.

Jacobs, C., and R.M. Helmut. 2002. Production and Characterization of a Budesonide Nanosuspension for Pulmonary Administration, Pharmaceutical Research 19: 189–194.

Jain, S. 2008. Guided by Dr. Mrs. Swarnlata Saraf Ph.D. Thesis, Design and Characterization of Biodegradable long circulation particulate some oral hypoglycemic drugs; Faculty of Technology, Pt.Ravishankar Shukla University, Raipur, Chhattisgarh, 492010, INDIA (Unpublished work).

Jain, S., and S. Swarnlata. 2008. Preparation and *in vitro* evaluation of glipizide loaded poly (D, Llactide- co-glycolide) nanoparticles for effective management of type II diabetes, Nano Science & Nano Technology: An Indian Journal 2: 1.

Jain, S., and S. Swarnlata. 2010.Type 2 diabetes mellitus its global prevalence and therapeutic strategies. Diabetes & Metabolic Syndrome: Clinical Research & Reviews 4: 48–56.

Jamzad, S., and R. Fassihi. 2007. Development of a robust once-a-day glipizide matrix system. J. Pharm. Pharmacol. 59: 769–75.

Kim, M.S., K.S. Seo, H. Hyun, et al. 2005. Sustained release of bovine serum albumin using implantable wafers prepared by mPEG-PLGA diblock copolymers. Int. J. Pharm. 304: 165–77.

Kitchell, J.P., and D.L. Wise. 1985. Poly (lactic/glycolic acid) biodegradable drug-polymer matrix systems. Meth Enzymol. 112: 436–48.

Melero, P.S, D. Antolinez and N. Abia. 2008. Oral anti-diabetes medicines, an update. Rev. Enferm 31: 7–10.

Negre-Salvayre, A., R. Salvayre, N. Auge, R. Pamplona and M. Portero-Otı. 2009. Hyperglycemia and Glycation in Diabetic Complications, Antioxidants & Redox Signaling. 11: 3071–3109.

Nwobu C.O., and C.C. Johnson 2007. Targeting obesity to reduce the risk for type 2 diabetes and other co-morbidities in African American youth: a review of the literature and recommendations for prevention. Diab. Vasc. Dis. Res. 4: 311–9.

Panyam, J., and V. Labhasetwar 2003. Biodegradable nanoparticles for drug and gene delivery to cells and tissue. Adv. Drug Deliv. Rev. 55: 329–47.

Rawat, M., D. Singh, S. Swarnlata and S. Saraf. 2006. Nanocarriers: promising vehicle for bioactive drugs. Biol. Pharm. Bull. 29: 1790–8.

Swarnlata, S. 2010a. Application of Colloidal properties in drug delivery, pp 55–70. *In:* Monzer Fenun edited.Surfactant science series Volume 148: Colloidal drug delivery. CRC Press, Tylor & Francis Group, USA.

Swarnlata, S. 2010b. Application of nano systems in cosmetics, pp380–397. *In:* R.S. Chaughule and R.V. Ramanujan, Nanoparticles: Synthesis, Characterization and applications. American Scientific publishers, USA.

Swarnlata, S. 2009. Process optimization for the production of nanoparticles for drug delivery applications, Expert Opin. Drug Deliv. 6 : 187–196.

Yadav, A.K., P. Mishra, A.K. Mishra, et al. 2007. Development and characterization of hyaluronic acid-anchored PLGA nanoparticulate carriers of doxorubicin. Nanomedicine 3: 246–57.

Yoo, H.S., J.E. Oh, K.H. Lee and T.G. Park. 1999. Biodegradable nanoparticles containing doxorubicin-PLGA conjugate for sustained release. Pharma. Res.16: 1114–8.

Youn, Y.S., S.Y. Chae, S. Lee, J.E. Jeon, H.G. Shin and K.C. Lee. 2007. Evaluation of therapeutic potentials of site-specific PEGylated glucagon-like peptide-1 isomers as a type 2 antidiabetic treatment: Insulinotropic activity, glucose-stabilizing capability, and proteolytic stability. Biochem. Pharmacol. 73: 84–93.

13

Immune Protection for Transplanted Pancreatic Islets by Nano-Encapsulation Strategies: A Chemist's Insight

Silke Krol,[1],* Ana Maria Waaga-Gasser[2] and Piero Marchetti[3]

ABSTRACT

Nanotechnology is gaining increasing importance in medicine. Researcher working on innovative treatments of diabetes, one of fastest expanding maladies in the developed and unfortunately also in the developing world use nanoparticulated formulation of insulin administration to improve the quality of life for the patients. One of the latest and perhaps most interesting involvements of nanotechnology in diabetes is the nano-encapsulation of isolated pancreatic islets for transplantation in patients with type 1 diabetes without immunosuppressant.

---

[1]Fondazione I.R.C.C.S. Istituto Neurologico "Carlo Besta", IFOM-IEO-campus, via Adamello 16, 20139 Milan, Italy; E-mail: silke.krol@ifom-ieo-campus.it

[2]University of Wuerzburg, Department of Surgery I, Molecular Oncology and Immunology, Oberduerrbacher Str. 6, D-97080 Wuerzburg, Germany;
E-mail: Waaga-Gasser@chirurgie.uni-wuerzburg.de

[3]Department of Endocrinology e Metabolism, Hospital of Cisanello, Via Paradisa 2, 56124 Pisa, Italy; E-mail: piero.marchetti@med.unipi.it

*Corresponding author

List of abbreviations after the text.

Due to the complexity of islets and the immune response or inflammatory reaction it looks that multifunctionality is the key to success, a feature unique to nanoparticles or nanolayers. The past 30 years of experience with microencapsulation of islets showed that the main problems are delayed response time to glucose stimulation, low biocompatibility to the recipient, low reproducibility of the coating, and graft rejection because of induction of immunologic or inflammatory pathways.

Nano-encapsulation means the deposition of polymeric multilayers anchored directly to the islets either by electrostatic or chemical interaction as well as lipid-incorporation. So far the nano-encapsulation showed promising results in improving the response time to stimulation especially with hydrophilic but neutral polymers which has been proven by a perifusion assay with different glucose concentrations in which no difference between coated and uncoated islets could be seen. Additional functionalities bound to the protective nano- coating and supposed to trigger insulin release proved successful and is promising in reducing the number of transplanted islets. The shortage of donor organs rather than the immune protection can be the real bottle neck in islet transplantation.

Further approaches included coagulation preventing moieties to the multilayer to manage the inflammatory response mainly for transplantation into the portal vein, the gold standard for islet transplantation.

The electrostatic assembled multilayers showed good protection against auto-antibodies from diabetes type 1 patients. Empty beads coated with the respective layers elicited low fibrosis after transplantation in the peritoneum along with neovascularization. All factors which can be crucial to prevent necrosis of the transplanted cell mass.

Nano-encapsulation is still in its infancy but the results *in vitro* are very promising so far. One critical point is the stability of the nanocoating, especially the lipid-anchored polymers as they disappear after only 24 to 48 h.

INTRODUCTION

Diabetes mellitus (DM) is one of the most common chronic diseases in the world, and continues to increase in prevalence and impact. Based on the data from 40 countries published by the WHO the total number of individuals with DM is expected to rise from 171 million in 2000 to prognosticated 366 million in 2030 (Shaw et al. 2010). It is associated with high morbidity and mortality, primarily mediated by the development of chronic vascular changes over time. This causes an excess in healthcare expenditure, and finally a large economic burden through loss of productivity and foregone economic growth. (American Diabetes Association [ADA] 2011)

DM is an heterogeneous disease (Table 1), and the two major forms are type 1 (T1DM) and type 2 diabetes (ADA 2011; Marchetti et al. 2008). In particular, T1DM is due primarily to autoimmune-mediated destruction of pancreatic islet beta-cells, resulting in dramatic insulin deficiency.

Table 1. Classification of diabetes (ADA 2011).

Class	
Type 1	• Immune mediated • Idiopathic
Type 2	• insulin resistance with relative insulin deficiency • secretory defect with insulin resistance
Gestational diabetes	• pregnancy
others	• Genetic defects of beta cells • Genetic defects in insulin action • Disease of the exocrine pancreas • Endocrinopathies • Drug or chemical induced • Infections • Uncommon forms of immune mediated • Other genetic syndromes sometimes associated with diabetes

An attractive alternative to daily insulin therapy is the replacement of a fully functional pancreatic β-cell to achieve a more physiological means for precise restoration of glucose homeostasis. Beta-cell replacement can be performed by either pancreas or islet allo-transplantation (Vardanyan et al. 2010). Although both procedures carry the burden of chronic immunosuppression, transplantation of isolated islets can be associated with aimed at selectively inhibiting undesired islet-specific or nonspecific immune responses, which may lead to-weaning/withdrawal of generalized immune suppression with preservation of long-lasting insulin independence.

Microencapsulation techniques of isolated pancreatic islets from the organ donors have be reported with different high viscous polymeric materials. After transplantation in diabetic patients they sometimes established normoglycemia for several years (Soon-Shiong et al. 1994). However, this success was not reproducible in all cases. This may be explained by the isolation technique as well as the origin and pretreatment (i.e., cold ischemia (Emamaullee and Shapiro 2007) of the islets prior to encapsulation which cast a cascade of alloantigen dependent events (Contreras et al. 2003; Laskowski et al. 2000) leading to chronic graft dysfunction.

In the past 10 years nanotechnology was explored in the field of encapsulation technology or more general in diabetes management. An overview about the different nanosystems can be found in a review by Krol (Krol 2008) or Pickup and collaborators (Pickup et al. 2008). Some developed nanoparticulated non-invasive insulin delivery systems to replace the daily multiple insulin injections. Others explored nanotechnological highly sensitive non-invasive glucose monitoring systems to substitute the multi-prick blood glucose testing.

While the technique of nano-encapsulation is still in its infancy it becomes already clear that this technique allows to develop more complex coatings able to respond to the manifold requirements in terms of islet functionality but also offers a flexible immune protection. The nano-encapsulation usually consist of multiple nanometer thick layers assembled directly of the surface of the islet and interconnected by high affinity and selective interactions.

Before developing novel protocols for nano-encapsulation of pancreatic islets a basic understanding of the function and requirements of islets as well as the complex innate and acquired immune responses at different times after transplantation.

The main requirements for immune protection and long-term function of transplanted pancreatic islets is a continuous coverage of the islet surface which allows for a fast response on external stimulation. High biocompatibility and low toxicity of the coating materials to either the recipient's tissue or the islets should be also given. Active or passive protection against immune responses and long-term stability of the coating are further important demands.

In the following the different approaches to reach those goals will be discussed controversially highlighting islet survival and immune modulation.

REQUIREMENTS FOR A SUCCESSFUL NANO-ENCAPSULATION

Pancreatic Islets and their Special Needs

It is important to understand that an islet is not a clump of cells but an organ in itself with well orchestrated communication between the different endocrine cells and other functional structures like the peri-insular basement membrane or ductal cells or acinar cells.

In a normal pancreas are approximately 1,000,000 islets of *Langerhans*, which contain several different types of endocrine cells (Fig. 1): the insulin secreting β–cells represent the majority of islet endocrine cells (60–80%), followed by glucagon containing cells (alpha cells, 20–30%) somatostatin (delta-) cells (5–15%), ghrelin (epsilon-) cells, and pancreatic polypeptide (PP-) cells (Cabrera et al. 2006). But islets require also non-endocrine cells such as intra-insular endothelial cells (Brissova et al. 2005), ductal epithelial cells (Ilieva et al. 1999) or the basement membrane for long-term survival and function. Moreover, remaining intrainsular endothelium after isolation contributes to neovascularization of the graft (Brissova et al. 2005). A peri- and intra-insular basement membrane separates endocrine (hormone producing islets) from exocrine tissue. The membrane consist of the typical extracellular matrix (ECM) proteins (i.e., laminin, collagen IV, and perlecan (Irving-Rodgers et al. 2008)). It differs between species in continuity and composition (van Deijnen et al. 1992). Collagen IV seems to play a particularly important role for proper insulin metabolism (Kaido et al. 2004). Moreover, the peri-insular capsule provides a barrier protective against diseases (Irving-Rodgers et al. 2008) but also against possible toxicity of the capsule material.

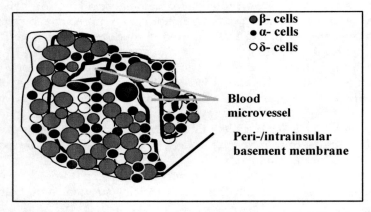

Fig. 1. Model of a pancreatic islet and connected functional units to it. S. Krol, unpublished data.

Immune Response

Beta-cell replacement can be performed by either pancreas or islet allotransplantation achieving consistent insulin independence and is usually associated with low morbidity. Nevertheless it has a limited success rates as the majority of islet recipients experience a decrease in graft function over time, with an insulin independence rate of only 10% at 5 years post-transplant (Ryan et al. 2005). Two events lead to decline in graft functionality: allo- or autoantigen-dependent rejection or alloantigen independent factors.

The main form of autoimmune destruction of β–cell death is apoptosis (Cnop et al. 2005) that, although not yet fully elucidated, seems to involve expression of Fas ligand (FasL) and its receptor Fas at the surface of the activated CD8+ T-cells and pancreatic β–cells respectively, release of perforin and granzyme by activated CD8+ T-lymphocytes, secretion of cytokines (including interleukin-1beta (IL-1β), tumor necrosis factor alpha (TNFα), and interferon-gamma (IFNγ) by the immune cells infiltrating the islets, and production of reactive oxygen species such as nitric oxide (NO) by macrophages, dendritic cells and the β–cells themselves (Fig. 2, Table 2).

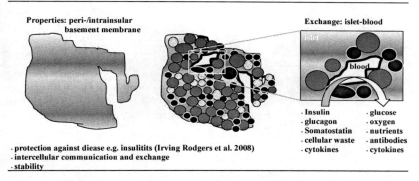

Fig. 2. Non-endocrine structures important for long-term islet survival and their functions. S. Krol, unpublished data.

The interaction between the T cell receptor and the MHC alloantigen in the presence of an appropriate co-stimulatory signal is now recognized as the central event that initiates alloimmune rejection (Waaga et al. 2000). Alloantigen is recognized either directly as intact allo-MHC on the surface of donor cells or indirectly as processed peptides derived from donor MHC and presented by recipient antigen-presenting cells (APCs) (Waaga et al. 2000). These in turn stimulate the production of a battery of cytokines, growth factors and alloantibodies.

After islet transplantation in T1DM, the decline in islet graft function is likely impacted by allograft rejection and recurrent autoimmunity. Clinical studies in autotransplantation show a gradual decrease in islet graft function after a sustained period of graft function, despite the absence of graft-specific immunity (Oberholzer et al. 1999, 2000; Robertson et al. 2001). Therefore, the gradual attrition in graft function can be partially attributed to nonimmunological factors.

Table 2. Physical properties of molecules interacting with the protective encapsulation.

Molecule	MW [Da]	Hydrodynamic size	charge of molecule
Insulin	5,733	(5x2.5) nm (6-mer)	Negative (pH7.4) pI: pH=5.3–5.35[1]
Glucagon	3,485		Neutral (pH7.4) pI: pH=7.0[2]
Pancreatic polypeptide			Negative (pH 7.4) pI: pH=4.5[3]
somatostatin	1,638		Positive (+1, pH 7.4) pI: pH=9.1[3]
ghrelin	3,314		positive (+4, pH 7.4)[4] pI: pH=11[3]
glucose	180		neutral (pH 7.4)
NO	30	136 pm	neutral (pH 7.4)
Cytokines	6,000–70,000		-
IL-1 β	17,000		pI: pH=7 [5]
TNF-α	25,000		pI: pH= 5.01[6]
IFN γ	17,146		pI: pH=9.54[7]
mAb	150–900 kD	10 nm	pI: pH=6.1[8]
macrophages	–	20–30 microns	

[1]Wintersteiner and Abramson 1933; [2]Andrade et al. 1997; [3]calculated from amino acid sequence by www.EMBL.de/cgi/pigrapper; [4]Bowers 2001; [5]http://www.grt.kyushu-u.ac.jp/spad/account/ligand/il1.html; [6]http://www.signaling-gateway.org/molecule/query?afcsid=A002291; [7]http://www.drugbank.ca/drugs/DB00033; [8]Vlasak and Ionescu 2008.

Whereas some antigen-independent events such as prolonged cold ischemia time may be capable of being minimized or even avoided, most other nonspecific and donor-associated factors cannot be influenced. One donor factor is brain death of the cadaveric donor. Animal models mimicking brain death had shown that this donor condition negatively affects islet yield as well as function due to the activation of pro-inflammatory cytokines (Contreras et al. 2003).

Another critical alloantigen-independent risk factor in chronic graft dysfunction is ischemia reperfusion (I/R) injury occurring during organ retrieval, storage and transplantation (Laskowski et al. 2000; Gasser et al. 2000).

In conclusion, auto- or alloimmune responses are caused either by immunological or non-immunological factors. During this process the capsule matrix will be exposed to a multitude of molecules with a broad range of physical properties. Some are listed in Table 2.

Key Facts of Diabetes and Islet Transplantation

- Pancreas consists of exocrine and endocrine tissue. Endocrine denominates the pancreatic islets or islets of *Langerhans*.

- Insulin is a hormone produced by β-cells in pancreatic islets. Deficiency leads to diabetes. The other cells are α- (glucagon), δ- (somatostatin), PP (pancreatic polypeptide), and ε–cells (ghrelin).
- Beta-cell mass is regulated by four major mechanisms: apoptosis (programmed cell death), size modification (hypo and hyperplasia), replication (mitotic division of differentiated β–cells) and neogenesis (development from precursor cells).
- Diabetes mellitus means persistent high blood glucose level (hyperglycemia). It is an heterogeneous disease, and one of its forms (type 1 or juvenile diabetes; 10% of new DM cases per year) is an autoimmune disease, mediated by the interplay between genetic and environmental factors, that leads to the destruction of the β–cells. Patients require life-long exogenous insulin administration.
- Insulin independence can be achieved by pancreas (>30,000 pancreas transplants worldwide) or isolated pancreatic islet transplantation.
- For islet transplantation, the first successful series was reported in 1990 in surgical diabetes. In 1999 insulin independence was reported in seven out of seven T1DM patients treated with intrahepatic islet transplantation and a glucocorticoid-free immunosuppressant.
- The gold standard for islet transplantation is the *Edmonton* protocol. It describes the injection of islets in the portal vein of the liver and associated regimen of generalized immunosuppression.

THE MICRO-PAST

Despite the encouraging outcome the advantages of both, pancreas and islet transplantation, have to be balanced against the unavoidable adverse effects of immunosuppression, that is needed to prevent allo-immunity and auto-immune recurrence of diabetes.

The story about the microencapsulation of transplanted pancreatic islets as immune protection began in the year 1980 (Lim and Sun 1980). The idea is to physically exclude immune-associated molecules (antibodies, cytokines) or cells (macrophages). Microencapsulation is mainly characterized by a random entrapment of isolated islets into droplets of neutral agarose or cellulose or negatively charged natural polymer, e.g., alginate cross-linked by bivalent cation (Wilson and Chaikof 2008a).

After purity of coating material and alginate composition were singled out as factors inducing fibrosis or immune activation the coating's thickness became the main problem. It became obvious that graft lost due to islet necrosis is caused by malnutrition and hypoxia. In this context Wilson and Chaikof (Wilson and Chaikof 2008a) raised the problem of general mass transport limitation through the microcapsule.

The micrometer thickness can seriously influencing the response time to glucose stimulation because both, glucose and insulin transport are diffusion driven, hence slow. Molecules crossing the microcapsule create a concentration gradient. Moreover if the molecules and the polymeric capsule material are interacting it is even possible that the glucose level is disconnected from the real blood level. Possible interactions can be either electrostatic or in neutral hydrophilic polymers hydrogen bonds.

Complexity of immune response as well as inflammatory processes caused by the islet or microcapsule material are broadly reviewed by Wilson and Chaikof (Wilson and Chaikof 2008a). The heterogeneity of molecules in size and charge (Table 2) indicate that a physical cut-off will hardly work. The most prominent example is cytotoxic NO released by, e.g., macrophages. It is much smaller than insulin and cannot be excluded by a microcapsule.

Recently the trend in microencapsulation moved towards more complex systems like alginate/poly-L-lysine (PLL)/alginate multilayers on the polymeric microbeads or other surface functionalization (Fig. 3).

The outcome of allo- and xeno-transplantation of microencapsulated islets in different species is reviewed by Wilson and Chaikof (Wilson and Chaikof 2008a). Predominantly the islets are transplanted in the peritoneum as the only site which can home the high islet/capsule volume. The low

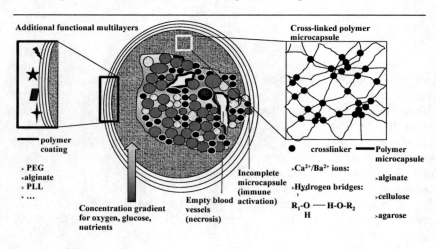

Fig. 3. Microencapsulation of pancreatic islet. Composition of the microcapsule and additional functional multiple nanolayers. The islet cells exposed by incomplete microencapsulation are rescued by the nanolayers. Neutral polymer like polyethyleneglycol (PEG) or positive polycation like poly-L-lysine (PLL) will allow free passage of Ca^{2+}-ions important for insulin release while the polyanion alginate will trap them and hamper proper islet function. S. Krol, unpublished data.

reproducibility of the microcoating was indicated by the fact that animals treated with the same experimental protocol varied strongly in duration of achieved normoglycemia. The differences are even more prominent for different microcoatings.

More sophisticated micro-approaches trying to solve hypoxia inducing fast neovascularization by surface coverage with VEGF (vascular endothelial growth factor) (Sigrist et al. 2003), improve of biocompatibility by attaching HEK cells via polyDNA hybridization (Teramura et al. 2010b), or reduce inflammatory response by heparin or other coagulation controlling drugs (Wilson and Chaikof 2008a).

Besides islet necrosis due to microcapsule thickness accompanied by hampered mass transport and a strong transplant mass increase which excludes the intrahepatic injection, the gold standard in islet transplantation. Nowadays the current clinical islet transplantation requires 600,000–700,000 islet equivalents (i.e.) with a volume of around 10 ml (Shapiro et al. 2000). The size of an islet is usually 200–300 micron but it increases to 500–1000 micron if a microcapsule is applied. This leads in the best case to a total transplant volume of 40 ml and in the worst case to 366 ml. These volumes can only be transplanted into the peritoneum. Diminution of microcapsule's thickness leads to a higher number of incompletely coated islets and hence graft failure by immune recognition (Wilson and Chaikof 2008a).

THE NANO-FUTURE

Initially the nanotechnology was introduced in diabetes therapy to monitor the blood glucose or administer insulin non-invasively. Reviews describing the different nano-approaches can be found from Pickup et al. (Pickup et al. 2008), Wilson and Chaikof (Wilson and Chaikof 2008a) and Krol (Krol 2008).

But the insolvable problems of necrosis and transplant mass increase in transplantation of microencapsulated islets were demanding for nano-strategies. Moreover the complexity of islets in terms of intra-insular communication as well as with exocrine structures crucial for long-term insulin secretion requires sophisticated coating strategies. That multifunctionality may be the key to success supported the latest developments in microencapsulation towards multilayers to improve biocompatibility or orchestrate the interaction with immune system or inflammatory mediators.

Nano-encapsulation, also called conformal encapsulation because it follows the shape of the islets, based on coatings which have a thickness of some nanometers up to several 100 nanometers, 1 nm being 10^{-9} or one billionth of a meter. Multiple nanometer thick polymer layers self-assembled directly on the islet surface by high affinity interaction of the building blocks can combine different functionalities and additionally guarantee completeness of the coating. The building blocks are polyelectrolytes, polymers which carry positively or negatively chargeable monomers or neutral hydrophilic polymers. Nano-encapsulation multilayers are assembled via biotin-SA recognition (Wilson et al. 2008b; highest binding constant in nature), nucleic acid base pairing (Teramura et al. 2010b), or electrostatic binding (Krol et al. 2006a).

Only recently a number of good and complete reviews (Teramura and Iwata 2010a; Pickup et al. 2008; Wilson and Chaikof 2008a) were published summarizing the latest developments in the field. In the following the different main developments in nano-encapsulation with their future perspective will be controversially discussed. An important aspect will be the capsule interaction with islet function and immune modulation.

As the first layer anchoring to the islet surface along with the nano-encapsulation material properties is the limiting factor for molecule exchange with the environment as well as capsule stability the different approaches are analyzed accordingly. Two main strategies can be distinguished: 1) Direct incorporation of lipid modified polymers such as polyethylenglycol (PEG) into the cell membrane which are suppose to prevent immune recognition (Miura et al. 2006). 2) Superficial interaction with the cell membrane or peri-insular capsule via electrostatic (Krol et al. 2006a), or chemical covalent binding (Kizilel et al. 2010).

Electrostatically Attached Polymer Multilayer Approach

The work of Krol and collaborators (Krol et al. 2006a) can mainly be considered as a proof-of-concept that layer-by-layer nano-encapsulation with the two synthetic polyelectrolytes polyallyamine (PAH) and polystyrenesulfonate (PSS) can be applied to sensitive cells like the pancreatic islets (Fig. 4). The first layer was the polycation which was anchored to the negatively charged peri-insular capsule or cell membrane. They noted that the intactness of the peri-insular basement membrane is crucial for a successful coating even with potentially toxic compounds like PAH. The islets were completely coated by up to 3 polyelectrolyte layers providing protection against the Glutamic Acid Decarboxylase (GAD)+ autoantibody found frequently in T1DM patients (Gronowski et al. 1995). The static insulin release assay showed small difference to the uncoated islets. Noteworthy in this context is that the net charge of the nano-capsule seems to influence the insulin secretion. With two polyelectrolyte layers (polydimethyldiammonium chloride/PSS) and a negative net charge the islets become non-responsive to a high glucose

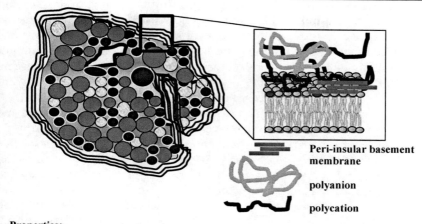

Fig. 4. Electrostatic anchoring to cell membrane or peri-insular basement membrane: Electrostatically assembled multilayers. S. Krol, unpublished data.

level (Krol et al. 2006a). A possible reason can be that the insulin release is triggered by a depolarization of the ATP-dependent K-channels in the membrane of β–cells controlling the membrane potential and which in turns lead to a Ca-ion influx (e.g., Cook et al. 1988). Hence, one can assume that any disturbance of the membrane potential of β–cells may cause an imbalance of insulin response to glucose stimulation. Especially because in human islets β–cells are localized close to the islet surface (Brissova et al. 2005). Electrostatic interactions can be mainly excluded if one considers that glucose is neutral and insulin is negatively charged at pH 7.4. The same is true for the cut-off as a capsule of 3 layers showed permeability for molecules with MW>150 kDa (Krol et al. 2006a).

Wilson et al. (Wilson et al. 2008b) developed a protective nano-coating for the islets anchoring the first layer by electrostatic interaction. The following layers were cross-linked by high affinity binding of streptavidin (SA) to PLL-PEG-biotin. PEG is well known and broadly used for its protein and cell repellent properties and hence to prevent immune recognition especially for nano-drug delivery systems in cancer treatment and antibody delivery. Wilson and his co-workers found that their coating is not influencing the insulin release of the nanocoated islets compared to uncoated cells in a static glucose stimulation assay.

Why to stress so much the charge of the polymeric material and the net charge of the nano-coating? Several "nanotoxicity" studies strongly indicate that the electrostatic interaction with the cell is the reason for polycation toxicity (e.g., Fischer et al. 2003). Moreover the net charge of the multilayers and of nanopores within the multilayer matrix may influence the transport of the different molecules listed in Table 2. Teramura and his co-workers (Teramura et al. 2008b) compared the electrostatic binding with a covalent binding of polymers and the lipid- or alkyl anchoring, both are discussed in the next paragraphs. They bound a single layer of synthetic polymers in the different modes to the surface and investigated its influence on cell viability of human embryonic kidney (HEK) cells and T-cell like cells as well as integrity and long-term stability of the resulting coating. PEG was covalently bound to membrane proteins of the cells and this layer was compared with a layer of electrostatically attached polyelectrolyte such as the cationic PEI (polyethyleneimine) or the anionic carboxylated PVA (polyvinylalcohol) and with layers of lipid-PEG and alkyl-PEG, both incorporating in the cell membrane by hydrophobic interactions. They found that the polycation leads to low cell viability. The polyanion does not bind the negative cell surface due to repulsive forces. For the other coatings they observe a good and homogenous coating but the material disappeared from the cell surface within 24 h after application.

Summarizing one can state that the coating with electrostatic anchoring gives homogenous layers and with a positive net charge of the nanocapsule good response in the static glucose response assay. The main drawback is the high toxicity of polycations used as first layer, even if an intact peri-insular capsule may provide some protection (Krol, unpublished data).

Covalently Attached Polymer Multilayer Approach

Polymers covalently attached to the membrane proteins as first layer for functionalized multilayer were explored by Kizilel et al. (Kizilel et al. 2010) for coating integrity and in a static glucose stimulation and a dynamic perifusion assay. Their first layer consist of biotin-PEG-NHS (N-hydroxysuccinimidyl) which covalently binds to free primary amine groups of membrane proteins. A scheme is sketched in Fig. 5.

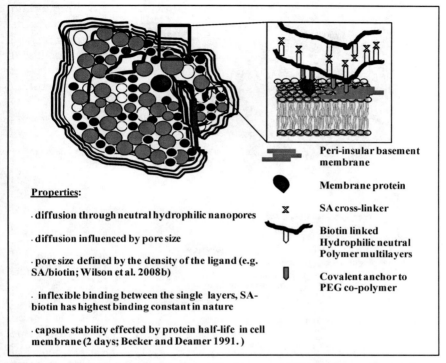

Fig. 5. Covalent anchoring to cell membrane or peri-insular basement membrane: e.g., hydrophilic streptavidin (SA)-biotin cross-linked multilayer nano-encapsulation. Polyethyleneglycol (PEG) prevents immune recognition. S. Krol, unpublished data.

In the following they cross-link the biotin via SA to a biotin-PEG linked insulinotropic truncated glucagon-like peptide 1 (GLP1, 7–37). The neutral hydrophilic coating does not induce toxicity to the enveloped cells. In both, the dynamic and the static glucose stimulation test the islets release more insulin compared to the uncoated control. In the perifusion the response time of the coated islets was minimally delayed compared to the uncoated ones. Moreover the coated islets showed the typical biphasic insulin secretion. This new promising approach can significantly reduce the number of transplanted islet equivalents as the release through the neutral coating and due to the stimulation by GLP-1 seem to be optimal. One critical point cane be the long-term stability of the membrane anchored coating (see: Teramura et al. 2008b). Here the limiting factor is the turnover half-life of a protein in the cell membrane which is around 2 days (Becker and Deamer 1991).

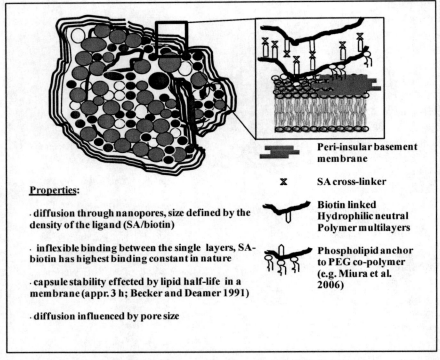

Fig. 6. Lipid anchoring to cell membrane: Biotin/SA assembled multilayers. S. Krol, unpublished data.

Membrane Anchored Lipid-based Approach

The pioneering work of Miura et al. (Miura et al. 2006) describes a multilayer coating starting with lipid-bound PEG molecules anchored into the membrane of islet cells via hydrophobic interactions. These PEG molecules serves as surface for additional layers of poly-L-lysine (PLL) and alginate. With 293 HEK cells and islets they showed that the coating is uniformly on the cell or islet surface and is not interfering with insulin release in static glucose stimulation assay. They replaced the electrostatically assembled and charged multilayers by a hydrophilic but neutral nano-capsule assembled via hydrophobic interactions as well as cysteine bonds (Teramura et al. 2007). This coating provides a slightly better responses to glucose stimulation than the charged multilayers.

The lipid-anchored multilayers were also functionalized in order to solve one of the problems related to intrahepatic transplantation which is that the occlusion induced by islets in the blood stream of the portal vein can cause coagulation and hence thrombosis and necrosis in the islet mass.

Several molecules were tested in the past for their properties to suppress blood-mediated inflammatory reactions such as Melagan and other. A short list of the investigated molecules can be found in the publication of Teramura and Iwata (Teramura and Iwata 2008a). They chose from the list of inflammation mediators heparin and urokinase and immobilized them on bovine serum albumin multilayers which were self-assembled via biotin-streptavidin interactions. Even so the described approach leads to a microcapsule it will be discussed here rather than in the micro-past paragraph as it begins with the above mentioned lipid-anchoring. Moreover the same group developed another nano-coating to modulate the coagulation by binding urokinase to the hydrophobically bound PVA-alkyl layers (Totani et al. 2008). For both, the nano-PVA capsule as well as the albumin microcapsule, the urokinase does not interfere with the insulin release in a static glucose stimulation assay. If islets coated with urokinase consisting multilayers are plated on a fibrin containing gel it was observed that after 13h of incubation the fibrin is dissolved. *In vivo* this effect can prevent the coagulation of fibrin in blood around the transplanted intraportal islets and early graft loss by thrombus formation. But the long-term study with the albumin microencapsulated islets showed that the coating is not stable. After 7 days the coating appears to be very patchy and the islets are no longer fully covered. This was supported by the findings of Inui et al. (Inui et al. 2010). Within 24h their lipid-anchored coating disappeared from the cell surface either by direct release to the medium or by endocytosis. An explanation can be that phospholipids have a quite short half-life of 170 mins in the cell membrane (Becker and Deamer 1991).

The nanocoatings with differently assembled multilayers show that the approach allows to combine in a very elegant way different functionalities such as cut-off, insulin stimulation (GLP-1), coagulation modulation (urokinase), and immune recognition (PEG). But so far the nano-coating is still in its infancy. The data from a static glucose stimulation assay, e.g., are not sufficient to conclude on response times. No immune protection assays such as, e.g., a GAD+ antibody recognition or others were performed except for the PAH/PSS/PAH polyelectrolyte coating. *In vivo* transplantation of the nano-coated islets will show if the materials cause fibrosis and if the nanocapsule is stable enough in a dynamic environment as the tissue. Very few nanocapsules, even without islets, were tested for their ability to induce neovascularization of the transplant. First attempts in that direction were the intraperitoneal transplantation of alginate beads coated with the same polyelectrolyte sequence (Chanana et al. 2005) as used for the nanocoated islets studied by Krol and co-workers (Krol et al. 2006a). Transplanted uncoated alginate beads showed a tight fibrotic capsule and no vessels. In contrast the polyelectrolyte multilayer coated alginate beads with a positive surface charge were embedded in a net of new grown vessels which connected them tightly to the peritoneum (Chanana et al. 2005).

Applications to other Areas of Health and Disease

The microencapsulation of pancreatic islet for immune protection and increased stability was already in the past the pioneer for immune protection of other types of cells. In 2003 this field of research was called cytomedicine (Yoshioka et al. 2003). Since then microcoated living cells or tissues such as hepatocytes, parathyroid cells, pituitary cells, and thymic epithelial cells were transplanted (for an overview: Krol et al. 2006b). Moreover the novel approach raised hope also for transplantation of stem cells, or so-called producer cells, cells genetically engineered to produce hormones or other proteins in patients with genetic deficiency diseases. But the success of the microencapsulation was somehow limited by the same problems discussed for the microencapsulated pancreatic islets. If the multifunctional nanocoating approach will succeed several severe disease like Parkinsonism or cancers can be treated with the immune protected engineered cells. Moreover the nanocoated cells can solve one of the main and most pressing problems in transplantation medicine which is the shortage of donor organs. If the coating is protective in case of allo-tranplantation the next rational step to make will be the xeno-transplantation.

Definitions of Key Terms

Auto-transplantation: donor and recipient are the same person.

Allo-transplantation: donor and recipient are within the same species.

Xeno-transplantation: across species.

Graft rejection: Allo-and xeno-transplantation requires immunosuppressive drugs otherwise the graft (islets) is destroyed by the immune system.

Microencapsulation: random entrapment of an undefined number of islets in high-viscose hydrophilic neutral polymers, e.g., agarose or cellulose or negatively charged alginate stabilized by divalent positively charged ions.

Conformal coating: assembled on the islet surface from single molecules. It is ultrathin and follows exactly the islet shape.

Multilayer nano-encapsulation: serial deposition of nanometer thick (1–100 nm) polymer layers connected by electrostatic or hydrophobic interactions as well as high affinity molecule recognition.

Static glucose stimulation assay: measures insulin concentration released by a defined number of pancreatic islets incubated for a defined time in medium containing usually two glucose concentrations: 3.3 and 16.8 mM.

Perifusion: dynamic glucose stimulation assay. Defined number of islets are in a chamber flushed by medium without or with different glucose concentrations. Response of isolated islets is a characteristic biphasic insulin secretion (for an example see: Kizilel et al. 2010). Insulin secretion can be delayed or modified by interaction with encapsulation material.

Summary Points

- Microencapsulation provides protection by physical cut-off molecules or cells. But the distance between islets and blood and the diffusion-driven molecule transport prolongs and unhinges the mass transport for glucose and insulin from the real blood levels.
- Islets are a complex system due to the interplay between the different islet cells and exocrine structures. The immune response caused by the allo-transplant or by inflammatory process are also complex and well orchestrated. If a complex nanocoating can support and maintain the communication between the islet cells and protects against several levels of immune response and inflammatory events islet survival and function will be prolonged.
- Nano-encapsulation creates nanometer thick multifunctional layers directly on the islet surface by phospholipid incorporation of modified polymers, electrostatic interaction of polyelectrolytes with charges on the cell or islet surface, or chemical covalent binding of polymers to proteins in the cell membrane or polymerization.
- Self-assembling via high affinity SA-biotin or base pairing as well as electrostatic binding and multilayers guarantees a complete coating of the islets and hence prevents islet exposure to the host immune system.
- Nanometer thickness of the coating allows for a fast molecule exchange between islets and blood. It can prevent graft failure due to necrosis and malnutrition.
- The multilayer approach offers the possibility to combine different functionalities such as improved biocompatibility (cells), suppression of coagulation (urokinase), reduced immune recognition (PEG), induction of neovascularization (VEGF), and enhanced insulin release (GLP-1).

Abbreviations

β-cell	:	beta cell
DM	:	Diabetes mellitus
ECM	:	extracellular matrix
GAD	:	Glutamic Acid Decarboxylase
GLP1	:	glucagon-like peptide 1
HEK	:	human embryonic kidney cells
i.e.	:	islet equivalents
IFNγ	:	interferon-gamma
IL-1β	:	interleukin-1 beta
NHS	:	N-hydroxysuccinimidyl
NO	:	nitric oxide
PEG	:	polyethylenglycol
PLL	:	poly-L-lysine
PVA	:	polyvinylalcohol
SA	:	streptavidin
TNFα	:	tumor necrosis factor alpha
T1DM	:	type 1 diabetes
VEGF	:	vascular endothelial growth factor
MW	:	molecular weight

References

American Diabetes Association. 2011. Diagnosis and Classification of Diabetes Mellitus. Diabetes Care 34: S62–S69.

Andrade, A.S.R., L. Vilela and H. Tunes. 1997. Purification of bovine pancreatic glucagon as a by-product of insulin production. Braz J. Med. Biol. Res. 30: 1421–1426.

Becker, W.M., and D.W. Deamer. 1991. The World of the Cell. Benjamin/Cummings Publishing Company, Redwood City (CA). USA.

Bowers, C.Y. 2001. Unnatural Growth Hormone-Releasing Peptide Begets Natural Ghrelin. J. Clin. Endocrin. Metab. 86: 1464–1469.

Brissova, M., M.J. Fowler, W.E. Nicholson, A. Chu, B. Hirshberg and D.M. Harlan, A.C. Powers. 2005. Assessment of Human Pancreatic Islet Architecture and Composition by Laser Scanning Confocal Microscopy. J. Histochem. Cytochem. 53: 1087–1097.

Cabrera, O., D.M. Berman, N.S. Kenyon, C. Ricordi, P.O. Berggren and A. Caicedo. 2006. The unique cytoarchitecture of human pancreatic islets has implications for islet cell function. Proc. Natl. Acad. Sci. USA 103: 2334–2339.

Cnop, M., N. Welsh, J.C. Jonas, A. Jörns, S. Lenzen and D.L. Eizirik. 2005. Mechanisms of pancreatic beta-cell death in type 1 and type 2 diabetes: many differences, few similarities. Diabetes 54: S97–S107.

Contreras, J.L., C. Eckstein, C.A. Smyth, M.T. Sellers, M. Vilatoba, G. Bilbao, F.G. Rahemtulla, C.J. Young, J.A. Thompson, I.H. Chaudry and D.E. Eckhoff. 2003.

Brain death significantly reduces isolated pancreatic islet yields and functionality *in vitro* and *in vivo* after transplantation in rats. Diabetes 52: 2935–2942.

Cook, D.L., L.S. Satin, M.L. Ashford and C.N. Hales. 1988. ATP-sensitive K-channels in pancreatic B-cells: The "spare channel" hypothesis. Diabetes 37: 495–498.

Emamaullee, J.A., and A.M. Shapiro. 2007. Factors influencing the loss of beta-cell mass in islet transplantation. Cell Transplant 16: 1–8.

Fischer, D., Y. Li, B. Ahlemeyer, J. Krieglstein and T. Kissel. 2003. *In vitro* cytotoxicity testing of polycations: Influence of polymer structure on cell viability and hemolysis. Biomaterials 24: 1121–1131.

Gasser, M., A.M. Gasser, I.A. Laskowski and N.L. Tilney. 2000. The influence of donor brain death on short and long-term outcome of solid organ allografts. Ann. Transplant 5: 61–67.

Gronowski, A.M., E.C.C. Wong, T.R. Wilhite, D.L. Martin, C.H. Smith, C.A. Parvin and M. Landt. 1995. Detection of Glutamic Acid Decarboxylase Autoantibodies with the *varelisa* ELISA. Clinical. Chem. 41: 1532–1534.

Ilieva, A., S. Yuan, R.N. Wang, D. Agapitos, D.J. Hill and I. Rosenberg. 1999. Pancreatic islet cell survival following islet isolation: the role of cellular interactions in the pancreas. J. Endocrinology 161: 357–364.

Inui, O., Y. Teramura and H. Iwata. 2010. Retention dynamics of amphiphilic polymers PEG-lipids and PVA-alkyl on the cell surface. Appl. Mat. Interf. 2: 1514–1520.

Irving-Rodgers, H.F., A.F. Ziolokowski, C.R. Parish, Y. Sado, Y. Ninomiya, C.J. Simeonovic and R.J. Rodgers. 2008. Molecular composition of the peri-islet basement membrane in NOD mice: a barrier against destructive insulitits. Diabetologia 51: 1680–1688.

Kaido, T., M. Yebra, V. Cirulli and A.M. Montgomery. 2004. Regulation of human β-cells adhesion, motility, and insulin secretion by collagen IV and its receptor α1β1. J. Biol. Chem. 279: 53762–53769.

Kizilel, S., A. Scavone, M.S. Xiang, J.-M. Nothias, D. Ostrega, P. Witkowski and M. Millis. 2010. Encapsulation of pancreatic islets within nano-thin functional polyethylene glycol coatings for enhanced insulin secretion 16: 2217–2227.

Krol, S. 2008. Nanomedicine in diabetes—Recent developments. Eur. J. Nanomedicine 1: 40–44.

Krol, S., S. del Guerra, M. Grupillo, A. Diaspro, A. Gliozzi and P. Marchetti. 2006a. Multilayer nanoencapsulation—New approach for immune protection of human pancreatic islets. Nano Letters 6: 1933–1939.

Krol, S., A. Gliozzi and A. Diaspro. 2006b. Polyelectrolyte Nanocapsules—Promising Progress in development of New Drugs and therapies. Frontiers in Drug Design & Discovery 2: 333–348.

Laskowski, I., J. Pratschke, M.J. Wilhelm, M. Gasser and N.L. Tilney. 2000. Molecular and cellular events associated with ischemia/reperfusion injury. Ann. Transplant 5: 29–35.

Marchetti, P., F. Dotta, D. Lauro and F. Purrello. 2008. An overview of pancreatic beta-cell defects in human type 2 diabetes: implications for treatment. Regul Pept. 146: 4–11.

Miura, S., Y. Teramura and H. Iwata. 2006. Encapsulation of islets with ultra-thin polyion complex membrane through poly(ethylene glycol)-phospholipids anchored to cell membrane. Biomaterials 27: 5828–5835.

Oberholzer, J., F. Triponez, J. Lou and P. Morel. 1999. Clinical islet transplantation: a review. Ann. N. Y. Acad. Sci. 875:189–199.

Oberholzer, J., F. Triponez, R. Mage, E. Andereggen, L. Bühler, N. Cretin, B. Fournier, C. Goumaz, J. Lou, J. Philippe and P. Morel. 2000. Human islet transplantation: lessons from 13 autologous and 13 allogeneic transplantation. Transplantation 69: 1115–1123.

Pickup, J.C., Z.L. Zhi, F. Khan, T. Saxl and D.J. Birch. 2008. Nanomedicine and its potential in diabetes research and practice. Diabetes Metab. Res. Rev. 24: 604–610.

Robertson, R.P., K.J. Lanz, D.E. Sutherland and D.M. Kendall. 2001. Prevention of diabetes for up to 13 years by autoislet transplantation after pancreatectomy for chronic pancreatitis. Diabetes 50: 47–50.

Ryan E.A., B.W. Paty, P.A. Senior, D. Bigam, E. Alfadhi, N.M. Kneteman, J.R. Lakey and A.M. Shapiro. 2005. Five-year follow-up after clinical islet transplantation. Diabetes 54: 2060–2069.

Shapiro, A.M., J.R. Lakey, E.A. Ryan, G.S. Korbutt, E. Toth, G.L. Warnock, N.M. Kneteman and R.V. Rajotte. 2000. Islet transplantation in seven patients with T1D mellitus using a glucocorticoid-free immunosuppressive regimen. N. Engl. J. Med. 343: 230–238.

Shaw, J.E., R.A. Sicree and P.Z. Zimmet 2010. Global estimates of the prevalence of diabetes for 2010 and 2030. Diabetes Res. Clin. Pract. 87: 4–14.

Sigrist, S., A. Mechine-Neuville, K. Mandes, V. Calenda, S. Braun, G. Legeay, J.P. Bellocq, M. Pinget and L. Kessler. 2003. Influence of VEGF on the viability of encapsulated pancreatic rat islets after transplantation in diabetic mice. Cell Transplant 12: 627–635.

Soon-Shiong, P., R.E. Heintz, N. Merideth, Q.X. Yao, Z. Yao, Z. Tianli, M. Murphy, M.K. Moloney, M. Schemehl, M. Harris, R. Mendez, R. Mendez and P.A. Sandford. 1994. Insulin independence in a type 1 diabetic patient after encapsulated islet transplantation. Lancet 343: 950–951.

Teramura, Y., Y. Kaneda and H. Iwata. 2007. Islet-encapsulation in ultra-thin layer-by-layer membranes of poly(vinyl alcohol) anchored to poly(ethylene glycol)-lipids in the cell membrane. Biomaterials 28: 4818–4825.

Teramura, Y., and H. Iwata. 2008a. Islets surface modification prevents blood-mediated inflammatory responses. Bioconjugate Chem. 19: 1389–1395.

Teramura, Y., Y. Kaneda, T. Totani and H. Iwata. 2008b Behavior of synthetic polymers immobilized on a cell membrane. Biomaterials 29: 1345–1355.

Teramura, Y., and H. Iwata. 2010a. Bioartificial pancreas Microencapsulation and conformal coating of islet of Langerhans. Adv. Drug Deliv. Rev. 62: 827–840.

Teramura, Y., L.N. Minh, T. Kawamoto and H. Iwata. 2010b. Microencapsulation of islets with living cells using PolyDNA-PEG-lipid conjugate. Bioconjugate Chem. 21: 792–796.

Totani, T., Y. Teramura and H. Iwata. 2008. Immobilization of urokinase on the islet surface by amphiphilic poly(vinyl alcohol) that carries alkyl side chains. Biomaterials 29: 2878–2883.

Van Deijnen, J.H.M., C.E. Hulstaert, G.H.J. Wolters and R. van Schilfgaarde. 1992. Significance of the peri-insular extracellular matrix for islet isolation from the pancreas of rat, dog, pig, and man. Cell Tissue Res. 267: 139–146.

Vardanyan, M., E. Parkin, C. Gruessner and H.L. Rodriguez Rilo. 2010. Pancreas vs. islet transplantation: a call on the future. Curr. Opin. Organ. Transplant 15: 124–30.

Vlasak, J., and R. Ionescu. 2008. Heterogeneity of Monoclonal Antibodies Revealed by Charge-Sensitive Methods. Curr. Pharma. Biotech. 9: 468–481.

Waaga, A.M., M. Gasser, I. Laskowski and N.L. Tilney. 2000. Mechanisms of chronic rejection. Curr. Opin. Immunol. 12: 517–521.

Wilson, J.T., and E.L. Chaikof. 2008a. Challenges and emerging technologies in the immunoisolation of cells and tissues. Adv. Drug Deliv. Rev. 60: 124–145.

Wilson, J.T., W. Cui and E.L. Chaikof. 2008b. Layer-by-layer assembly of a conformal nanothin PEG coating for intraportal islet transplantation. Nano Letters 8: 1940–1948.

Wintersteiner, O., and H.A. Abramson. 1933. The isoelectric point of insulin: electrical properties of adsorbed and crystalline insulin. J. Biol. Chem. 99: 741–753.

Yoshioka, Y., R. Suzuki, H. Oka, N. Okada, T. Okamoto, T. Yoshioka, Y. Mukai, H. Shibata, Y. Tsutsumi, S. Nakagawa, J.I. Miyazaki and T. Mayumi. 2003. A novel cytomedical vehicle capable of protecting cells against complement. Biochem. Biophys. Res. Comm. 305: 353–358.

14

Nanoparticles, Interleukin-10 and Autoimmune Diabetes

Rhishikesh Mandke,[1,a] *Ashwin Basarkar*[2] *and Jagdish Singh*[1,b,*]

ABSTRACT

Type 1A diabetes, an immunological endocrine disorder, is one of the most common chronic disorders accounting for 5–10% of total diagnosed diabetes cases. The requirement of a lifelong therapy, which might prove inadequate, results in deterioration in quality of life and is a huge burden on the healthcare systems. The pathogenesis of autoimmune diabetes involves generation of markers such as anti-islet antibodies and subsequent inflammatory immune reactions. The resulting generation of inflammatory cytokines is implicated in destruction of pancreatic β-cells and loss of insulin production. Interleukin-10 (IL-10), an anti-inflammatory cytokine, is produced by T_H2 population

---

[1]Department of Pharmaceutical Sciences, NDSU Dept 2665, PO Box 6050, College of Pharmacy, Nursing, and Allied Sciences, North Dakota State University, Fargo, ND 58105, United States.
[a]E-mail: Rhishikesh.Mandke@ndsu.edu
[b]E-mail: Jagdish.Singh@ndsu.edu
[2]Scientist, Formulations, Acceleron Pharma, Inc., 128, Sydney Street, Cambridge, MA 02139, United States; E-mail: Ashwin.Basarkar@gmail.com; Ashwin.Basarkar@merck.com
Present affiliation: Senior Formulation Development Scientist, Sterile Product Development, Merck Research Laboratories, 556 Morris Avenue, Summit, NJ 07901, United States
*Corresponding author
List of abbreviations after the text.

of helper T cells. Administration of IL-10 is known to alleviate the symptoms of various inflammatory and algesic reactions by downregulating the inflammatory cytokine production. Systemic administration of IL-10 is known to prevent autoimmune insulitis and subsequent loss of insulin production ability of the pancreas. However, owing to very short plasma half life (~2 min), sustaining the plasma levels of IL-10 is challenging. In this scenario, gene therapy has the potential to offer a solution, whereby the genes encoding IL-10 can be administered using a viral or nonviral vector. A successful transfection by these systems can result in a sustained expression of IL-10 endogenously and assuage the need of multiple IL-10 injections. Nanoparticles have been reported to successfully deliver plasmids encoding IL-10 with high tissue compatibility and greater safety. This chapter discusses the prevalence, pathogenesis and treatment options of autoimmune or type 1A diabetes. The newer, experimental options involving the applications of IL-10 are elaborated. Finally, applications of nanoparticulate delivery systems for effective administration of therapy involving IL-10 are briefly reviewed here.

INTRODUCTION

Autoimmune diabetes (known as type 1A diabetes) is an autoimmune disorder, characterized by selective destruction of pancreatic β-cells resulting in gradual loss of insulin production. Annually, it affects more than 11,000 new cases in US alone and accounts for 5–10% of total diagnosed cases of diabetes (Devendra and Eisenbarth 2003). Though children and adolescents are the major populations affected by this disorder, it has been reported to occur as late as 90 years of age. Hence, the term "juvenile-onset" is not used anymore to describe this type of diabetes. Diabetes and associated groups of diseases put a tremendous burden on the healthcare system in US (Jacobson et al. 1997. The expert committee on the diagnosis and classification of Diabetes mellitus 2002).

The treatment of type 1A diabetes is essentially dependant on the delivery of insulin and focuses on the use of short- and long-acting insulins to simulate the physiological insulin secretion profile, which often prove to be inadequate to prevent various diabetes-related complications. Various novel therapies are currently being investigated to prevent or treat type 1A diabetes. The major treatment options under the investigations include islet cells transplantation and cytokine gene therapy. Currently, more than 300 clinical trials involving cytokine genes are underway (http://www.wiley.com/legacy/wileychi/genmed/clinical/). This fact underlines the level of interest generated by these gene therapies. Type 1A diabetes is one of the candidate diseases which can potentially be prevented by cytokine gene therapy. Though viral vectors play an important role in delivering the genes to their target cells, their immunogenic and mutagenic potential usually provides a cause of concern and has led scientists to search for safer nonviral vectors. The number of nonviral vectors used in various studies involving gene therapies is steadily increasing over the period of time (Gao et al. 2007; Kawakami et al. 2008).

Nanoparticles have been studied for various drug delivery applications including gene delivery. These delivery systems offer various advantages over the viruses as vectors for gene delivery. The nanoparticles have a large surface area, which can be modified using various functionalities for diverse applications such as pH buffering, electrostatic binding with DNA, targeting of specific tissues and stealth from the components of the immune system. The nanoparticles can demonstrate high tissue penetrability and, rapid escape from degradative endosomal/lysosomal compartments which are essential features of a nonviral vector. The availability of diverse fabrication materials ranging from biological and synthetic polymers to various metals has made the nanoparticulate delivery system an attractive carrier for delivery of various biotherapeutics (Basarkar and Singh 2007).

PATHOGENESIS OF AUTOIMMUNE DIABETES

Autoimmune diabetes is a multipronged disorder which may lead to cardiovascular complications, retinopathy, neuropathy, and nephropathy. Cell-mediated autoimmunity is primarily implicated in the pathogenesis of type 1A diabetes. Auto-reactive T cells may exist normally but are restrained by immunoregulatory mechanisms leading to a self tolerant state. However, when these mechanisms fail, auto-reactive T cells are activated, expand clonally, and cause a series of autoimmune reactions which results in the destruction of β-cells in the pancreas (Fig. 1, Kawasaki et al. 2004).

Primarily, the inflammation in the pancreas is a T cell-mediated response directed against one or more β-cell markers such as islet cells, insulin, GAD and tyrosine phosphatase. The presence of auto-antibodies for one or more of these markers has been observed in more than 85% of diagnosed cases with elevated fasting blood glucose levels and can be used to identify target population for suitable interventions. Though diabetogenic seroconversion can occur at the age < 10 years, there is no limiting age factor for development of autoimmunity. Interestingly, more than 30% of the young population with type 2 diabetes are known to express anti-islet autoantibodies and would eventually shift to active insulin therapy within 36 months (Devendra and Eisenbarth 2003).

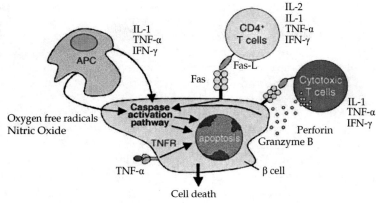

Fig. 1. Mechanisms of β cell destruction in type 1 diabetes. (With permission from Elsevier, Kawasaki et al. 2004).

Color image of this figure appears in the color plate section at the end of the book.

Newly diagnosed diabetics with insulin deficiency usually exhibit invasion of pancreatic islets by CD8+ T cells and associated β-cell destruction. Pancreatic autopsy in such patients indicate mononuclear cell infiltration into the pancreatic islets and a gradual destruction of β-cells (Kawasaki et al. 2004).

The mechanism of autoimmunity in type 1A diabetes has been extensively investigated using animal models such as NOD mice, BB rats, and multiple low-dose streptozotocin (MLD-STZ)-treated mice enhancing our understanding of the pathogenesis of human type 1A diabetes. Although the exact mechanism of initiation and progression of autoimmunity leading to the destruction of pancreatic β-cells is not yet known, a broad array of immune cells have been identified as contributors towards this response. In the NOD mouse, a lymphocytic infiltration of pancreas (insulitis) is followed by β-cell destruction with gradual insulin deficiency and resultant hyperglycemia. In response to the auto-antigens presence on the surface of pancreatic β-cells, antigen presenting cells such as macrophages and dendritic cells migrate towards the pancreas. Studies done in NOD mice and BB rats have suggested the critical role of macrophages in the progression of autoimmune response. Inactivation of macrophages in these animal models resulted in significant protection against development of diabetes. T cells in these animals lose their ability to differentiate into cytotoxic cells. Macrophage-depleted NOD mice also showed an upregulation of T_H2 response and a downregulation of T_H1 response. CD4+ and CD8+ T cells and B cells usually follow the dendritic cells and macrophages. This insulitis progresses from the vicinity of the islets and eventually invades and destroys the islets.

Proinflammatory cytokines such as IL-1, IL-2, TNF-α, and IFN-γ, produced primarily by T_H1 cells play an important role in the pathogenesis of type 1A diabetes by causing activation and migration of more inflammatory cells into the pancreas. Additionally, effector T cells also cause β-cell destruction through direct contact with surface ligands such as FasL and membrane-bound TNF-α which induces apoptosis. Cytotoxic T cells also release perforins and granzymes leading to activation of nucleases. All these mechanisms together, either directly or through the induction of caspase pathway, lead to apoptosis of pancreatic β-cells. This β-cell destruction is enhanced by the T_H1 subset of CD4+ T cells and the type 1 cytokines, such as INF-γ, TNF-α and IL-2 which can be inhibited by T_H2 and T_H3 cytokines, such as IL-4, IL-5, and IL-10 (Carter et al. 2005; Cox et al. 2001; Devendra and Eisenbarth 2003).

DIAGNOSIS AND PREDICTION OF AUTOIMMUNE DIABETES

Expression of two or more autoantibodies (for β-cell markers such as islet cells, insulin, GAD and tyrosine phosphatase) has a strong positive correlation (>85%) with type 1A diabetes while the presence of a single autoantibody is usually useful for prediction in >65% of cases (Devendra and Eisenbarth 2003).

STRATEGIES FOR PREVENTION/TREATMENT OF TYPE 1A DIABETES

Type 1A diabetes gradually reduces the body capacity to produce insulin. So far, the only available treatment of type 1A diabetes is insulin delivery, either through injection or insulin pump to mimic the basal and simulated insulin levels. Delivered insulin replaces body's own insulin resulting in restoration of normal or near normal levels of glucose in the blood. However, this therapy is aimed towards management rather than the treatment of the condition. Insulin therapy has to be continued for the life of the patient and must be complimented with strict diet control and exercise routine. Moreover, insulin requirement keeps changing over the lifetime of the patient. Insulin therapy is expensive and has patient compliance issues in case of daily injections. Use of insulin pump may lead to problems such as pump-malfunction and infection at the implantation site. These limitations associated with insulin therapy have led to an intensive research in the development of alternative therapies for type 1 diabetes. Despite the advances in research for novel insulin delivery methods, management of diabetes and associated multiple acute and chronic complications is difficult and expensive. It is reported that there was no impact of low-dose parenteral insulin therapy in delaying the onset of diabetes (Diabetes Prevention Trial–Type 1 Diabetes Study Group 2002). Therefore, there is a need to develop alternatives of insulin injections for delaying or preventing the onset of type 1A diabetes.

ISLET TRANSPLANTATION

Pancreatic islet transplantation has been pursued as a promising treatment strategy for type 1A diabetes. Several clinical trials have been performed using islet transplantation with immunosuppressive therapy. The goal of islet transplantation is to achieve good glycemic control with minimal side effects. However, so far the results have been disappointing in terms of insulin dependence with only about 10% of subjects being off insulin at one year (Brendel et al. 1999). Moreover, since the transplanted islets are rejected by patient's immune system, they have to be maintained on immunosuppressive medication such as tacrolimus, sirolimus, and cyclosporine A for long durations after transplant, making them more susceptible to infections. Long-term immunosuppressive therapy may also cause serious side effects such as nephrotoxicity, hepatotoxicity, and neurological complications. Recently, investigators have proposed that surface modification of islets using poly (ethylene glycol) can lead to prolonged survival of islets by prevention of immunogenic reactions. This study showed that PEGylated islets could survive for a long time in rats with low dose of immunosuppressive therapy with cyclosporine A. However, after the immunosuppressive treatment was discontinued, the islets could not survive for a long time which underlines the need for continuous immunosuppression with islet transplantation.

GLUTAMIC ACID DECARBOXYLASE (GAD)

GAD is one of the most commonly studied β-cell autoantigen. It is a 65kd protein which was precipitated from the sera of type 1A diabetic patients. Antibodies to GAD65 have been found in 70–75% of type 1A diabetics compare to 1–2% in healthy individuals (Sanjeevi et al. 1996). Administration of purified GAD protein has been reported to enhance the tolerance of T cell mediated immune response against pancreatic β-cells causing prevention or delay in the development of insulitis and diabetes. Induction of tolerance with the use of GAD may also prevent the development of immune reaction against other autoantigens on the β-cells. Autoimmune reaction against β-cells may also be prevented by suppressing the expression of GAD in transgenic mice using antisense therapy.

HEAT SHOCK PROTEIN (HSP)

Heat shock proteins are intracellular molecular chaperones that are upregulated in response to stress. HSP is also an important autoantigen involved in the pathogenesis of type 1A diabetes. Hsp60 was found in NOD mice that were developing diabetes. Transplantation of anti-Hsp60 T cells in healthy mice was also found to result in induction of diabetes. A 24 amino acid peptide on Hsp60 was identified as the epitope recognized by autoreactive T cells and was termed p277. Vaccination with p277 in mice has been shown to prevent both the onset of autoimmune diabetes and the deterioration of diabetes after onset in NOD mice (Elias and Cohen 1994).

CYTOKINE GENE THERAPY

Cytokines are small peptide molecules synthesized and secreted inside the body by activated lymphocytes (lymphokines), macrophages/monocytes (monokines) and cells outside the immune system and mediate and regulate immunity, inflammation, and hematopoiesis. The cytokine action can be autocrine (on the producing cell), paracrine (on neighboring cells), or endocrine (on distant organs). A complex interplay between cytokines is responsible for the pathology of autoimmune diabetes (Tarner and Fathman 2001; Hill and Sarvetnick 2002).

Type 1A diabetes is generally explained by T_H1/T_H2 balance model which concludes that autoimmunity is caused due to dominance of T_H1 cytokines. T_H1 cells with their cytokine effectors such as IFN-γ, IL-2 and TNF-α elicit cell-mediated responses; whereas T_H2 cells through their cytokines such as IL-4 and IL-10 elicit humoral responses. Restoration of balance between T_H1 and T_H2 cytokines by upregulation of T_H2 cytokine expression may lead to prevention of autoimmune diabetes. Delivery of cytokines, however, is not possible due to high cost and very short plasma half life. Delivery of genes encoding T_H2 cytokines has the potential to eliminate these shortcomings by facilitating in situ expression of cytokines. It may also circumvent the problems associated with immune reaction against foreign cytokines. Genes encoding anti-inflammatory cytokines, IL-4 and IL-10 have been delivered to suppress autoimmunity in type 1A diabetes (Chernajovsky et al. 2004; Li et al. 2008; Basarkar and Singh 2009).

INTERLEUKIN-4 GENE DELIVERY

IL-4, a prototypical T_H2 cytokine, has been widely used for gene therapy in experimental models of autoimmune disorders such as experimental allergic encephalomyelitis (EAE), and collagen-induced arthritis (CIA). Gene encoding IL-4 has been delivered using both viral and non-viral vectors. Expressed IL-4 inhibits the production of proinflammatory cytokines thereby suppressing autoimmunity. IL-4 gene therapy has achieved mixed success in treatment of type 1A diabetes. Delivery of gene encoding IL-4 had been only partially successful in spontaneously developing Insulin Dependent Diabetes Mellitus (IDDM) in NOD mice. Studies showed that *i.v.* injection of IL-4 plasmid was effective only when administered in combination with IL-10 plasmid. Delivery of IL-4 plasmid complexed with polymer, on the other hand led to complete amelioration of the disease (Lee et al. 2002).

INTERLEUKIN-10 GENE DELIVERY

IL-10 was first described in 1989 by Mosmann and coworkers as a cytokine that is produced by T_H2 cell clones and inhibited interferon-γ synthesis in T_H1 clones. IL-10 is a pleiotropic cytokine produced primarily by the T_H2 subset of helper T cells, B cells and macrophages (Fig. 2). It also induces proliferation and differentiation of B cells and mast cells. IL-10 demonstrates potent anti-inflammatory activity via downregulation of the pro-inflammatory cytokines released by activated immune competent cells. In addition, IL-10 inhibits monocyte MHC class II molecule, CD23, ICAM-1, and B7 expression leading to inhibition of the ability of the APC to activate the helper T cells. IL-10 also inhibits eosinophil survival and IL-4-induced IgE synthesis while in B lymphocytes, it stimulates cell proliferation and Ig secretion. TNF-α and other cytokines stimulate IL-10 secretion, suggesting a homeostatic mechanism whereby an inflammatory stimulus induces TNF-α secretion, which in turn stimulates IL-10 secretion leading to suppression of TNF-α synthesis. Among the T_H2 cytokines, IL-10 uniquely inhibits cytokine production by mononuclear cells (Borish and Steinke 2003). It has been previously used for treatment of pain, inflammation and autoimmune diseases such as psoriasis and Crohn's disease (Milligan et al. 2005; Sloane et al. 2009).

These studies indicate that IL-10 might be a valuable agent for treatment of T cell-mediated autoimmune diseases such as type 1A diabetes. Production and action of IL-10 is found to be low in both human and animal models of type 1A diabetes which suggests that administration of IL-10 can lead to correction of diabetes (Schloot et al. 2002). The plasma

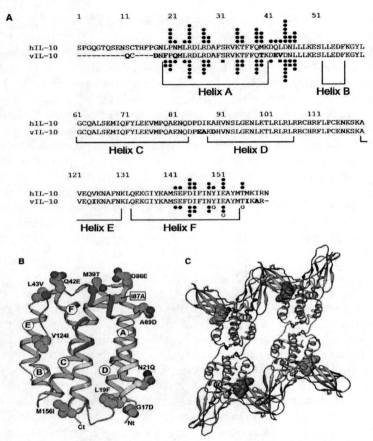

Fig. 2. Sequence and Structure of vIL-10. (A) Structure-based sequence alignment based on vIL-10/sIL-10R1 and hIL-10/sIL-10R1 complexes. vIL-10 amino acids that differ from the hIL-10 sequence are shown in bold. The approximate amount of surface area buried by residues in viral (vIL-10 and vIL-10$_{A87I}$) and human IL-10 is shown with black circles according to the following code; (5Å^2 < one circle ≤ 20Å^2), (20Å^2 < two circles ≤ 40Å^2), (40Å^2 < three circles ≤ 60Å^2), (60Å^2 < four circles ≤ 80Å^2), (80Å^2 < five circles ≤ 100Å^2), (>100Å^2 = six circles). Residues that bury surface area only in vIL-10 or only in vIL-10$_{A87I}$ are shown by black squares and open circles, respectively. If two markers are shown, the greatest position of each marker reflects the amount of surface area buried in the respective complexes. (B) Ribbon diagram of one vIL-10 domain. The side chains of vIL-10 residues that differ from hIL-10 are shown. vIL-10 residues that bury surface area in the site Ia and Ib interfaces are shown in purple and gold, respectively. (C) 2:4 vIL-10/sIL-10R1 complex. The site II interface is located between the cyan/green and gold/purple 1:2 vIL-10/sIL-10R1 complexes. Residues that form the putative IL-10R2 binding site are shown on the vIL-10s in space filling representation. IL-10R2 binding sites are colored green if they are accessible to the IL-10R2 chain or red if the site is occluded in the site II interface. (With Permission from Elsevier, Yoon et al. 2005).

Color image of this figure appears in the color plate section at the end of the book.

half life of IL-10, however, is very short (~2 min) which severely limits its use as a therapeutic agent. IL-10 gene has been delivered in a number of studies using both viral and non-viral vectors and has been found to prevent or slow down the progression of autoimmune diabetes. IL-10 gene delivered using adeno-associated viral vector resulted in protection of pancreatic islets along with suppression of T cell activation (Goudy et al. 2001). Intramuscular injection of naked plasmid DNA encoding IL-10 also led to prevention of STZ-induced diabetes (Zhang et al. 2003). Additionally, IL-10 gene delivery has also been used for the suppression of autoimmune reaction after pancreatic islet transplantation (Carter et al. 2005).

ROLE OF NANOPARTICULATE DELIVERY SYSTEMS IN IL-10 GENE THERAPY FOR PREVENTION OF AUTOIMMUNE DIABETES

Nanotechnology has been critical for gene delivery in last few years. The nanoparticles as non-viral gene therapy vectors have been used with mixed degree of success. The advantages of these delivery systems include small size, high stability, high payload capacity, availability of various fabrication materials and techniques, greater cellular uptake and ability to cross BBB, amenability for modifications that can render them target specific and availability of variable routes of administration, including oral and parenteral. The schematics of the barriers encountered by nanoparticle-based vector at cellular level are shown in Fig. 3. The major drawbacks of nanoparticulate gene delivery systems include their lower transfection efficiency as compared to the viral vectors. However, in terms of safety and lack of immunogenicity and mutagenicity, the nanoparticulate delivery systems are found to be superior to their viral counterparts. Various types of nanoparticles have been successfully applied for gene therapy and a great number of them are already in various phases of clinical trials (Basarkar and Singh 2007).

Studies have showed that administration of chimeric plasmid encoding both IL-4 and IL-10 resulted in synergistic effect in prevention of autoimmune insulitis when administered in the form of polymeric complexes with poly[α-(4-aminobutyl)-L-glycolic acid] (Lee et al. 2003). These complexes were nanoscale in size, non-toxic, biodegradable and exhibited superior transfection efficiency than frequently used gene carrier, Poly-L-Lysine (Lim et al. 2000). Basarkar and Singh have successfully demonstrated the ability of IL-10 gene therapy for prevention of autoimmune diabetes using cationic nanoparticles in mice (Basarkar and Singh 2009). The cationic nanoparticles prepared by double emulsion/

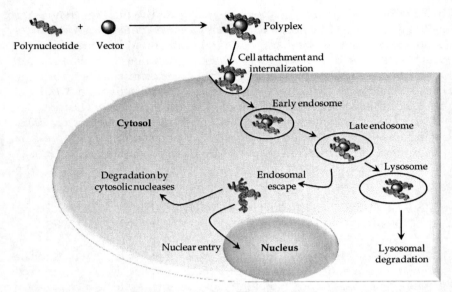

Fig. 3. Barriers to the nanoparticle-based gene delivery. Schematic diagram of the barriers encountered by nanoparticle-based vector at cellular level towards efficient gene delivery.

solvent evaporation technique were non-toxic and biocompatible (Fig. 4, Basarkar et al. 2007). The ability of these nanoparticles to deliver the genes of interest to the target cells was demonstrated both, *in vitro* and *in vivo*. Most importantly, when injected intramuscularly, these nanoparticles could deliver the IL-10 genes to myocytes, leading to sustained production of IL-10 thereby preventing the autoimmune insulitis and associated hyperglycemia (Fig. 5 and 6). Our studies with chitosan-based polymeric nanoparticles/nanomicelles indicate that these polymeric systems are biocompatible, non-toxic and exhibit excellent transfection efficiencies. A bicistronic plasmid DNA encoding IL-4 and IL-10 complexed with these nanoparticulate delivery systems, when administered intramuscularly could prevent insulitis and hyperglycemia in MLD-STZ induced diabetic mice model (Mandke and Singh, Submitted manuscript).

Key Facts of Immune-mediated Diabetes

- Autoimmune destruction of pancreatic β-cells which gradually hampers the insulin-production capacity of the pancreas is known as autoimmune or type 1A diabetes.
- It is estimated that in US alone, more than 11,000 patients are affected by type 1A diabetes annually and require lifelong therapy.

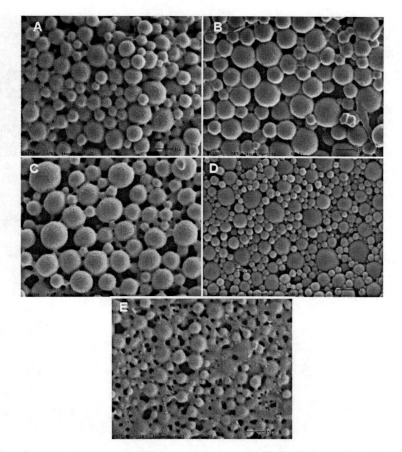

Fig. 4. Morphology of cationic PLGA nanoparticles used for IL-10 gene delivery. Scanning electron micrographs of PLGA nanoparticles containing 25% (A), 40% (B), 50% (C), 60% (D), and 75% (E), E100, a cationic surfactant. Progressive deterioration in particle morphology was observed with increasing concentrations of E100 in the formulation. (Reproduced from Basarkar 2007).

- The pathogenesis of immune-mediated diabetes involves a complex interplay of various genetic, immunological and environmental factors.
- IL-10 gene therapy, either alone or combination with other cytokine gene therapy, can successfully prevent autoimmune destruction of pancreatic islets in murine models.

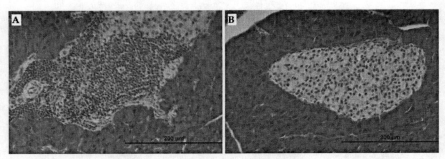

Fig. 5. Protective ability of IL-10 gene therapy in pancreas of Streptozotocin-treated mice. Light micrographs of histological section of pancreas from (A) streptozotocin-treated animal indicating infiltration by the immune cells and (B) streptozotocin-treated animal, also treated with IL-10 plasmid loaded on PLGA/50%E100 nanoparticles at a dose of 50μg of DNA exhibiting protection of pancreatic islets from immune destruction (Reproduced from Basarkar 2007).

Color image of this figure appears in the color plate section at the end of the book.

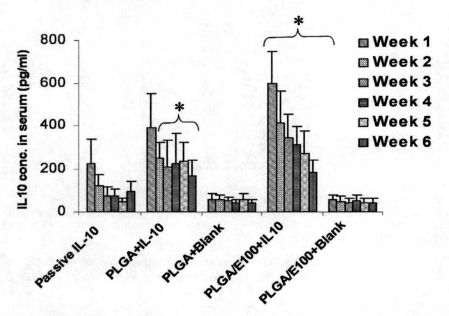

Fig. 6. Serum Interleukin-10 levels in mice treated with IL-10 plasmid using PLGA nanoparticles. Animals were injected with IL-10 plasmid passively or loaded onto nanoparticles at a dose of 50μg. Mean ± SD (n=6) are presented. A * indicates that the values are statistically greater than those of passive group for same time point (Reproduced from Basarkar 2007).

Key Facts of Interleukin-10

- It is a pleiotropic cytokine produced by a variety of cells associated with immunological functionalities, most importantly T_H2 type of helper T cells.
- It has dose-dependent immunosuppressive properties and can downregulate cell mediated immune and inflammatory response.
- IL-4 is known to have an additive effect on the immunosuppressive action of IL-10.
- Due to its short plasma half life (~2 min), a sustained delivery or a gene therapy option is most suitable.

Key Facts of Nanoparticulate Delivery Systems

- For various delivery applications, nanoparticles offer distinct formulation advantages such as small size, large surface area, greater cellular uptake and ability to cross BBB, availability of various fabrication materials and amenability for modifications that can render them target specific.
- Nanoparticles can be used for delivery of proteins and genes with better therapeutic outcome.
- Nanoparticulate delivery systems can offer distinct advantages over viral vectors for gene delivery due to their overall superior safety profile.
- In general, IL-10 gene therapy, administered using nanoparticulate delivery systems can be an attractive approach in preventing immune-mediated diabetes.

Applications to Area of Health and Disease

Autoimmune diabetes is characterized by β-cell-specific immune response leading to tissue destruction and loss of insulin production ability of the body. The associated diseases such as celiac disease, Addison's disease, thyroid disease, and pernicious anemia; that have been correlated with incidences of autoimmune diabetes; are some of the major chronic diseases in the world. These complications require long-term therapies and have major socio-economic impact.

Insulin resistance and allergy are among the most widely observed complications in type 1A diabetes. It is known that irrespective of the type of insulin used, long term insulin therapy leads to production of autoantibodies against insulin. High titer of these antibodies necessitates the use of higher insulin doses. Such complications coupled with high (and increasing) prevalence of type 1A diabetes make it a major candidate for newer therapies. Delivery of genes encoding IL-10 can be one such candidate therapy for prevention of autoimmune diabetes. Such gene therapy, administered with nanoparticulate delivery systems can be effective, especially when combined with other cytokine gene therapies.

Summary Points

- Type 1A diabetes is an autoimmune disorder that presently requires lifelong therapy.
- Parenteral insulin delivery currently forms the mainstay of diabetes management and is the only approved form of therapy.
- The cost, efficacy and compliance issues in the insulin therapy necessitate the search for alternate therapies for type 1A diabetes.
- Various strategies including islet transplantation, GAD therapy and cytokine gene therapy have been investigated for prevention/ management of type 1A diabetes.
- IL-10 is a potent anti-inflammatory cytokine that has been reported to prevent/slow down the progression of several autoimmune disorders including type 1A diabetes.
- The short plasma half life of IL-10 warrants the use of plasmid DNA for in situ production of IL-10.
- Nanoparticles have been shown to be safe and efficient vehicles for plasmid DNA delivery through various routes.
- The IL-10 gene delivery using nanoparticles has shown promise of preventing autoimmune diabetes in several preclinical studies.

Abbreviations

APC	:	Antigen presenting cells
BB	:	Bio-breeding
BBB	:	Blood-brain barrier
GAD	:	Glutamic acid decarboxylase
HLA	:	Human leukocyte antigen
HSP	:	Heat shock protein
ICAM	:	Inter-cellular adhesion molecule
IDDM	:	Insulin dependent diabetes mellitus
IFN	:	Interferon
Ig	:	Immunoglobulin
IL	:	Interleukin
MHC	:	Major histocompatibility complex
NOD	:	Non-obese diabetic
PEG	:	Polyethylene glycol
STZ	:	Streptozotocin
TNF	:	Tumor necrosis factor

References

Basarkar, A., D. Devineni, R. Palaniappan and J. Singh. 2007. Preparation, characterization, cytotoxicity and transfection efficiency of poly(DL-lactide-co-glycolide) and poly(DL-lactic acid) cationic nanoparticles for controlled delivery of plasmid DNA. Int. J. Pharm. 343: 247–254.

Basarkar, A. 2007. In vitro and in vivo gene delivery using cationic nanoparticles. Ph.D. Thesis, North Dakota State University, Fargo, North Dakota, USA.

Basarkar, A., and J. Singh. 2007. Nanoparticulate systems for polynucleotide delivery. Int. J. Nanomed. 2: 353–360.

Basarkar, A., and J. Singh. 2009. Poly (lactide-co-glycolide)-polymethacrylate nanoparticles for intramuscular delivery of plasmid encoding interleukin-10 to prevent autoimmune diabetes in mice. Pharm. Res. 26: 72–81.

Borish, L.C., and J.W. Steinke. 2003. Cytokines and chemokines. J. Allergy Clin. Immunol. 111: S460–S475

Brendel, M., B. Hering, A. Schulz and R. Bretzel. 1999. International Islet Transplant Registry Report. Justus-Liebis University of Giessen. 1–20.

Carter, J.D., J.D. Ellett, M. Chen, K.M. Smith, L.B. Fialkow, M.J. McDuffie, K.S. Tung, J.L. Nadler and Z. Yang. 2005. Viral IL-10-mediated immune regulation in pancreatic islet transplantation. Mol. Ther. 12: 360–368.

Chernajovsky, Y., D.J. Gould and O.L. Podhajcer. 2004. Gene therapy for autoimmune diseases: Quo vadis? Nat. Rev. Immunol. 4: 800–811.

Cox, N.J., B. Wapelhorst, V.A. Morrison, L. Johnson, L. Pinchuk, R.S. Spielman, J.A. Todd and P. Concannon. 2001. Seven regions of the genome show evidence of linkage to type 1 diabetes in a consensus analysis of 767 multiplex families. Am. J. Hum. Genet. 69: 820–830.

Devendra, D., and G.S. Eisenbarth. 2003. Immunologic endocrine disorders J. Allergy Clin. Immunol. 111: S624–S636.

Diabetes Prevention Trial–Type 1 Diabetes Study Group. 2002. Effects of insulin in relatives of patients with type 1 diabetes mellitus. N. Engl. J. Med. 346: 1685–1691.

Elias, D., and I.R. Cohen. 1994. Peptide therapy for diabetes in NOD mice. Lancet 343: 704–706.

Gao, X., K. Kim and D. Liu. 2007. Nonviral gene delivery: what we know and what is next. AAPS J. 9: E92–E104.

Goudy, K., S. Song, C. Wasserfall, Y.C. Zhang, M. Kapturczak, A. Muir, M. Powers, M. Scott-Jorgensen, M. Campbell-Thompson, J.M. Crawford, T.M. Ellis, T.R. Flotte and M.A. Atkinson. 2001. Adeno-associated virus vector-mediated IL-10 gene delivery prevents type 1 diabetes in NOD mice. Proc. Natl. Acad. Sci. USA 98: 13913–13918.

Hill N., and N. Sarvetnick. 2002. Cytokines: promoters and dampeners of autoimmunity. Curr. Opin. Immunol. 14: 791–797.

Jacobson D.L., S.J. Gange, N.R. Rose and N.M.H. Graham. 1997. Epidemiology and estimated population burden of selected autoimmune diseases in the united states. Clin. Immunol. Immunop. 84: 223–243.

Kawakami, S., Y. Higuchi and M. Hashida. 2008. Nonviral approaches for targeted delivery of plasmid DNA and oligonucleotide. J. Pharm. Sci. 97: 726–745.

Kawasaki, E., N. Abiru and K. Eguchi. 2004. Prevention of type 1 diabetes: from the view point of β cell damage. Diabetes Res. Clin. Pr. 66: S27–S32.

Lee, M., J.J. Koh, S.O. Han, K.S. Ko and S.W. Kim. 2002. Prevention of autoimmune insulitis by delivery of interleukin-4 plasmid using a soluble and biodegradable polymeric carrier. Pharm. Res. 19: 246–249.

Lee, M., K.S. Ko, S. Oh and S.W. Kim. 2003. Prevention of autoimmune insulitis by delivery of a chimeric plasmid encoding interleukin-4 and interleukin-10. J. Control Release 88: 333–342.

Li, L., Z. Yi, R. Tisch and B. Wang. 2008. Immunotherapy of type 1 diabetes. Arch. Immunol. Ther. Exp. 56: 227–236.

Lim, Y.B., S.O. Han, H.U. Kong, Y. Lee, J.S. Park, B. Jeong and S.W. Kim. 2000. Biodegradable polyester, poly[alpha-(4-aminobutyl)-L-glycolic acid], as a nontoxic gene carrier. Pharm. Res. 17: 811–816.

Milligan, E.D., S.J. Langer, E.M. Sloane, L. He, J. Wieseler-Frank, K. O'Connor, D. Martin, J.R. Forsayeth, S.F. Maier, K. Johnson, R.A. Chavez, L.A. Leinwand, and L.R. Watkins. 2005. Controlling pathological pain by adenovirally driven spinal production of the anti-inflammatory cytokine, interleukin-10. Eur. J. Neurosci. 21: 2136–2148.

Sanjeevi, C.B. and A. Falorni, I. Kockum, W.A. Hagopian and A. Lernmark. 1996. HLA and glutamic acid decarboxylase in human insulin-dependent diabetes mellitus. Diabet. Med. 13: 209–217.

Schloot, N.C., P. Hanifi-Moghaddam, C. Goebel, S.V. Shatavi, S. Flohe´, H. Kolb and H. Rothe. 2002. Serum IFN-γ and IL-10 levels are associated with disease progression in non-obese diabetic mice. Diabetes Metab. Res. Rev. 18: 64–70.

Sloane, E.M., R.G. Soderquist, S.F. Maier, M.J. Mahoney, L.R. Watkins and E.D. Milligan. 2009. Long-term control of neuropathic pain in a non-viral gene therapy paradigm. Gene Ther. 16: 470–475.

Tarner, I.H., and C.G. Fathman. 2001. Gene therapy in autoimmune disease. Curr. Opin. Immunol. 13: 676–682.

The expert committee on the diagnosis and classification of Diabetes mellitus. 2002. Report of the expert committee on the diagnosis and classification of diabetes mellitus. Diabetes Care. 25: S5–S20.

Vehik, K., and D. Dabelea. 2011. The changing epidemiology of type 1 diabetes: why is it going through the roof? Diabetes Metab. Res. Rev. 27: 3–13.

Yoon, S.I., B.C. Jones, N.J. Logsdon and M.R. Walter. 2005. Same structure, different function: Crystal structure of the Epstein-Barr virus IL-10 bound to the soluble IL-10 R1 chain. Structure. 13: 551–564.

Zhang, Z.L., S.X. Shen, B. Lin, L.Y. Yu, L.H. Zhu, W.P. Wang, F.H. Luo and L.H. Guo. 2003. Intramuscular injection of interleukin-10 plasmid DNA prevented autoimmune diabetes in mice. Acta. Pharmacol. Sin. 24: 751–756.

15

In vivo MR Imaging of Nanolabeled Diabetogenic T cells

Amol Kavishwar,[1,a] *Zdravka Medarova*[1,b] *and Anna Moore*[1,c,*]

ABSTRACT

Chronic infiltration of pancreatic islets by autoreactive CD4+ and CD8+ T cells (also termed insulitis) results in specific and complete destruction of insulin producing beta cells, marking the onset of Type 1 diabetes. Because of the severity of the disease, the ability to track autoreactive lymphocytes *in vivo* is a long-sought goal of diabetes researchs. Recent advancements in MRI and the introduction of higher magnetic strength small animal imaging platforms has made it possible to track single cells in deep organs in rodents (10–100 microm voxel resolution). This chapter describes the use of MRI and magnetic nanoparticles to track the migration and homing of diabetogenic T cells.

---

[1]Molecular Imaging Laboratory, MGH/MIT/HMS Athinoula A. Martinos Center for Biomedical Imaging, Massachusetts General Hospital/Harvard Medical School, Building 75, 13th Street, Charlestown, MA 02129.
[a]E-mail: amol@nmr.mgh.harvard.edu
[b]E-mail: zmedarova@partners.org
[c]E-mail: amoore@helix.mgh.harvard.edu
*Corresponding author
List of abbreviations after the text.

Application to Areas of Health and Disease

Diabetes mellitus represent a metabolic disorder, which originates from defects in insulin secretion and/or action. Type 1 diabetes results in autoimmune destruction of insulin-producing beta-cells. Type 2 diabetes is characterized by inadequate/impaired insulin production. Both conditions lead to elevated blood glucose (hyperglycemia). Diabetes is a chronic disease for which there is no cure. During its progression, diabetes can lead to devastating cardiovascular, retinal, kidney, and neural complications.

Molecular imaging is a fast growing field of non-invasive methods for detection and monitoring of various pathological conditions. It offers the exceptional potential to resolve the complex natural history of the disease and to permit diagnosis at the earliest contributory stages, characterized by the first signs of metabolic or molecular abnormality. Furthermore, molecular imaging permits the combination of the data on anatomical/physiologic level obtained by currently available modalities with the detailed molecular/cellular data from biochemistry and cell and molecular biology. In the context of diabetes it can provide answers to many of the questions related to the natural history of the disease and its pathophysiology. Furthermore, it allows the noninvasive monitoring of diabetes progression in real time as well as response to therapy non-invasively and in natural physiologic environments. Currently, molecular imaging is being explored in many areas of diabetes research including assessment of beta-cell mass, imaging of autoimmune attack in type 1 diabetes, evaluation of diabetes-associated beta-cell death, imaging of islet vasculature and imaging of islet transplantation among others.

Key Facts of Magnetic Nanoparticle Labeled Cells

- The development of noninvasive methods to image the diabetic pancreas can dramatically improve the diagnosis and characterization of diabetes mellitus. MRI provides high spatial resolution, unlimited depth penetration through tissues, high tissue contrast and tomographic capability. Thus it is capable of providing high-quality three dimensional anatomical images of the body.
- For monitoring cell migration by MRI, cells can be labeled with magnetic nanoparticles by simple incubation prior to transplantation.
- For monitoring cells already present in the body (such as in the case of insulitis) nanoparticles can be injected systemically with the purpose of labeling cells at the site of their accumulation and/or in circulation.

- Labeling can be greatly increased by modifying the surface of the particle with a targeting moiety. For example, magnetic nanoparticles decorated with membrane-translocation peptides significantly increase their uptake by the cells.
- Magnetic nanoparticles represent negative contrast agents, which produce a void of signal intensity on MR images upon their accumulation in the cells/tissue of interest.
- Diabetogenic T-cells can be labeled with magnetic nanoparticles prior to transplantation and monitored for their migration to pancreatic islets in an adoptive transfer model.
- Diabetogenic CD8+ T cells that have already accumulated in the islets can be detected after systemic administration of magnetic nanoparticles coated with a peptide/MHC complex, which is recognized by the T cell receptor.

THE PATHOGENESIS OF TYPE 1 DIABETES

Type 1 diabetes is characterized by the loss of tolerance to islet beta cells causing autoreactive CD4+ and CD8+ T-cells to attack and ultimately destroy insulin producing beta cells. Evidence for involvement of T cells came from histological studies showing presence of CD4+, CD8+ T-cells and macrophages in islets (Tsai et al. 2008). Infiltration, migration and homing of T-lymphocytes to islet beta cells is a multistep process that includes recirculation from blood to tissue, to lymph nodes, and back to blood (Springer 1994). In terms of immune recognition, the most likely sequence of events involves initial acquirement of beta-cell autoantigens by antigen presenting cells (APCs) in the islets of Langerhans, followed by activation and migration of APCs to pancreatic lymph nodes. Next, activated APCs encounter naïve T lymphocytes present in the immediately draining pancreatic lymph nodes, which leads to T lymphocyte activation and their subsequent migration to the islets where they re-encounter beta-cell derived autoantigen and are retained (Mathis et al. 2001). This infiltration is a gradual process that begins long before the onset of diabetes, persists for several years in humans and then declines when most of the beta cell mass is destroyed (Gepts 1965; Bottazzo et al. 1985). Testing for autoantibody in combination with genetic testing can identify individuals at risk, but it is not yet possible to clinically identify insulitis as the first step. Biopsy does not necessarily provide the required information (Imagawa et al. 2001) and due to its invasive nature, has poor patient compliance. Therefore, the early detection with subsequent monitoring of mononuclear cell infiltration of the pancreas non-invasively in real-time would represent a significant step towards identifying initial events leading to beta-cell destruction.

Furthermore, non-invasive methods would allow for investigation of the mechanism behind the pathology and would ultimately lead to the means to monitor patients and establish individualized therapeutic regimens.

CELL TRACKING BY MRI

In vivo imaging techniques have gained significant interest in recent years from the research and clinical community due to their non-invasiveness, versatility and ability to monitor information on the cellular and molecular level. In that context, cell tracking has been used in various pathologies included cancer, neurologic disorders and others. Magnetic resonance imaging has been the most used modality for cell tracking in intact organisms. With the introduction of high-field strength small animal MR scanners it is now possible to achieve high spatial resolution on the order of 10–100 μm voxel. This permits tracking single cells in deep organs in rodents provided that cells are labeled with MR contrast agent (Heyn et al. 2005; Heyn et al. 2006; Shapiro et al. 2006).

Superparamagnetic iron oxide nanoparticles (SPIO) and their modifications are probably the most promising contrast for cell tracking. Nanoparticles are usually composed of a superparamagnetic core of maghemite and/or magnetite ranging in size from 2–3 nm in diameter to well over 300 nm. Particles in the range of 10–50 nm are ideal for extravasation and cellular uptake studies and owing to their small size are rapidly cleared by the kidneys. When placed in a magnetic field, the magnetic moment of iron oxide nanoparticles aligns with the direction of the external field and enhances the magnetic flux eliciting disturbance that promotes rapid dephasing of surrounding protons. This reduction of spin-spin relaxation time (T2) is used to generate contrast, which appears as a dark area on MR images. These bare SPIO's are often coated with starch, dextran, peptides, fluorescent dyes etc that bestow them with properties such as fluorescence, selectivity towards particular cell or tissue and make them more biocompatible. Besides adding specificity, these coatings also increase the size of monocrystalline SPIO, preventing their rapid renal clearance and in turn increasing their targeting-capacity (Thorek et al. 2006).

Cells can be labeled with iron oxides for *in vivo* tracking over prolonged periods of time, due to the persistence of the label inside cells and the stability of the resultant signal (Bulte and Kraitchman 2004). Various iron oxides have been used for cell labeling by fluid phase endocytosis (Schulze et al. 1995; Moore et al. 1997; Weissleder et al. 1997; Schoepf et al. 1998) and/or receptor mediated endocytosis (Shen et al. 1996; Moore et al. 1998). The labeling efficiency of the nanoparticles can be increased by the addition of a targeting or membrane translocation moiety such as Tat peptide, a cell-penetrating peptide (GRKKRRQRRRPQ) adopted from the HIV-1 transactivator of transcription (TAT) protein. The peptide has been shown to translocate a variety of fluorescent probes, peptides, proteins and pathogenic epitopes across the cell membrane. A variety of cells have been labeled and detected with Tat-labeled iron oxides (CLIO-Tat; crosslinked iron oxide nanoparticles modified with Tat) including normal splenocytes, neuroprogenitor cells and CD34+ stem cells (Lewin et al. 2000). Further, superparamagnetic iron oxide nanoparticles have been successfully used to label neural progenitor cells (Bulte et al. 2001; Magnitsky et al. 2005), embryonic stem cells (Arai et al. 2006), dendritic cells (de Vries et al. 2005), bone marrow-derived endothelial precursors (Anderson et al. 2005), and mesenchymal stem cells (Hill et al. 2003; Kraitchman et al. 2003; Bulte et al. 2005).

Cell migration is another area where MRI has been employed for tracking of labeled lymphocytes. Specifically, MRI has been used to detect the migration of autoreactive lymphocytes in models of multiple sclerosis (Anderson et al. 2004), AIDS (Sundstrom et al. 2004), and cancer (Kircher et al. 2003). In this chapter we will concentrate on recent advances in the application of superparamagnetic iron oxide nanoparticles and their derivatives for visualization of diabetogenic T-cells using MR imaging.

CELL TRACKING OF DIABETOGENIC LYMPHOCYTES

Other Modalities

Previous limited attempts to track diabetogenic lymphocytes were accomplished by different modalities. As such, accumulation of radiolabeled autologous lymphocytes was demonstrated in pancreata of diabetic patients by nuclear imaging (Kaldany et al. 1982). Indirect methods were used by Signore et al. where radiolabeled cytokines such as IL-2 were used for tracking of autoimmune infiltration in diabetes by scintigraphy (Signore et al. 1994; Signore et al. 2003). Serious drawbacks of nuclear imaging methods include the short half-life of radioisotopes and potential toxic effects of radiation if long-half-life radiotracers are used.

Optical imaging was also used for lymphocyte detection and monitoring and involved fluorescent labeling of T-lymphocytes and the visualization of their homing to islet grafts through a body window device (Bertera et al. 2003). However, this modality does not allow for clinical translation since no optical imaging systems are available for patients at this time.

MRI

Magnetic resonance imaging with its high spatial resolution and availability of imaging probes is ideally suited for the purposes of visualization of diabetogenic lymphocytes. Diabetic lesions in islets of humans and non-obese diabetic (NOD) mice are populated by CD8+ T-cells that play a critical role in the pathogenesis of type 1 diabetes (Tsai et al. 2008). In our initial experiments we attempted to show that lymphocytes could be labeled with contrast agent with enough efficiency for MRI detection. To demonstrate that, lymphocytes from 18–20 week old diabetic NOD mice were labeled with CLIO-Tat. In this preparation, Tat peptide was also conjugated to a fluorescent dye for correlative microscopy. These labeled cells were then adoptively transferred to NOD.scid mice by tail vein injection (Moore et al. 2002). NOD.scid mice were selected as recipients because these animals do not develop diabetes due to the *scid* mutation, which prevents maturation of endogenous lymphocytes. The autoimmune response observed in the islets of these mice are only due to the transferred lymphocytes.

Twenty-four hours after transfer the pancreas was removed and processed for *ex vivo* MRI and histology. MR imaging of excised pancreas revealed signal voids in pancreatic islets signifying the presence of lymphocytes labeled with superparamagnetic contrast agent. Dithizone staining, which is used to reveal pancreatic islets in the tissue co-localized with the signal voids in side-to-side comparison images (Fig. 1) (Moore et al. 2002). This study confirmed that labeling of diabetogenic lymphocytes with iron oxide nanoparticles modified with a membrane translocation peptide resulted in highly efficient uptake of magnetic particles without compromising cell viability and function. Labeling of these cells allowed for the visualization of their infiltration in the islets with high-resolution MRI systems. These first results set the path to future studies showing that direct visualization of insulitis as the first stage in diabetes development could be possible and that it could serve as a new surrogate endpoint in the clinical management of diabetes onset.

The first *in vivo* studies on tracking lymphocytes to the pancreas were performed by Billotey et al. (Billotey et al. 2005). This group used anionic magnetic nanoparticle (AMNP) for cell labeling. The negative charge on the surface of these particles allowed for rapid uptake of the particles by cells. Transmission electron microscopy confirmed the presence of nanoparticles in the cells. Similar to our previously mentioned study, this study also involved isolation of diabetogenic T-cells from NOD mice and their *ex vivo* labeling with magnetic nanoparticles. The labeled cells were then adoptively transferred to irradiated NOD mice. Irradiation of the NOD mouse reduces the number of regulatory T-cells, permitting adoptively

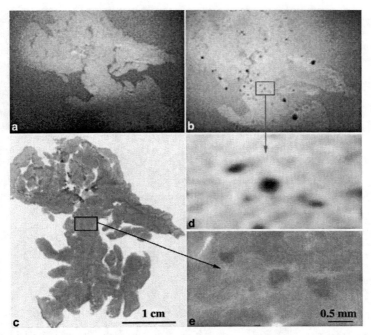

Fig. 1. MRI of the pancreas after adoptive transfer of diabetic splenocytes labeled with CLIO-Tat into NOD.scid mice. T_1-weighted image showing physiological details of the pancreas (a), T_2-weighted image, showing infiltration of labeled cells into pancreatic islets (b). Dithizone stained pancreas (c). Correlation between MR images (d) and staining for DTZ (e) shows the presence of infiltrated cells labeled with CLIO-Tat (dark dots on d) in the pancreatic islets (purple DTZ staining on e). Magnification bars: a,b,c, 1 cm; d,e, 0.5 mm. Reproduced by permission of John Wiley and Sons from Magnetic Resonance in Medicine, 2002, 47(4): 751–758.

Color image of this figure appears in the color plate section at the end of the book.

transferred cells to migrate and home to the pancreatic islets. Diabetes in these animals was detected 20 days later indicating that labeling did not cause any impairment in cell function. MRI studies performed on the 11th day after intravenous transfer of labeled T-cells revealed dark spots in the caudal region of the pancreas in recipient animals confirming migration and homing of the cells before the onset of diabetes (Fig. 2). This study demonstrates that similar to CLIO-Tat, AMNP can also be used to label T-cells followed by their detection by *in vivo* MRI (Billotey et al. 2005).

The two proof-of-principle studies listed above utilized non-specific labeling of diabetogenic lymphocytes with the purpose of demonstrating the possibility of tracking cells to the pancreas *in vivo*. At the same time, the pool of diabetogenic lymphocytes consists of multiple specificities. These specificities target autoantigens presented on the surface of beta-cells in

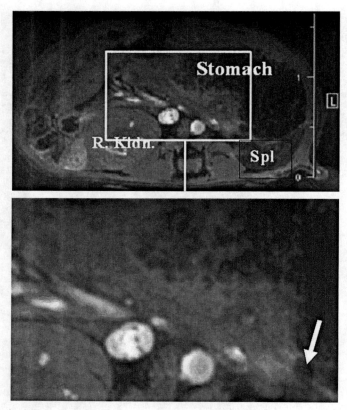

Fig. 2. *In vivo* MR images obtained in a mouse 11 days after transfer of labeled T cells (top). Transverse view (13/4) centered on pancreas (white box) with corresponding magnified view (bottom) 13/4. Arrow corresponds to lymph nodes close to the caudal part of the pancreas. Reproduced by permission of the Radiological Society of North America from Radiology, 2005, 236: 579–587.

the context of MHC Class I proteins during autoimmune attack. Many autoantigens have been already identified in mice and humans (Tsai et al. 2008). Clearly, the ability to label T cells involved in autoantigen recognition could discriminate between autoreactive and irrelevant lymphocyte specificities as well as between specificities within the diabetogenic pool itself.

Recent studies in mice and humans have identified an islet specific glucose-6-phosphatase catalytic subunit-related protein (IGRP) as an autoantigen that is processed and presented in the context of class I MHC as a peptide$_{206-214}$. This peptide-MHC complex is recognized by H-2Kd restricted CD8+ T cells carrying a Vα17-Jα42 T-cell receptor in NOD mice (Tsai et al. 2008). Based on this knowledge our group has designed a magnetic nanoparticle that carries this NOD-related peptide (NRP-V7, KYNKANVFL (Lieberman et al. 2003)) complexed on H-2Kd class 1 MHC (Moore et al. 2004). In this case iron oxide nanoparticles were conjugated to FITC-avidin to provide linkage to biotinylated MHC-I-NRP-V7 peptide (Fig. 3).

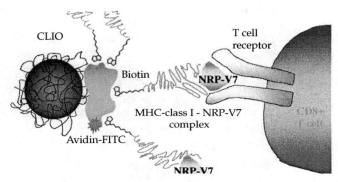

Fig. 3. Schematic representation of CLIO-NRP-V7 probe. Biotinylated NRP-V7/H-2Kd complex was coupled to CLIO particles modified with avidin. CLIO-NRP-V7 probe is recognized by the TCR on NRP-V7–reactive CD8+ T-cells. Copyright 2004 American Diabetes Association. From Diabetes, Vol. 53, 2004; 1459–1466. Reproduced by permission of The American Diabetes Association.

Color image of this figure appears in the color plate section at the end of the book.

The specificity of the CLIO-NRP-V7 probe was tested by incubating CD8+ T cells from NOD mice and 8.3-NOD mice (transgenic mice expressing NRP-V7 specific TCR on NOD background) with CLIO-NRP-V7 and analyzing them by flow cytometry. The probe labeled almost 92% of CD8+ T-cells in 8.3-NOD mice whereas only 1.83% CD8+ T-cells from NOD mice took the label. This showed that the probe was specific for NRP-V7 reactive CD8+ T-cells. Almost 90% of the peripheral blood CD8+ T-lymphocytes were labeled with the probe confirming that the avidin-biotin conjugation did not compromise its avidity for CD8+ T-cells. Fluorescence microscopy studies revealed that most of the probe was located inside the cells. Internalization was specific because no uptake was observed with CLIO-avidin alone or with an irrelevant conjugate (Moore et al. 2004).

For *in vivo* MRI studies, the NOD mouse model of diabetes was selected. First, CD8+ T-cells from 8.3-NOD mice were isolated and incubated with CLIO-NRP-V7 followed by adoptive transfer into 5 week old NOD. scid mice. The changes in signal intensity on T2-weighted images were already noticeable on day 2 after the transfer. The tissue became darker over time as more and more cells migrated and accumulated in pancreas.

The decrease in signal intensity continued up until the 16th day after the transfer. Fluorescence microscopy performed on pancreatic sections from these mice confirmed the presence of CD8+ T-cells in the pancreas of the recipient mice. This study served as the first demonstration of imaging autoimmune response by MRI in live animals.

The above-described study was performed using diabetogenic CD8+ T cells labeled *in vitro* prior to adoptive transfer. However, for clinical translation of these studies it is necessary to deliver the probe to CD8+ T cells that have already accumulated in the pancreas. Therefore, as the next step we performed studies with intravenously injected probe. We expected that magnetic nanoparticle coated with NRP-V7-peptide would accumulate in pancreatic islets (and potentially in pancreatic lymph nodes) and produce a drop in regional T2 relaxation time with a corresponding loss of signal on MR images. As a result, by monitoring changes in signal intensity in the expected regions, it would be possible to derive quantitative information on the accumulation of IGRP-reactive CD8+ T-cells.

To this end we injected the probe intravenously in 5, 8, 15 and 24 weeks NOD old mice followed by MRI to determine the recruitment of diabetogenic CD8+ T-cells carrying TCR specific for NRP-V7 peptide (Medarova et al. 2008). Control mice of the same age groups were injected with unmodified parental magnetic nanoparticles. A sufficient contrast-to-noise ratio (difference in T2 relaxation before and after injection of contrast agent) was observed on T2 weighted MR images allowing for semi-quantitative estimation of probe accumulation. Animals injected with unmodified parental magnetic nanoparticles did not show any significant difference between pre and post-contrast T2 values. This suggested that the drop in signal observed with magnetic nanoparticles coated with NRP-V7-peptide is a function of their targeted nature. The progression of diabetes is a gradual process. In NOD mice, insulitis begins as early as 4 wks of age and continues until more than 90% of the beta cell mass is destroyed. At this point (10–12 wks of age) animals become overtly diabetic. This progression was captured on MR images of mice of different ages. A small change of 1±1.2% in T2 was observed for 5 wks old NOD mice consistent with lower levels of insulitis. The drop in T2 was pronounced the most in 8 week old mice amounting to more than 18% and consistent with the highest insulitic score at this age. As animals grew older the differences in pre and post-contrast T2 were ~6% and 4% respectively. The smaller difference in the 15 and 24 week old groups suggests that after 15 weeks of age, when more than 90% of beta cell mass is already destroyed, there is an overall reduction in the recruitment of $IGRP_{206-214}$ reactive CD8+ T-cells to the islets and hence less reduction of the signal on T2 weighted MR images (Fig. 4 (Medarova et al. 2008)).

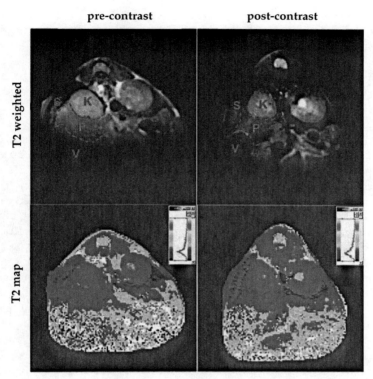

Fig. 4. MRI of NRP-V7-specific CD8+ T lymphocyte infiltration of the pancreas. Representative T_2-weighted image (top) and its corresponding multiecho T_2 map (bottom) before (left) and 24 h after (right) intravenous injection of MN-NRP-V7. The pancreas is outlined in red. V = stomach, S = spleen, K = kidney, P = pancreas. Reproduced by permission of John Wiley and Sons from Magnetic Resonance in Medicine, 2008, 59(4): 712–720.

Color image of this figure appears in the color plate section at the end of the book.

Given that "insulitis" lesions consist of various different types of autoreactive CD4+, CD8+ T-cells, macrophages and other mononuclear cells, we next investigated what percentage of $IGRP_{206-214}$ reactive CD8+ T-cells were in the lesions of wild type NOD mice. Thus mice after MRI imaging were sacrificed and islet-associated CD8+ T-cells were analyzed by flow cytometry. At 5 weeks of age, only 0.5% of the islet-associated CD8+ T-cells were $IGRP_{206-214}$ reactive that grew to 4% in 8 week old mice and then was reduced to 1.5% in 15 week old mice and to less than 0.5% by 24 weeks of age. Remarkably the percentage of islet associated $IGRP_{206-214}$ reactive CD8+ T-cells from mice of the different age groups correlated very well with the drop in T2 signal on MR images in the same animals. A similar analysis of $IGRP_{206-214}$ reactive splenic and blood CD8+ T-cells was performed by flow cytometry. Interestingly, NRP-V7-peptide reactive CD8+ T-cells peaked in the blood of 7 to 8 week old mice but not in the spleen indicating that the relative frequency of these cells in circulation parallels their frequency in islet lesions. These experiments allow one to calculate the rate of recruitment of autoreactive T-cells in real time after intravenous injection of T cell-reactive magnetic nanoparticles (Medarova et al. 2008).

It is noteworthy magnetic nanoparticles decorated with peptide-MHC complexes could be used not only for imaging but for therapy as well. Recent studies from the University of Calgary (Dr. Pere Santamaria laboratory) showed that nanoparticles coated with disease-relevant peptide-major histocompatibility complexes expanded cognate autoregulatory T cells, suppressed the recruitment of noncognate specificities, prevented disease in prediabetic mice, and restored normoglycemia in diabetic animals (Tsai et al. 2010). Importantly, while these nanoparticles were first developed to label diabetogenic CD8+ T cells during diabetes development, they can also be used for other pathologies. For example, a recent study showed that magnetic nanoparticles conjugated to a specific MHC-peptide complex were capable of selectively labeling and imaging CTL cells expressing T-cell receptors (TCR) that recognize cognate MHC-peptide complexes on the surface of antigen-presenting tumor cells (Gunn et al. 2008).

Imaging of autoimmune attack in diabetes is a fast developing area with multiple outgoing studies. Various approaches are considered including the use of ^{19}F based polyfluorine nanoparticle instead of iron oxides for studying the recruitment of CD8+ T-cells in the pancreas (Srinivas et al. 2007). The advantage of ^{19}F –labeled contrast agents is in the absence of any background in the body. However its low sensitivity due to poor molecular mobility, association of fluorinated segments and low fluorine content could present potential problems. In this study cells were labeled with PFPE (perfluoropolyether) nanoparticles and adoptively transferred to NOD.scid mice. *In vivo* MR imaging demonstrated accumulation of labeled cells in the pancreas and therefore the possibility of using ^{19}F-based nanoparticles for monitoring of recruitment of autoreactive T-cells (Srinivas et al. 2007).

Conclusions

In conclusion, MR imaging of the recruitment of T cells in diabetes could serve as a prototype for further diagnostic and therapeutic studies. It is possible, for example, to create nanoparticle-MHC-peptide conjugates with multiple specificities that would allow not only imaging pancreatic inflammation in real time, but also the delivery of tolerogenic doses of many epitopes simultaneously, enabling the depletion of different autoreactive T-cell pools (such as those recognizing most IGRP epitopes) below the threshold required for diabetogenesis (Medarova and Moore 2009).

Summary Points

- Type 1 diabetes is characterized by the loss of tolerance to islet beta cells causing autoreactive CD4+ and CD8+ T-cells to attack and ultimately destroy insulin producing beta cells.
- T-cells can be loaded with magnetic nanoparticles and can be visualized by MR imaging.
- Diabetogenic T-cells localize to the pancreas and can be used to study progression of diabetogenensis.
- Tat-peptide can significantly enhance labeling of magnetic, isolated T-cells by nanoparticles.
- Magnetic nanoparticles decorated with NRP-V7 peptide can label diabetogenic T-cells *in vivo*.
- Magnetic nanoparticles are already in the clinic, which makes it easier to apply these methods to humans.

Definitions

MRI: Magnetic Resonance Imaging is an imaging technique that uses magnetic fields and radio frequency to create an image of internal organs. Because different organs and tissue have different water content, a contrast is generated that helps discern various organs and tissue, including any abnormality like tumor growth, edema, clot formation etc.

Magnetic nanoparticle: Magnetic nanoparticles consist of an iron core and stabilizing coating (usually dextran). When present in the cells/tissue of interest they create a negative contrast on MR images thus marking their position.

SPIO: Superparamagnetic iron oxide nanoparticles are composed of cores of iron, magnetite and maghemite, mostly coated with dextran that is sometimes crosslinked for stabilization. Their size can range from 2–4 nm to well over a micrometer in orally ingested contrast agents.

IDDM: Type 1 Diabetes is an autoimmune condition in which the immune cells of one's body start to attack its own cells. In this case they attack the cells that produce insulin, ultimately leading to insulin deficiency and requiring insulin injections. This condition is called insulin dependent diabetes mellitus (IDDM).

Autoimmune T cells: Autoimmune T-cells are white blood cells that mature in the thymus and hence the name T-cells. These cells sometimes attack the body's own cells and hence are called autoimmune T-cells.

CLIO-Tat: Magnetic nanoparticle coated with crosslinked dextran and labeled with a cell penetrating peptide derived from the AIDS virus "transactivator of transcription" protein. This protein has been shown to penetrate cell membranes and deliver a cargo (in this case a magnetic nanoparticle) into the cell.

PFPE nanoparticles: Contrast provided by these nanoparticles depends upon the paramagnetic nature of the fluorine atom. Because the concentration of fluorine in animals is very low, clear images are obtained with fluorine nanoparticles that are co-registered with regular proton MRI to localize their position.

H-2Kd restricted CD8+ T cells: T-cells recognize antigen through T-cell receptors (TCR). These TCR's bind to antigenic peptides that are presented by members of the major histocompatibility complex proteins or MHC. There are several groups or haplotypes of MHC in every species. One such haplotype found in mice that spontaneously develop diabetes is H-2Kd. T cells from these mice, when transferred to any other mouse strain, will induce diabetes if that strain has MHC class I H-2Kd haplotype and hence the name H-2Kd restricted T cells.

Abbreviations

IDDM	:	insulin dependent diabetes mellitus
APC	:	antigen presenting cell
MR	:	magnetic resonance
MRI	:	magnetic resonance imaging
SPIO	:	superparamagnetic iron oxide nanoparticle
Tat	:	transactivator of transcription
CLIO	:	crosslinked iron oxide nanoparticle
NOD mice	:	non obese diabetic mice
scid	:	severe combined immunodeficiency
AMNP	:	anionic magnetic nanoparticle
IGRP	:	islet specific glucose-6-phosphatase catalytic subunit-related protein
CTL	:	cytotoxic T-lymphocyte
PFPE	:	perfluoropolyether

References

Anderson, S.A., J. Glod, A.S. Arbab, M. Noel, P. Ashari, H.A. Fine and J.A. Frank. 2005. Noninvasive MR imaging of magnetically labeled stem cells to directly identify neovasculature in a glioma model. Blood 105: 420–425.

Anderson, S.A., J. Shukaliak-Quandt, E.K. Jordan, A.S. Arbab, R. Martin, H. McFarland and J.A. Frank. 2004. Magnetic resonance imaging of labeled T-cells in a mouse model of multiple sclerosis. Ann. Neurol. 55: 654–659.

Arai, T., T. Kofidis, J.W. Bulte, J. de Bruin, R.D. Venook, G.J. Berry, M.V. McConnell, T. Quertermous, R.C. Robbins and P.C. Yang. 2006. Dual in vivo magnetic resonance evaluation of magnetically labeled mouse embryonic stem cells and cardiac function at 1.5 t. Magn. Reson Med. 55: 203–209.

Bertera, S., X. Geng, Z. Tawadrous, R. Bottino, A.N. Balamurugan, W.A. Rudert, P. Drain, S.C. Watkins and M. Trucco. 2003. Body window-enabled in vivo multicolor imaging of transplanted mouse islets expressing as insulin-Timer fusion protein. Bio. Techniques 35: 718–722.

Billotey, C., C. Aspord, O. Beuf, E. Piaggio, F. Gazeau, M.F. Janier and C. Thivolet. 2005. T-cell homing to the pancreas in autoimmune mouse models of diabetes: in vivo MR imaging. Radiology 236: 579–587.

Bottazzo, G.F., B.M. Dean, J.M. McNally, E.H. MacKay, P.G. Swift and D.R. Gamble. 1985. In situ characterization of autoimmune phenomena and expression of HLA molecules in the pancreas in diabetic insulitis. N. Engl. J. Med. 313: 353–360.

Bulte, J.W., T. Douglas, B. Witwer, S.C. Zhang, E. Strable, B.K. Lewis, H. Zywicke, B. Miller, P. van Gelderen, B.M. Moskowitz, I.D. Duncan and J.A. Frank. 2001. Magnetodendrimers allow endosomal magnetic labeling and in vivo tracking of stem cells. Nat. Biotechnol. 19: 1141–1147.

Bulte, J.W., L. Kostura, A. Mackay, P.V. Karmarkar, I. Izbudak, E. Atalar, D. Fritzges, E.R. Rodriguez, R.G. Young, M. Marcelino, M.F. Pittenger and D.L. Kraitchman. 2005. Feridex-labeled mesenchymal stem cells: cellular differentiation and MR assessment in a canine myocardial infarction model. Acad. Radiol. 12 Suppl. 1: S2–S6.

Bulte, J.W., and D.L. Kraitchman. 2004. Monitoring cell therapy using iron oxide MR contrast agents. Curr. Pharm. Biotechnol. 5: 567–584.

de Vries, I.J., W.J. Lesterhuis, J.O. Barentsz, P. Verdijk, J.H. van Krieken, O.C. Boerman, W.J. Oyen, J.J. Bonenkamp, J.B. Boezeman, G.J. Adema, J.W. Bulte, T.W. Scheenen, C.J. Punt, A. Heerschap and C.G. Figdor. 2005. Magnetic resonance tracking of dendritic cells in melanoma patients for monitoring of cellular therapy. Nat. Biotechnol. 23: 1407–1413.

Gepts, W. 1965. Pathologic anatomy of the pancreas in juvenile diabetes mellitus. Diabetes 10: 619–633.

Gunn, J., H. Wallen, O. Veiseh, C. Sun, C. Fang, J. Cao, C. Yee and M. Zhang. 2008. A multimodal targeting nanoparticle for selectively labeling T cells. Small 4: 712–715.

Heyn, C., C.V. Bowen, B.K. Rutt and P.J. Foster. 2005. Detection threshold of single SPIO-labeled cells with FIESTA. Magn. Reson Med. 53: 312–320.

Heyn, C., J. Ronald, S. Ramadan, L. MacKenzie, D. Mikulis, D. Palmieri, J. Bronder, P. Steeg, T. Yoneda, I. MacDonald, A. Chambers, B. Rutt and P. Foster. 2006.

In vivo tracking of growth and dormancy of solitary cells in a mouse model of breast cancer metastasis to the brain using MRI. Magn. Reson. Med. 56: 1001–1010.

Hill, J.M., A.J. Dick, V.K. Raman, R.B. Thompson, Z.X. Yu, K.A. Hinds, B.S. Pessanha, M.A. Guttman, T.R. Varney, B.J. Martin, C.E. Dunbar, E.R. McVeigh and R.J. Lederman. 2003. Serial cardiac magnetic resonance imaging of injected mesenchymal stem cells. Circulation 108: 1009–1014.

Imagawa, A., T. Hanafusa, S. Tamura, M. Moriwaki, N. Itoh, K. Yamamoto, H. Iwahashi, K. Yamagata, M. Waguri, T. Nanmo, S. Uno, H. Nakajima, M. Namba, S. Kawata, J. I. Miyagawa and Y. Matsuzawa. 2001. Pancreatic biopsy as a procedure for detecting *in situ* autoimmune phenomena in type 1 diabetes: close correlation between serological markers and histological evidence of cellular autoimmunity. Diabetes 50: 1269–1273.

Kaldany, A., T. Hill, S. Wentworth, S.J. Brink, J.A. D'Elia, M. Clouse and J.S. Soeldner. 1982. Trapping of peripheral blood lymphocytes in the pancreas of patients with acute-onset insulin-dependent diabetes mellitus. Diabetes 31: 463–466.

Kircher, M.F., J.R. Allport, E.E. Graves, V. Love, L. Josephson, A.H. Lichtman and R. Weissleder. 2003. *In vivo* high resolution three-dimensional imaging of antigen-specific cytotoxic T-lymphocyte trafficking to tumors. Cancer Res. 63: 6838–6846.

Kraitchman, D.L., A.W. Heldman, E. Atalar, L.C. Amado, B.J. Martin, M.F. Pittenger, J.M. Hare and J.W. Bulte. 2003. *In vivo* magnetic resonance imaging of mesenchymal stem cells in myocardial infarction. Circulation 107: 2290–2293.

Lewin, M., N. Carlesso, C. Tung, X. Tang, D. Cory, D. Scadden and R. Weissleder. 2000. Tat peptide-derivatized magnetic nanoparticles allow *in vivo* tracking and recovery of progenitor cells. Nature Biotechnology 18: 410–414.

Lieberman, S.M., A.M. Evans, B. Han, T. Takaki, Y. Vinnitskaya, J.A. Caldwell, D.V. Serreze, J. Shabanowitz, D.F. Hunt, S.G. Nathenson, P. Santamaria and T.P. DiLorenzo. 2003. Identification of the beta cell antigen targeted by a prevalent population of pathogenic CD8+ T cells in autoimmune diabetes. Proc. Natl. Acad. Sci. USA 100: 8384–8388.

Mathis, D., L. Vence and C. Benoist. 2001. Beta-Cell death during progression to diabetes. Nature 414: 792–798.

Magnitsky, S., D.J. Watson, R.M. Walton, S. Pickup, J.W. Bulte, J.H. Wolfe and H. Poptani. 2005. *In vivo* and *ex vivo* MRI detection of localized and disseminated neural stem cell grafts in the mouse brain. Neuroimage 26: 744–754.

Medarova, Z., and A. Moore. 2009. MRI in diabetes: first results. AJR Am. J. Roentgenol. 193: 295–303.

Medarova, Z., S. Tsai, N. Evgenov, P. Santamaria and A. Moore. 2008. *In vivo* imaging of a diabetogenic CD8+ T cell response during type 1 diabetes progression. Magn. Reson. Med. 59: 712–720.

Moore, A., J. Basilion, E. Chiocca and R. Weissleder. 1998. Measuring transferrin receptor gene expression by NMR imaging. Biochim. Biophys. Acta. 1402: 239–249.

Moore, A., J. Grimm, B. Han and P. Santamaria. 2004. Tracking the recruitment of diabetogenic CD8+ T-cells to the pancreas in real time. Diabetes 53: 1459–1466.

Moore, A., P. Z. Sun, D. Cory, D. Hogemann, R. Weissleder and M.A. Lipes. 2002. MRI of insulitis in autoimmune diabetes. Magn. Reson. Med. 47: 751–758.

Moore, A., R. Weissleder and A. Bodganov. 1997. Uptake of dextran-coated monocrystalline iron oxides in tumor cells and macrophages. Journal of Magnetic Resonance Imaging 7: 1140–1145.

Schoepf, U., E.M. Marecos, R.J. Melder, R.K. Jain and R. Weissleder. 1998. Intracellular magnetic labeling of lymphocytes for *in vivo* trafficking studies. Biotechniques 24: 642–646.

Schulze, E., J. Ferrucci, K. Poss, L. Lapointe, A. Bogdanova and R. Weissleder. 1995. Cellular uptake and trafficking of a prototypical magnetic iron oxide label *In vitro*. Investigative Radiology 30: 604–610.

Shapiro, E.M., K. Sharer, S. Skrtic and A.P. Koretsky. 2006. *In vivo* detection of single cells by MRI. Magn. Reson. Med. 55: 242–249.

Shen, T., A. Bogdanov, A. Bogdanova, K. Poss, T. Brady and R. Weissleder. 1996. Magnetically labeled secretin retains receptor affinity to pancreas acinar cells. Bioconjugate Chemistry 7: 311–316.

Signore, A., M. Chianelli, E. Ferretti, A. Toscano, K.E. Britton, D. Andreani, E.A. Gale and P. Pozzilli. 1994. New approach for *in vivo* detection of insulitis in type I diabetes: activated lymphocyte targeting with 123I-labelled interleukin 2. Eur. J. Endocrinol. 131: 431–437.

Signore, A., A. Picarelli, A. Annovazzi, K.E. Britton, A.B. Grossman, E. Bonanno, B. Maras, D. Barra and P. Pozzilli. 2003. 123I-Interleukin-2: biochemical characterization and *in vivo* use for imaging autoimmune diseases. Nucl. Med. Commun. 24: 305–316.

Springer, T.A. 1994. Traffic signals for lymphocyte recirculation and leukocyte emigration: the multistep paradigm. Cell 76: 301–314.

Srinivas, M., P. A. Morel, L.A. Ernst, D.H. Laidlaw and E.T. Ahrens. 2007. Fluorine-19 MRI for visualization and quantification of cell migration in a diabetes model. Magn. Reson. Med. 58: 725–734.

Sundstrom, J.B., H. Mao, R. Santoianni, F. Villinger, D.M. Little, T.T. Huynh, A.E. Mayne, E. Hao and A.A. Ansari. 2004. Magnetic resonance imaging of activated proliferating rhesus macaque T cells labeled with superparamagnetic monocrystalline iron oxide nanoparticles. J. Acquir. Immune Defic. Syndr. 35: 9–21.

Thorek, D.L., A.K. Chen, J. Czupryna and A. Tsourkas. 2006. Superparamagnetic iron oxide nanoparticle probes for molecular imaging. Ann. Biomed. Eng. 34: 23–38.

Tsai, S., A. Shameli and P. Santamaria. 2008. CD8+ T cells in type 1 diabetes. Adv. Immunol. 100: 79–124.

Tsai, S., A. Shameli, J. Yamanouchi, X. Clemente-Casares, J. Wang, P. Serra, Y. Yang, Z. Medarova, A. Moore and P. Santamaria. 2010. Reversal of autoimmunity by boosting memory-like autoregulatory T cells. Immunity 32: 568–580.

Weissleder, R., C. Cheng, A. Bogdanova and A. j. Bogdanov. 1997. Magnetically labeled cells can be detected by MR imaging. Journal of Magnetic Resonance Imaging 7: 258–263.

16

Nanoparticle-Mediated Delivery of Angiogenic Inhibitors in Diabetic Retinopathy

Krysten M. Farjo,[1,a] Rafal Farjo,[2,a] Ronald Wassel[2,b] and Jian-xing Ma[1,b,]*

ABSTRACT

Diabetes is a progressive disease that may lead to complications associated with defects in angiogenesis. In the case of diabetic retinopathy (DR), retinal capillaries are damaged, causing retinal inflammation. Eventually, impaired retinal capillary function may cause retinal ischemia, which induces retinal neovascularization (RNV), a process in which new blood vessels sprout from retinal capillaries and invade retinal tissue. If left untreated, RNV causes

---

[1]Department of Physiology, University of Oklahoma Health Sciences Center, Oklahoma City, OK 73104 USA.
[a]E-mail: Krysten-farjo@ouhsc.edu
[b]E-mail: Jian-xing-ma@ouhsc.edu
[2]Charlesson LLC, Oklahoma City, OK 73104 USA.
[a]E-mail: rfarjo@charlessonllc.com
[b]E-mail: dwassel@charlessonllc.com
*Corresponding author
List of abbreviations after the text.

vascular leakage, retinal inflammation, and retinal scarring, which may result in permanent vision loss, including blindness. Thus, inhibiting RNV is essential for preserving vision in patients with DR. The pathogenesis of RNV is characterized by increased expression of angiogenic stimulators and decreased expression of angiogenic inhibitors. Vascular endothelial growth factor (VEGF) is the primary angiogenic stimulator that is up-regulated during RNV. Over the past decade, several anti-VEGF therapies have shown success in clinical trials for treating DR-induced RNV. However, in order to maintain therapeutic benefits these therapies must be administered by intravitreal (IVT) injection every 4–12 weeks. This can be problematic, since repeated IVT injections can lead to severe complications, such as endophthalmitis, cataract, retinal tears, and retinal detachment, which may delay or prevent continued treatment. Thus, there is a strong need to develop alternative drug delivery systems to eliminate or at least reduce the frequency of intravitreal injections. A variety of both natural and chemical angiogenic inhibitors can effectively inhibit RNV in animal models, and some angiogenic inhibitors have also achieved measures of success in clinical trials. In many cases, nanotechnology is being utilized to create nanoparticle formulations that improve and prolong the bioavailability and efficacy of angiogenic inhibitor therapies. This chapter reviews recent studies which utilize nanoparticles to deliver different types of angiogenic inhibitors for the treatment of DR-induced RNV.

INTRODUCTION

Angiogenesis is the process by which new blood vessels are generated from existing vasculature (Risau 1997). During diabetes, tissue-specific and systemic changes in vascular biology may occur and alter the "angiogenic capacity" of the vasculature to either enhance or inhibit angiogenesis (Frank 2004). Research studies have demonstrated that such angiogenic changes play a role in the development and progression of different types of diabetic complications. For instance, decreased angiogenic capacity contributes to impaired wound healing in diabetic patients. In contrast, excessive angiogenesis is a significant component in the pathogenesis of diabetic retinopathy (DR) and nephropathy (DN). Thus, drugs which inhibit angiogenesis could be effective to limit the progression of DR and DN, although a local drug delivery approach is most desirable in order to avoid systemically inhibiting angiogenesis and worsening wound healing problems in diabetic patients. This chapter focuses on how nanotechnology is being utilized to develop nanoparticle-based systems to deliver angiogenic inhibitors to target tissues, with a specific emphasis on delivery to the retina for the treatment of DR.

ANGIOGENESIS IN THE PATHOGENESIS OF DIABETIC RETINOPATHY

The retina is a light-sensitive neuronal tissue lining the back of the eye. It contains several types of neuronal cells, including the photoreceptor cells that respond to light stimuli by generating electrical signals that travel through other retinal neurons to the optic nerve, which then carries the signal to the brain (Fig. 1). DR is a disease that is initiated and perpetuated by aberrant changes in the retinal capillaries (Frank 2004). In early preclinical stages of DR, hyperglycemia induces oxidative stress in retinal capillary endothelial cells, which alters cell metabolism and causes thickening of the basement membrane layer that lies between the endothelial cells and their surrounding support cells, the pericytes. At this stage, there is no symptom. As the disease progresses to clinical non-proliferative DR (NPDR), retinal capillary pericytes begin to die, and microaneurysms and localized hemorrhaging occur. In some cases, blood and fluid from hemorrhages can accumulate and cause macular edema, and this may cause noticeably blurred vision. In late stages of NPDR, retinal capillaries become occluded and non-perfusion occurs, causing retinal ischemia. This strongly promotes angiogenesis to produce new blood vessels in a process called retinal neovascularization (RNV), which marks the beginning of proliferative diabetic retinopathy (PDR).

Table 1. Key Facts of Diabetic Retinopathy.

- After 20 years with diabetes, more than 75% of patients will develop some form of diabetic retinopathy (Barcelo et al. 2003).
- DR is leading cause of blindness in working-age adults, and DR accounts for 4.8% of all cases of blindness worldwide (Resnikoff et al. 2004).
- The incidence of DR will likely increase in coming years, since the number of people with diabetes is expected to double by the year 2030 (Resnikoff et al. 2004).
- Blindness due to DR is highly preventable, as patients with DR have a 90% chance of maintaining vision if properly treated (Resnikoff et al. 2004).
- Current treatment options for DR include laser photocoagulation and a limited number of angiogenic inhibitor drugs, all of which are anti-VEGF therapies that specifically target and inhibit VEGF (Abdallah and Fawzi 2009).

This table lists the key facts of the prevalence of diabetic retinopathy and current treatment options.

RNV generates abnormal and fragile blood vessels which leak blood into the retina and vitreous, perturbing light from reaching the retina (Figs. 1 and 2). This causes spotty or cloudy vision. If RNV progresses it will cause increased retinal inflammation, macular edema, retinal scarring, and in some cases retinal detachment. This may result in permanent vision loss up to the point of legal blindness. Thus, stopping RNV is crucial for preserving vision.

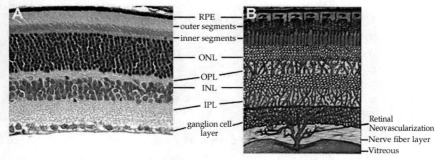

Fig. 1. Schematic representation of the retina and RNV. (A) Hematoxylin and eosin-stained transverse section of mouse retina (top: posterior, bottom: anterior). (B) Diagram of the retina which includes an illustration of RNV. RNV occurs when retinal capillaries invade the retinal tissue, primarily in the ganglion cell layer. (RPE: retinal pigment epithelium, outer segments: of photoreceptor cells, inner segments: of photoreceptor cells, ONL: outer nuclear layer of photoreceptor cell nuclei, OPL: outer plexiform layer, INL: inner nuclear layer of other retina cell types, IPL: inner plexiform layer).
Source: Authors' unpublished images.

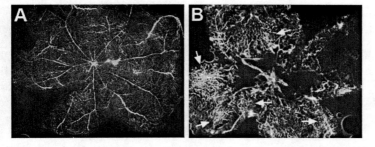

Fig. 2. Visualization of retinal neovascularization in the oxygen-induced retinopathy (OIR) mouse model. (A and B) Fluorescein angiography was performed using high molecular weight FITC-dextran, and then retinas were collected and whole-mounted onto slides for visualization using fluorescent microscopy. (A) P18 retina from a mouse that was not exposed to OIR shows the normal pattern of retinal capillaries extending from the center of the retina in a highly organized pattern. (B) P18 retina from a mouse which underwent OIR. In the OIR model, mouse pups are transferred at P7 into a hyperoxic chamber. The mice are transferred back to normoxic room air at P12, which creates relative hypoxia in the retinal tissue, leading to ischemia-induced RNV. The white arrows point out sites of RNV, and there is a large area of non-perfusion in the center of the retina due to impaired retinal capillary function.
Source: Authors' unpublished data.

In DR, pathogenic RNV results from retinal ischemia-induced activation of angiogenesis, which increases expression of angiogenic stimulators and decreases expression of angiogenic inhibitors (Gao et al. 2001). This causes the unfettered formation of abnormal blood vessels which lack integrity and leak. Vascular endothelial growth factor (VEGF) is the primary angiogenic stimulator during normal blood vessel development as well as during RNV (Penn et al. 2008). VEGF is a secreted glycoprotein which binds to VEGF receptors on endothelial cells to stimulate their proliferation and migration. VEGF expression is minimal in normal retina, but expression is increased during RNV. Clinical trials have already demonstrated that anti-VEGF therapies can effectively treat DR-induced RNV, although these therapies require repetitive IVT injections in order to sustain therapeutic benefits (Abdallah and Fawzi 2009).

The normal retina expresses high levels of the angiogenic inhibitor pigment epithelium-derived factor (PEDF) (Gao et al. 2001). PEDF is a secreted glycoprotein which has been shown to counteract the effects of VEGF, at least in part by inhibiting VEGF binding to VEGF receptors on endothelial cells (Zhang et al. 2006). PEDF expression is decreased during RNV, and the resulting increase in the VEGF/PEDF ratio is thought to play a significant role in the pathogenesis of RNV. Decreasing the VEGF/PEDF ratio by increasing PEDF expression can inhibit RNV progression. In addition, angiogenic inhibitors other than PEDF as well as pharmacological compounds, can also be beneficial for the treatment of RNV (Gao et al. 2003; Zhang et al. 2001).

THE CAVEATS OF DRUG DELIVERY TO THE RETINA

Drug delivery to the retina has always presented a significant challenge for pharmaceutical drug development. This is largely due to the blood-retinal-barrier (BRB), which functions to reduce the passage of macromolecules and inflammatory cells from the blood into the retinal tissue in order to isolate and protect the retina, similar to the blood-brain-barrier which protects the brain (Cunha-Vaz 2004). The BRB severely limits the delivery of therapeutics to the retina via systemic routes of administration. Likewise, topical drug administration via conventional eye drops or ointments is also very inefficient for targeting drugs to the retina. Hence, local delivery by intravitreal (IVT) injection is the standard procedure for delivering drugs to treat retinal disease. In the case of drug delivery by IVT injection, the BRB becomes an ally instead of an adversary, by preventing the diffusion of the drug out of the retina and into the systemic circulation, which prolongs drug bioavailability in the retina. Nevertheless, prolonged treatment of chronic retinal diseases, such as DR, necessitates that drug be redelivered by IVT injections every 1 to 3 months to sustain therapeutic effectiveness. This can become problematic, since numerous IVT injections can cause severe complications, including endophthalmitis, cataract, retinal tears, and retinal detachment (Sampat and Garg 2010), which may delay or prevent prolonged treatment regimens. Thus, more effective drug delivery systems are needed to reduce the frequency of IVT injections, and thereby reduce complications and improve treatment outcome.

UTILIZING NANOTECHNOLOGY TO IMPROVE DRUG DELIVERY

Nanotechnology is believed to have the potential to revolutionize ocular drug delivery. Nanotechnology classically refers to the development of materials in the size range of 1–100 nm, but is often extended to include any materials less than 1 µm in diameter (Zhang et al. 2008a). Biologically compatible nanosystems promise several advantages over conventional drug formulations, including the ability to enhance the solubility and bioavailability of hydrophobic and unstable drug molecules by encapsulating drugs inside nanoparticle systems that impede physical, chemical, and biological degradation mechanisms. This often leads to improved pharmacokinetics of drug metabolism and prolonged drug half-life, resulting in a net increase in therapeutic activity in the retina.

Nanoparticle Materials and Formulations

Nanoparticles (NPs) consist of 1 µm or smaller particles which may be composed of either natural or synthetic materials, including lipids, polymers, polypeptides, polysaccharides, metals, ceramics, and semiconductor materials (Fig. 3) (Pinto Reis et al. 2006). The most common lipid-based NPs are liposomes, which consist of a phospholipid bilayer membrane that encapsulates cargo molecules (Torchilin 2005). Several liposome-based drugs have been FDA-approved for clinical use. Liposomes are generally biocompatible, non-toxic, and non-immunogenic, since they are often made using naturally-occurring phospholipids. Drugs and other therapeutic agents, whether hydrophobic or hydrophilic, can be encapsulated within liposome NPs with high efficiency. Liposomes are often unstable *in vivo*, but stability and bioavailability can be improved by generating hybrid liposome-polymer NPs. Polyethylene glycol (PEG) is commonly used for this purpose.

More recently, solid lipid NPs (SLN) have been tested for ocular drug delivery applications (Viola et al. 2009). SLNs can be generated from different lipids, including triglycerides, aliphatic alcohols, polyalcohol esters, cholesterol, and cholesterol esters. Lipid NPs typically have a mean

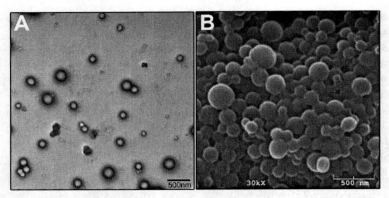

Fig. 3. Electron microscopy images of nanoparticles loaded with a small molecule angiogenic inhibitor (A) Transmission and (B) scanning electron microscopy images showing drug-loaded NPs composed of PLGA in the size range of 100–350 nm, with an average size of 200 nm. (PLGA: Poly(D,L-lactide-co-glycolide)).
Source: Authors' unpublished images.

diameter between 50–400 nm, a sufficiently small size for enhancing ocular delivery. Topical administration of SLNs in rabbits resulted in detectable delivery to the retina with no evidence of toxicity (Viola et al. 2009).

Polymer-based NPs are often generated using polylactide (PLA) and/or polyglycolide (PGA), which are often mixed to generate the copolymer Poly(D,L-lactide-co-glycolide) (PLGA) (Panyam et al. 2003). PLA and PGA can be combined in different ratios to generate PLGA-NPs which have distinct and well-characterized rates of degradation (see Figs. 3, 4, 5, and 6 for studies using PLGA-NPs). PLGA is biocompatible, biodegradable, non-toxic, and non-immunogenic, and thus, numerous PLGA-containing therapeutic agents have been approved by the FDA. Moreover, studies have already demonstrated that PLGA-NPs can be used to deliver therapeutic agents to the retina. Other natural polymers, such as polypeptides and polysaccharides can also be used to generate NPs. Polypeptide-based nanoparticles are most commonly generated using either albumin or poly-L-lysine. Polysaccharides, such as hyaluronic acid, heparin, chitosan, and cyclodextrin, can be used alone or in combination with lipids or PLGA to generate NPs (Liu et al. 2008).

Nanoparticles can also be generated from metals, such as gold, silver, and platinum. Gold is most commonly used, as it is inert, non-toxic, and non-immunogenic. A recent study showed that gold-NPs of 20 nm can pass through the BRB and exhibit no retinal toxicity (Kim et al. 2009). Interestingly, naked gold-NPs have intrinsic anti-angiogenic activity, and gold NPs conjugated with glycosaminoglycans have enhanced anti-angiogenic activity (Kemp et al. 2009). This inherent anti-angiogenic

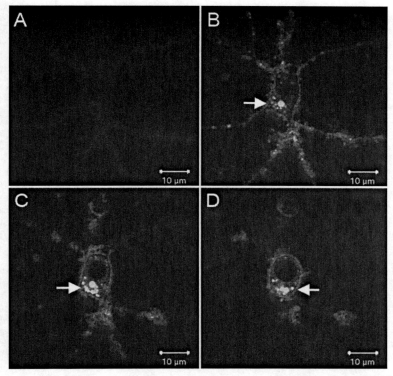

Fig. 4. Confocal fluorescent microscopy images of cells treated with fluorescently-labeled PLGA nanoparticles. Mouse neuronal cells were incubated for 1 hr in the presence of Fluorescently-labeled PLGA nanoparticles (PLGA-NP). Then cells were fixed and imaged by confocal microscopy at 63x magnification. (A-D) The images represent 1.5 μm-thick sections through a single cell, beginning in the plane just at the cell surface (A), and extending inside the cell (B-D). The fluorescent spheres (white arrows) inside the cell in panels B-D illustrate that fluorescent PLGA-NPs have been taken up inside the cell. (PLGA: Poly(D,L-lactide-co-glycolide))
Source: Authors' unpublished data.

activity has also been observed in chitosan NPs and sixth generation poly-L-lysine dendrimers (Xu et al. 2009). These observations suggest some types of NPs may be particularly useful for treating DR-induced RNV.

Therapeutic agents can be incorporated into NPs by either encapsulation, adsorption to the exterior surface, or covalent attachment (Pinto Reis et al. 2006; Zhang et al. 2008a). Regardless of the materials used to generate NPs, chemical formulation methods can be used to alter the size, charge, shape, and molecular weight of NPs to suit the drug delivery application. Nanoparticle formulations can greatly impact drug loading and subsequent release (Fig. 5) and directly affect drug pharmacokinetics, biodistribution, and bioavailability.

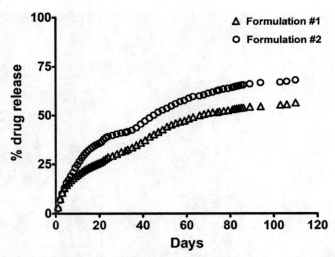

Fig. 5. Nanoparticle drug release kinetics can be affected by altering the chemical methods of nanoparticle formulation. The drug release kinetics of two different formulations of PLGA-NPs containing a small molecule angiogenic inhibitor molecule (CLT-003) were evaluated as follows: Nanoparticles were dispersed in 500 µl of PBS in dialysis tubing with a molecular weight cut-off of 25 kDa. The tubing was then stored in 60 ml of PBS and stirred at 37°C. Each day the 60 ml of PBS was replaced with fresh PBS and the PBS from the previous day's release was evaluated by HPLC to quantify the amount of CLT-003 compound released from NPs. The graph shows that CLT-003 NPs continuously released drug for up to 110 days, and the rate of release was altered by different PLGA-NP formulations. (PLGA: Poly(*D,L*-lactide-*co*-glycolide))
Source: Authors' unpublished data.

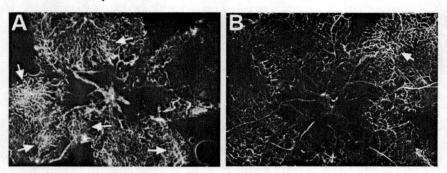

Fig. 6. K5-NP significantly reduces RNV in the OIR mouse model. (A and B) Fluorescein angiography was performed using high molecular weight FITC-dextran, and then the retinas were collected and whole-mounted onto slides for visualization using fluorescent microscopy. (A) P18 retina from an OIR mouse which had no therapeutic treatment. (B) P18 retina from an OIR mouse which received a single IVT injection of K5-NP (8.8 µg) at P12. The K5-NP significantly reduced both the central area of non-perfusion and sites of RNV (white arrows). (K5: plasminogen kringle 5)
Source: Authors' unpublished data.

NANOPARTICLE-BASED DELIVERY OF ANGIOGENIC INHIBITORS

Therapeutic agents which could be used to treat DR-induced RNV include gene therapy vectors, peptide-based inhibitors, antibodies, oligonucleotide aptamers, and small molecules. In preclinical studies, some of these therapeutic agents have been packaged into NPs in an effort to create superior "nanotherapeutics" that have enhanced biodistribution to target cells, increased bioavailability, overall increased therapeutic activity, and reduced adverse side effects compared to conventional drug delivery methods in animal models. Such studies which are applicable to the development of specialized nanoparticles for the delivery of angiogenic inhibitors to treat RNV are highlighted in the following sections.

Nanoparticle-Mediated Delivery of Gene Therapy Vectors to Express Angiogenic Inhibitors

Gene therapy-based treatments have the potential to generate significantly prolonged therapeutic activity after a single injection. Gene therapy refers to the delivery of DNA which is engineered to express a specific gene within the host individual's cells. Gene therapy theoretically reduces the need for repeated dosing, since as long as the DNA remains stable and accessible in the host cell, it can serve as a reusable template for the transcription and translation of the gene product.

Over the past two decades, there have been noted failures as well as promising achievements in gene therapy (Thomas et al. 2003). Virus-derived (viral) DNA vectors have been most commonly tested for gene therapy. However, human clinical trials have raised serious concerns about the safety of using viral vectors for gene therapy, as viral vectors have caused oncogenesis and fatal systemic inflammation in some cases. In addition, rAAV vectors are relatively small, which limits the size of insert DNA they can carry to less than 5 kilobases. Viral vectors may also have limited cell tropism, which restricts the cell types that uptake and express DNA form viral vectors.

As an alternative to viral vectors, non-viral DNA vectors are non-immunogenic and non-toxic and can be engineered to contain relatively large insert DNA sequences (Fink et al. 2006). However, the application of non-viral vectors has been hindered by their low transfection efficiency and high susceptibility to degradation. Recent advances in nanotechnology have enhanced the potential for using non-viral vectors for gene therapy applications. For instance, encapsulating non-viral vectors into NPs can protect the DNA from nuclease-mediated degradation, and in some cases specialized NPs have been developed to enhance cellular uptake and delivery to the nucleus for gene expression.

As a potential therapeutic strategy for treating DR, we recently developed plasminogen kringle 5 (K5) nanoparticles (K5-NP) composed of a PLGA: Chitosan shell which encapsulated a non-viral K5 expression plasmid (Park et al. 2009). K5 is a 80 kDa peptide fragment produced by the proteolytic cleavage of plasminogen. K5 potently inhibits endothelial cell proliferation (ED_{50} = 50 nM), and IVT injection of recombinant K5 peptide in rodent models significantly decreases VEGF expression, increases PEDF expression, and reduces RNV (Gao et al. 2002). However, the effects of K5 peptide injection are very transient, and repeated injections are necessary to maintain the effect of angiogenic inhibition during ischemia-induced RNV. In order to provide sustained expression of K5, we developed a non-viral DNA plasmid to constitutively express K5, and packaged the K5 DNA into nanoparticles to enhance delivery of the plasmid to retinal cells. We chose to develop K5-NP using PLGA since it is biocompatible, biodegradable, and FDA-approved for use in humans. PLGA-NPs are particularly well-suited for gene therapy applications, since PLGA nanoparticles interact with the endo-lysosomal membrane to escape from the endocytic pathway into the cell cytosol, which is thought to increase delivery of PLGA-NPs and their DNA cargo to the nuclear compartment (Panyam et al. 2002).

In our studies we tested the effect of K5-NP in two different rat models: the oxygen-induced retinopathy (OIR) model, which undergoes ischemia-induced RNV, and the streptozotocin (STZ)-induced diabetic model, which undergoes hyperglycemia-induced retinal vascular leakage (Park et al. 2009). The K5-NP was administered by IVT injection, which resulted in persistent expression of K5 in the retina for up to 4 weeks following a single injection. We observed a high level of K5-NP transfection efficiency, which was primarily in the ganglion cell layer of the retina. The K5-NP injection resulted in a significant decrease in retinal vascular leakage in both STZ-induced diabetic and OIR rats. In addition, K5-NP reduced the severity of ischemia-induced RNV in the OIR model (Fig. 6). The effects of the K5-NP were mediated in part by blocking the up-regulation of VEGF and ICAM1 in diabetic retinas, and this K5-dependent angiogenic inhibitory activity persisted for up to 4 weeks post-injection of K5-NP. Furthermore, we detected no toxicity associated with the K5-NP injection, as retinal structure and function were unaffected. This study demonstrates how nanoparticle technology can be utilized to enhance the delivery of non-viral gene therapy vectors to the retina as a potential therapy for DR.

DNA vectors must not only reach the target cell and transverse the plasma membrane to enter the cell, but once inside the cell they must also transverse the nuclear membrane. Some nanoparticle materials, such as PLGA, can facilitate the release of DNA nanoparticles from the endo-lysosomal pathway, which passively increases the chance of the DNA vector reaching the nuclear compartment, but does not actively target the DNA vector to the nucleus. New strategies for nanoparticle-mediated gene therapy focus on circumventing the endocytic pathway and directly targeting the nanoparticle-DNA vector to the nucleus. In some studies, peptide carriers, which have the ability to transverse cell membranes without the use of transporters, cell surface receptors, or the endocytic pathway (Gros et al. 2006), have been incorporated into NPs to enhance cellular uptake and increase nuclear targeting of gene therapy vectors (de la Fuente and Berry 2005; Munoz-Morris et al. 2007).

A novel nanoparticle formulation uses a 30-mer polylysine peptide that terminates with a single cysteine moiety (CK30) (Ziady et al. 2003). The terminal cysteine facilitates covalent bond formation with 10 kDa PEG to generate "PEGylated" CK30 (CK30-PEG). Plasmid DNA is then mixed with the CK30-PEG to generate CK30-PEG-DNA nanoparticles that contain a single DNA plasmid. The size and shape of CK30-PEG-DNA nanoparticles can be altered by using different lysine amine counterions, and the minor diameter can be restricted to less than 25 nm, which increases DNA delivery to post-mitotic cell types that do not undergo nuclear envelope breakdown, including neuronal cells in the retina (Farjo et al. 2006). An additional benefit of using CK30-PEG nanoparticles is that cellular uptake and nuclear targeting does not involve the endocytic pathway, but appears to be mediated at least in part by binding to nucleolin. Nucleolin is selectively expressed on the plasma membrane of specific cell types, including post-mitotic retinal cells (Chen et al. 2008). Nucleolin is otherwise largely concentrated in the nucleolus, and it is thought that nucleolin on the plasma membrane traffics back to the nucleolus to transport CK30-PEG-DNA NPs to the nucleus. CK30-PEG-DNA nanoparticles have been tested for use in retinal gene therapy. IVT injection of CK30-PEG-DNA nanoparticles containing a GFP reporter plasmid in mice generated significant GFP expression in the retina, mainly within the ganglion cell layer, which is invaded by RNV during DR (Farjo et al. 2006). These NPs elicited no detectable abnormalities in retinal structure or function.

Nanoparticle-Mediated Delivery of Drugs and Biologics to Inhibit Angiogenesis

There are a variety of small molecule drugs as well as biologics, such as peptides, antibodies, and aptamers which have angiogenic inhibitory activity and could be developed as useful therapeutics for DR. However, these compounds often have poor solubility, short half-life, or elicit cytotoxicity, which may produce an unfavorable pharmacokinetic profile, limited bioavailability, or adverse side effects that prevent the development of such molecules into viable drugs for the treatment of DR. Recent advancements in nanoparticle formulations offer the opportunity to overcome these obstacles and improve drug solubility (Fig. 7), prolong drug half-life, and even reduce cytotoxicity to increase bioavailability and reduce side effects.

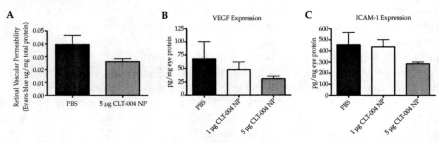

Fig. 7. An insoluble small molecule angiogenic inhibitor is solubilized with PLGA-nanoparticles and effectively reduces retinal vascular leakage and VEGF and ICAM-1 expression in STZ-induced diabetic rats. (A and B) STZ was administered by intraperitoneal injection to induce diabetes. Blood glucose levels were monitored to ensure the onset of diabetes. After two weeks, STZ-induced diabetic rats received an IVT injection of CLT-004 PLGA-NP in one eye and an equal volume of PBS vehicle in the contralateral eye. At 14 days following intravitreal injection, (A) retinal vascular permeability was quantified with the Evans blue extravasation method and retinal tissue levels of (B) VEGF and (C) ICAM-1 were quantified by ELISA. The data demonstrated that CLT-004 NPs caused a significant reduction in retinal vascular permeability (n = 8, p < 0.05) and reduced levels of VEGF and ICAM-1 in the retina. Graphs represent the mean +/− the SEM. These effects were sustained for 14 days following a single IVT injection of CLT-004 NPs. This data is a significant improvement over previous studies in which the relatively insoluble CLT-004 was administered without NPs, and lost therapeutic activity between 2 and 4 days post-IVT injection. Therefore, the nanoparticle formulation appears to confer sustained slow release of the drug.
Source: Authors' unpublished data.

Nanoparticle formulations can be especially helpful for drug molecules which have significant cytotoxic effects, such as the small molecule drug TNP-470 (Benny et al. 2008), a very potent and effective angiogenic inhibitor. In human clinical trials, TNP-470 appeared to be an effective therapy for several types of cancer; however, clinical trials were terminated when TNP-470 elicited neurotoxic effects, including short-term memory loss, seizures, dizziness, and decreased motor coordination. TNP-470 is so small that it could easily penetrate the blood-brain-barrier (BBB) to elicit these effects. Initial attempts to reformulate TNP-470 to block BBB penetration dramatically reduced overall bioavailability and efficacy. Recently, TNP-470 was formulated into monomethoxy-PEG-PLA nanomicilles of approximately 20 nm in diameter. This new nano formulation, named Lodamin, can now be orally administered to effectively treat cancer in animal models, with no evidence of BBB penetration or neurotoxicity. In addition, Lodamin is undergoing preclinical tests in a mouse model of choroidal neovascularization, and appears to be safe and effective at reducing retinal VEGF levels (Benny et al. 2010).

A peptide fragment of PEDF can act on its own as an angiogenic inhibitor, although it has limited bioavailability and activity in mice due to the protease-mediated degradation of the peptide (Li et al. 2006). Formulation of the PEDF peptide into PLGA nanoparticles (PLGA-PEDF) significantly increased the efficacy of the peptide therapy and prolonged therapeutic activity compared to PEDF peptide alone for the treatment of retinal ischemia in mice (Li et al. 2006). This study shows that IVT injection of PLGA-PEDF can effectively target retinal ganglion cells, the cells that are invaded by RNV during DR.

ENGINEERING SPECIALIZED NANOPARTICLES FOR TARGETED DELIVERY

Simply improving the overall bioavailability and cellular uptake of drugs, biologics, and DNA vectors with NP formulations is not always sufficient to improve therapeutic activity, because non-target cells, such as cells adjacent to the target cell population or engulfing macrophages, may uptake NPs more efficiently than the target cell population. This can result in decreased NP delivery to target cells and increased adverse side effects. The next sections will review recent efforts to develop multi-component NPs for targeting the delivery of angiogenic inhibitors to the retina, and specifically to neovascular endothelial cells for the treatment RNV in DR.

Engineering Nanoparticles to Target Neovascular Endothelial Cells

Neovascular endothelial cells express cell surface markers that are not expressed on normal, non-proliferative endothelial cells. These unique markers can be used to target NP delivery specifically to neovascular endothelial cells without affecting nearby normal endothelial cells or other cell types. These cell surface markers include intercellular adhesion molecule 1 (ICAM1) and two specific types of integrins, $\alpha_v\beta_3$ and $\alpha_v\beta_5$ (Silva et al. 2008).

Naturally-occurring extracellular matrix proteins normally bind to different integrins via arginine-glycine-apartic acid (RGD) motifs, and based on these sequences, synthetic cyclic and linear RGD peptides have been developed which specifically bind to only $\alpha_v\beta_3$ and/or $\alpha_v\beta_5$ integrins. Several RGD peptides are undergoing preclinical and clinical testing to target neovascularization (Silva et al. 2008). Highly specific antibodies against $\alpha_v\beta_3$ and/or $\alpha_v\beta_5$ integrins are also being developed to target and inhibit neovascular endothelial cells. In the future, RGD peptides and integrin antibodies could be utilized in NP-based therapeutics to target and inhibit RNV in DR.

Peptides and antibodies that bind to ICAM1 are also being developed for targeting neovascular endothelial cells. A peptide fragment derived from leukocyte function-associated antigen-1, known as cLABL, binds to ICAM1 on neovascular endothelial cells. Coating PLGA-PEG nanoparticles with cLABL significantly increased cellular uptake of NPs by neovascular endothelial cells compared to uncoated PLGA-PEG NPs (Zhang et al. 2008b). Likewise, conjugation of anti-ICAM1 antibodies to liposomes created immunoliposomes with enhanced endothelial cell uptake *in vitro* versus naked liposomes (Hua et al. 2010).

An ongoing preclinical study uses quantum dot nanocrystals (QDs) that have ICAM1 antibodies conjugated to their external surface to generate ICAM1-targeted nanocarriers (ITNs) (Jayagopal et al. 2007). ITNs are smaller than 200 nm, and bind to neovascular endothelial cells via ICAM1, which leads to clathrin-mediated endocytosis of the ITNs. Various therapeutic agents, including siRNA, peptides, small molecule drugs, and DNA vectors can be encapsulated in ITNs for selective delivery to neovascular endothelial cells. ITNs carrying different cargoes are being tested in animal models of neovascularization.

Enhancing Ocular Delivery of Nanoparticles

Systemic administration routes are desirable for treating DR in order to avoid complications from IVT injections. However, systemic therapies designed to specifically target neovascular endothelial cells may still encounter difficulty in reaching the retinal vasculature due to the isolated ocular environment and the BRB. In addition, systemic routes of administration could increase adverse side effects. Thus, strategies to deliver angiogenic inhibitors to the retina for the treatment of DR should consider how to overcome these difficulties and optimize ocular delivery.

There has been one notable success in using systemic administration to deliver an angiogenic inhibitor to ocular neovascular lesions. Although this study focused on choroidal neovascularization which occurs in the subretinal space, unlike RNV which occurs in the retinal layers near the vitreous, the results suggest this methodology could be used to treat RNV lesions as well. This study utilized the Flt23K DNA plasmid, which encodes the anti-VEGF intraceptor, a recombinant protein that includes VEGF-binding domains 2 and 3 of VEGFR-1 coupled to the endoplasmic reticulum retention signal sequence Lys-Asp-Glu-Leu (KDEL) (Singh et al. 2009). The anti-VEGF intraceptor is designed to bind to VEGF as it is synthesized to sequester VEGF in the endoplasmic reticulum and inhibit VEGF secretion. The Flt23K plasmid was encapsulated into PLGA nanoparticles, which were conjugated with either transferrin (TF), RGD peptide, or both in order to facilitate delivery to retinal CNV lesions following intravenous injection (Singh et al. 2009). Transferrin was chosen for as a targeting peptide because the retina expresses transferrin receptors. Within 24 hours of intravenous administration, TF/RGD-targeted nanoparticles were delivered specifically to choroidal neovascular lesions in the retina, and were not present in the contralateral healthy control retina. A much smaller amount of the non-targeted nanoparticle was also delivered to neovascular lesions, likely due to the non-specific effect of vascular leakage. Importantly, intravenous administration did not lead to any nanoparticle detection in the brain. Only TF/RGD-functionalized nanoparticles, and not unconjugated nanoparticles, resulted in expression of the anti-VEGF intraceptor in the RPE cell layer. RGD conjugation also produced significant gene delivery to retinal endothelial cells, whereas TF-conjugated nanoparticles were targeted more generally to the retina than to the retinal endothelial cells. Impressively, the intravenous administration of either TF- or RGD-functionalized nanoparticles delivered enough Flt23K-NP to the neovascular lesions to block up-regulation of VEGF protein in the retina and significantly reduce the neovascular lesion size.

Applications to Areas of Health and Disease

Nanotechnology has greatly impacted commercial pharmaceutical development and spawned a high demand for skilled formulation scientists with nanotech expertise. Whereas solubility used to play a major role in early drug screening, it can now be circumvented. Many small molecules often consist of large hydrophobic moieties which encumber drug delivery and shelf-stability. By using nanotechnology-based formulations, this drawback can be addressed. As small molecules tend to have an inherently lower cost of production than biologics, the development of small molecule therapies can drive down the long-term costs of healthcare associated with treating chronic diseases, such as DR.

Definitions

Diabetic Retinopathy (DR): A disease in which the retina is damaged as a consequence of pre-existing diabetes.

Proliferative DR: In the proliferative stage of diabetic retinopathy, retinal capillaries become occluded and cause retinal ischemia, which induces new blood vessels to sprout from damaged capillaries and invade the retina in a process termed retinal neovascularization (RNV).

Retinal Neovascularization (RNV): The process by which compromised retinal capillaries undergo abnormal angiogenesis to generate new, abnormally leaky blood vessels that invade retinal tissue and induce inflammation and may cause spotty or cloudy vision. In worst cases, retinal scarring leads to permanent vision loss and even blindness.

Angiogenic Inhibitor: A molecule which acts to inhibit or prevent mechanisms of angiogenesis, such as occurs during RNV.

VEGF: Vascular endothelial growth factor (VEGF) is a secreted glycoprotein which binds to VEGF-receptors on endothelial cells to strongly activate angiogenic signaling pathways and induce angiogenesis.

PEDF: Pigment epithelium-derived factor (PEDF) is a secreted glycoprotein which strongly inhibits angiogenic signaling pathways to inhibit angiogenesis. At least one mechanism of PEDF inhibition involves binding to VEGF-receptors to block VEGF binding.

K5: Plasminogen Kringle 5 (K5) is a peptide fragment, produced from the proteolytic cleavage of plasminogen, which has inherent angiogenic inhibitory activity.

Gene Therapy: The delivery of recombinant DNA molecules encoding a specific gene to a target host cell, in which the gene product is anticipated to provide a therapeutic benefit to the host.

Nanoparticle: A particle of less than 1 µm which is may be composed of various natural or synthetic materials, including polymers, lipids, polysaccharides, metals, and ceramic. Drugs may either be packaged inside or attached to the outside of NPs by adsorption or covalent linkage.

Summary
- Proliferative diabetic retinopathy (PDR) is characterized by retinal neovascularization, a process in which new blood vessels sprout from retinal capillaries and invade retinal tissue. RNV may ultimately result in permanent vision loss, including blindness if left untreated.
- RNV can be treated by delivering angiogenic inhibitors, which may be in the form of plasmid DNA, recombinant protein or peptide, antibodies, siRNA, oligonucleotide aptamers, and small molecule drugs.
- Conventional drug delivery utilizes intravitreal (IVT) injections, which must be repeated every 1 to 3 months to sustain therapeutic concentrations of drugs in the retina. Repetitive IVT injections can cause severe complications, which may delay or prevent prolonged treatment regimens.
- Nanoparticles (NPs) consist of 1 µm or smaller particles which may be generated from a variety of natural and synthetic materials. Therapeutic agents may either be packaged inside or attached to the outside of NPs.

- Nanoparticles are being utilized to deliver angiogenic inhibitors to the retina as a potential treatment for DR-induced RNV.
- Nanoparticles often improve pharmacokinetics of drug metabolism and prolonged drug half-life, resulting in a net increase in therapeutic activity in the retina.
- Multi-component NPs are also being designed using peptide and antibody tags that bind to cell surface markers expressed exclusively by neovascular endothelial cells to target drug delivery to the retina and specifically to sites of RNV.

Abbreviations

BRB	:	blood-retinal-barrier
DN	:	diabetic nephropathy
DR	:	diabetic retinopathy
ICAM1	:	intercellular adhesion molecule 1
IVT	:	intravitreal
K5	:	plasminogen kringle 5
NPDR	:	non-proliferative diabetic retinopathy
NPs	:	nanoparticles
PDR	:	proliferative diabetic retinopathy
PEDF	:	pigment epithelium-derived factor
PEG	:	polyethylene glycol
PGA	:	polyglycolide
PLA	:	polylactide
PLGA	:	Poly(*D,L*-lactide-*co*-glycolide)
POD	:	peptide for ocular delivery
QD	:	quantum dot
RGD	:	arginine-glycine-apartic acid
RNV	:	retinal neovascularization
RPE	:	retinal pigment epithelium
RPE65	:	RPE-specific protein 65 kDa
SLN	:	solid lipid nanoparticle
TF	:	transferrin
VEGF	:	vascular endothelial growth factor

References

Abdallah, W., and A.A. Fawzi. 2009. Anti-VEGF therapy in proliferative diabetic retinopathy. Int. Ophthalmol. Clin. 49: 95–107.

Barcelo, A., C. Aedo, S. Rajpathak and S. Robles. 2003. The cost of diabetes in Latin America and the Caribbean. Bull World Health Organ 81: 19–27.

Benny, O., O. Fainaru, A. Adini, F. Cassiola, L. Bazinet, I. Adini, E. Pravda, Y. Nahmias, S. Koirala, G. Corfas, R.J. D'Amato and J. Folkman. 2008. An orally delivered small-molecule formulation with antiangiogenic and anticancer activity. Nat. Biotechnol. 26: 799–807.

Benny, O., K. Nakai, T. Yoshimura, L. Bazinet, J.D. Akula and R.J. D'Amato. 2010. Broad Spectrum Anitangiogenic Therapy for Ocular Neovascularization [abstract]. IOVS, ARVO E-Abstract 2983.

Chen, X., D.M. Kube, M.J. Cooper and P.B. Davis. 2008. Cell surface nucleolin serves as receptor for DNA nanoparticles composed of pegylated polylysine and DNA. Mol. Ther. 16: 333–342.

Cunha-Vaz, J.G. 2004. The blood-retinal barriers system. Basic concepts and clinical evaluation. Exp. Eye Res. 78: 715–721.

de la Fuente, J.M., and C.C. Berry. 2005. Tat peptide as an efficient molecule to translocate gold nanoparticles into the cell nucleus. Bioconjugate chemistry 16: 1176–1180.

Farjo, R., J. Skaggs, A.B. Quiambao, M.J. Cooper and M.I. Naash. 2006. Efficient non-viral ocular gene transfer with compacted DNA nanoparticles. PLoS One 1: e38.

Fink, T.L., P.J. Klepcyk, S.M. Oette, C.R. Gedeon, S.L. Hyatt, T.H. Kowalczyk, R.C. Moen and M.J. Cooper. 2006. Plasmid size up to 20 kbp does not limit effective *in vivo* lung gene transfer using compacted DNA nanoparticles. Gene Ther. 13: 1048–1051.

Frank, R.N. 2004. Diabetic retinopathy. The New England journal of medicine 350: 48–58.

Gao, G., Y. Li, S. Gee, A. Dudley, J. Fant, C. Crosson and J.X. Ma. 2002. Down-regulation of vascular endothelial growth factor and up-regulation of pigment epithelium-derived factor: a possible mechanism for the anti-angiogenic activity of plasminogen kringle 5. J. Biol. Chem. 277: 9492–9497.

Gao, G., Y. Li, D. Zhang, S. Gee, C. Crosson and J. Ma. 2001. Unbalanced expression of VEGF and PEDF in ischemia-induced retinal neovascularization. FEBS Lett 489: 270–276.

Gao, G., C. Shao, S.X. Zhang, A. Dudley, J. Fant and J.X. Ma. 2003. Kallikrein-binding protein inhibits retinal neovascularization and decreases vascular leakage. Diabetologia 46: 689–698.

Gros, E., S. Deshayes, M.C. Morris, G. Aldrian-Herrada, J. Depollier, F. Heitz and G. Divita. 2006. A non-covalent peptide-based strategy for protein and peptide nucleic acid transduction. Biochim. Biophys. Acta. 1758: 384–393.

Hua, S., H.I. Chang, N.M. Davies and P.J. Cabot. 2010. Targeting of ICAM-1-directed immunoliposomes specifically to activated endothelial cells with low cellular uptake: use of an optimized procedure for the coupling of low concentrations of antibody to liposomes. J. Liposome Res.

Jayagopal, A., P.K. Russ and F.R. Haselton. 2007. Surface engineering of quantum dots for *in vivo* vascular imaging. Bioconjugate chemistry 18: 1424–1433.

Kemp, M.M., A. Kumar, S. Mousa, E. Dyskin, M. Yalcin, P. Ajayan, R.J. Linhardt and S.A. Mousa. 2009. Gold and silver nanoparticles conjugated with heparin derivative possess anti-angiogenesis properties. Nanotechnology 20: 455104.

Kim, J.H., J.H. Kim, K.W. Kim, M.H. Kim and Y.S. Yu. 2009. Intravenously administered gold nanoparticles pass through the blood-retinal barrier depending on the particle size, and induce no retinal toxicity. Nanotechnology 20: 505101.

Latinovic, S. 2006. Global initiative for the prevention of blindness: Vision 2020—the Right to Sight. Medicinski pregled 59: 207–212.

Li, H., V.V. Tran, Y. Hu, W. Mark Saltzman, C.J. Barnstable and J. Tombran-Tink. 2006. A PEDF N-terminal peptide protects the retina from ischemic injury when delivered in PLGA nanospheres. Exp. Eye Res. 83: 824–833.

Liu, Z., Y. Jiao, Y. Wang, C. Zhou and Z. Zhang. 2008. Polysaccharides-based nanoparticles as drug delivery systems. Adv. Drug Deliv. Rev. 60: 1650–1662.

Munoz-Morris, M.A., F. Heitz, G. Divita and M.C. Morris. 2007. The peptide carrier Pep-1 forms biologically efficient nanoparticle complexes. Biochem. Biophys. Res. Commun. 355: 877–882.

Panyam, J., M.M. Dali, S.K. Sahoo, W. Ma, S.S. Chakravarthi, G.L. Amidon, R.J. Levy and V. Labhasetwar. 2003. Polymer degradation and *in vitro* release of a model protein from poly(D,L-lactide-co-glycolide) nano- and microparticles. J. Control Release 92: 173–187.

Panyam, J., W.Z. Zhou, S. Prabha, S.K. Sahoo and V. Labhasetwar. 2002. Rapid endo-lysosomal escape of poly(DL-lactide-co-glycolide) nanoparticles: implications for drug and gene delivery. Faseb. J. 16: 1217–1226.

Park, K., Y. Chen, Y. Hu, A.S. Mayo, U.B. Kompella, R. Longeras and J.X. Ma. 2009. Nanoparticle-mediated expression of an angiogenic inhibitor ameliorates ischemia-induced retinal neovascularization and diabetes-induced retinal vascular leakage. Diabetes 58: 1902–1913.

Penn, J.S., A. Madan, R.B. Caldwell, M. Bartoli, R.W. Caldwell and M.E. Hartnett. 2008. Vascular endothelial growth factor in eye disease. Prog. Retin. Eye Res. 27: 331–371.

Pinto Reis, C., R.J. Neufeld, A.J. Ribeiro and F. Veiga. 2006. Nanoencapsulation I. Methods for preparation of drug-loaded polymeric nanoparticles. Nanomedicine 2: 8–21.

Resnikoff, S., D. Pascolini, D. Etya'ale, I. Kocur, R. Pararajasegaram, G.P. Pokharel and S.P. Mariotti. 2004. Global data on visual impairment in the year 2002. Bull World Health Organ 82: 844–851.

Risau, W. 1997. Mechanisms of angiogenesis. Nature 386: 671–674.

Sampat, K.M., and S.J. Garg. 2010. Complications of intravitreal injections. Curr. Opin. Ophthalmol. 21: 178–183.

Silva, R., G. D'Amico, K.M. Hodivala-Dilke and L.E. Reynolds. 2008. Integrins: the keys to unlocking angiogenesis. Arterioscler Thromb. Vasc. Biol. 28: 1703–1713.

Singh, S.R., H.E. Grossniklaus, S.J. Kang, H.F. Edelhauser, B.K. Ambati and U.B. Kompella. 2009. Intravenous transferrin, RGD peptide and dual-targeted nanoparticles enhance anti-VEGF intraceptor gene delivery to laser-induced CNV. Gene Ther. 16: 645–659.

Thomas, C.E., A. Ehrhardt and M.A. Kay. 2003. Progress and problems with the use of viral vectors for gene therapy. Nat. Rev. Genet. 4: 346–358.

Torchilin, V.P. 2005. Recent advances with liposomes as pharmaceutical carriers. Nat. Rev. Drug Discov. 4: 145–160.

Viola, F., C. Mapelli, D. Galimberti, G. De Martini, R. Esposti, L. Moneghini, P. Braidotti, M. Cresta, M.R. Gasco and R. Ratiglia. 2009. Solid Lipid Nanoparticles Topically Administered in Rabbits as New Drug Delivery System: A Preliminary Study of Safety and Bioavailability [abstract]. IOVS, ARVO E-abstract 2423.

Xu, Y., Z. Wen and Z. Xu. 2009. Chitosan nanoparticles inhibit the growth of human hepatocellular carcinoma xenografts through an antiangiogenic mechanism. Anticancer Res. 29: 5103–5109.

Zhang, D., P.L. Kaufman, G. Gao, R.A. Saunders and J.X. Ma. 2001. Intravitreal injection of plasminogen kringle 5, an endogenous angiogenic inhibitor, arrests retinal neovascularization in rats. Diabetologia 44: 757–765.

Zhang, L., F.X. Gu, J.M. Chan, A.Z. Wang, R.S. Langer and O.C. Farokhzad. 2008a. Nanoparticles in medicine: therapeutic applications and developments. Clin. Pharmacol. Ther. 83: 761–769.

Zhang, N., C. Chittasupho, C. Duangrat, T.J. Siahaan and C. Berkland. 2008b. PLGA nanoparticle—peptide conjugate effectively targets intercellular cell-adhesion molecule-1. Bioconjugate chemistry 19: 145–152.

Zhang, S.X., J.J. Wang, G. Gao, K. Parke and J.X. Ma. 2006. Pigment epithelium-derived factor downregulates vascular endothelial growth factor (VEGF) expression and inhibits VEGF-VEGF receptor 2 binding in diabetic retinopathy. J. Mol. Endocrinol. 37: 1–12.

Ziady, A.G., C.R. Gedeon, T. Miller, W. Quan, J.M. Payne, S.L. Hyatt, T.L. Fink, O. Muhammad, S. Oette, T. Kowalczyk, M.K. Pasumarthy, R.C. Moen, M.J. Cooper and P.B. Davis. 2003. Transfection of airway epithelium by stable PEGylated poly-L-lysine DNA nanoparticles *in vivo*. Mol. Ther. 8: 936–947.

17

Drug Loaded Nanofiber Matrices as Diabetic Wound Dressings

William Gionfriddo[1] and Lakshmi S. Nair[2,*]

ABSTRACT

The pathological nature of persistent diabetic and pressure ulcers, consisting of uncontrolled inflammation, immune dysfunction, and infection, require the use of multifaceted biomedical strategies in treatment models. This chapter highlights the production and employment of polymeric nanofibers capable of a diverse array of bioactive functions. A discussion of the electrospinning process, followed by mention of the various biocompatible materials utilized in the development of fibrous meshworks, underscores the importance of the ability of this technology to produce biomimetic scaffolds on a nanoscale. Furthermore, this chapter provides an in-depth analysis of recent research on nanofiber modalities and the mechanisms behind bioactive wound dressings capable of releasing molecules for the purposes of modulating the healing process in the diabetic environment.

---

[1] School of Medicine, University of Connecticut Health Center, E-7041, MC-3711, 263 Farmington Avenue, Farmington, CT – 06030; E-mail: wgionfriddo@student.uchc.edu

[2] Department of Orthopaedic Surgery, Department of Chemical, Materials and Biomolecular Engineering, University of Connecticut Health Center, E-7041, MC-3711, 263 Farmington Avenue, Farmington, CT-06030; E-mail: nair@uchc.edu

*Corresponding author

List of abbreviations after the text.

Key Terms

- **Polymer:** Large molecules consisting of repeating patterns of structural subunits.
- **Electrospinning:** The process of using electrical energy to make long fibers of nano size from solutions of polymers.
- **Nanofiber Matrix:** A meshwork of many nanofibers created by electrospinning that are chemically linked together, and potentially used as a wound dressing.
- **Hemostasis:** The hardening of blood following an injury, consisting of coagulation through platelet aggregation, adhesion, and fibrin clot formation.
- **Angiogenesis:** A physiological process by which new blood vessels are generated.
- **Stem and Progenitor Cells:** Cells with long rates of survival that can be activated to divide, migrate, and differentiate into different cell types with more specific functions.
- **Vinculin:** A prominent component of the complexes and focal adhesions that links cytoskeleton, plasma membrane and the extracellular matrix.
- **Curcumin:** A dye that confers yellow color to the popular spice, Turmeric (*Curcuma longa*), a common ingredient in the diet of many cultures, and known for its anti-inflammatory properties.

INTRODUCTION

Cutaneous wound healing is characterized by an intricate interplay of cell-signaling, migration, differentiation, and proliferation. *Hemostasis*, including the subsequent secretion of cytokines and chemotactic factors, is the first of four stages of wound healing (Pradhan et al. 2010). The second stage involves the significant role of an immune response via the infiltration of monocytes, neutrophils, and leukocytes. Phagocytic cells clear debris, and perpetuate *inflammatory* mechanisms. Healing is further characterized by rapid *proliferation* of cells involved in the repair and restoration of the damaged region. Among others: epithelia, endothelia, keratinocytes and fibroblasts migrate, differentiate, divide, and interact to replace the extracellular matrix (ECM) and injured tissue. Deposition of the ECM further promotes angiogenesis, migration, re-epithelialization, and wound closure. The final *remodeling* stage is characterized by ECM modification and scar tissue formation (Pradhan et al. 2009). The typical human internal environment is conducive to this appropriate response and resolution of disturbances. However, the hyperglycemic environment in diabetic individuals prevents the appropriate execution of the aforementioned stages, thus contributing to persistent ulceration. Several studies have noted the pathogenesis of impaired ECM formation; abnormal cellular proliferation, migration, and infiltration; insufficient macrophage activity; and impaired re-epithelialization and angiogenesis (Duraisamy et al. 2001; Khanna et al. 2010; Loots et al. 1998). In such instances pro-inflammatory cytokines such as TNFα (Tumor Necrosis Factor α), IL-6 (Interleukin-6), and IL-8 (Interleukin-8, also known as CXCL8), as well as MCP-1 (Monocyte Chemo-attractant Protein-1) and MMPs (Matrix Metalloproteinases) are over-expressed (Landis et al. 2010). Wound persistence, perpetuated by chronic inflammation and inappropriate immune responses in the hyperglycemic environment, leads to increased infection susceptibility. Therefore, treatment of diabetic wounds must involve attenuation of uncontrolled inflammation by promoting appropriate cytokine expression and self-limitation, as well as, eliminating infection and the promotion of appropriate re-epithelialization and wound repair mechanisms.

Research underwent in the past two decades in the area of wound dressings led to the development of a wide array of products ranging from gauzes, films, foams, hydrogels and hydrocolloids (Ovington and Pierce 2001). With a newer understanding of the mechanism of wound repair, including the role of inflammation in wound healing, novel drugs and biologicals are also considered along with the various wound dressings. Currently, hydrogels are considered to be the most advanced class of wound dressings, particularly for treating burn injuries and chronic wounds such as diabetic and pressure ulcers (Lay-Flurrie 2004). However, most of these dressings are biologically inert and their primary functions are to provide biocompatible, moist and permeable barriers to assist healing. The emergence of tissue engineering/regenerative medicine raises possibilities of developing novel wound healing strategies mainly by modulating the biological environment at the wound site. Also, recent research in cellular biology led to a deeper understanding of stem cells and progenitor cells and ways to control their functions using biological cues such as low molecular weight drugs, peptides, hormones, growth factors along with biomimetic structural, mechanical and biophysical environments. Among these, the biomimetic structural component has been shown to play a very important role in modulating cellular functions to promote optimal healing.

The extracellular matrix (ECM) in natural tissue provides the structured environment with mechanical, biochemical and physical cues that enable the cells to interact with each other and with the ECM to allow optimal cell function. The ECM is mainly composed of a cross-linked network of fibrous proteins and hydrated proteoglycans arranged in a unique tissue-specific three-dimensional architecture (collectively called structural/physical signals) in which other biomolecules (growth factors, chemokines and cytokines) and ions are bound (Nair et al. 2006). The natural ECM serves to organize cells in space to give them form, provide environmental signals, direct site specific cellular regulation and separate one tissue space from another. In short, the multifunctional nature of ECM is achieved through chemical cues such as insoluble signals or factors which interact with the soluble signals of cells along with the unique tissue-specific hierarchical structural features. It is therefore pertinent to consider the structural features of ECM along with the biological principles that govern cell-cell and cell-matrix interactions, while developing new therapeutic strategies to promote tissue regeneration. The hierarchical structure of the ECM has length scales varying from a few nanometers (nm) to millimeters (mm) to control cellular functions and tissue properties. Since the role of scaffolds in tissue engineering is to present a transient ECM mimic environment, it is important that they closely resemble the structural and topographical features of the ECM for optimal cell performance.

Table 1. Key Features of the Extracellular Matrix.

1. Fibroblasts are the chief cells involved in the synthesis and maintenance of the extracellular matrix.
2. The Extracellular Matrix composition is variable but usually consists of a network of fibers: Collagen, Elastin, and Reticular; as well as other proteins and the interstitial matrix.
3. It provides structural support and influences growth and survival of cells such as epithelia.
4. The ECM is important in wound healing due to its role as a scaffold for cells to adhere, migrate, differentiate and divide.
5. The ECM holds several different types of biologically active molecules such as: growth factors, proteases, and cytokines important in cell signaling.
6. Central to healing, the ECM also provides a dense barrier to the spread of infection.

Key features of the Extracellular Matrix and its function in wound healing and cellular support. ECM: Extracellular Matrix (Loots et al. 1998).

In the search for effective therapies in treating diabetic wounds, nanofibers have emerged as an encouraging modality due to their physical resemblance of the ECM and their capacity to deliver bioactive molecules, proteins and antibiotics over sustained periods of time. Before delving into small molecule release mechanisms, it is first beneficial to discuss the direct role of physical nanofiber characteristics and ECM-associated protein conjugation in facilitating wound repair.

Recent developments in nanotechnology provide us with tools to develop 3-dimensional structures with nanoscale features as well as model substrates to gain insights on the sensitivity of cells towards subtle ECM features on the nanometer scale. Using *in vitro* cell culture studies on nanopatterned surfaces fabricated by electron beam lithographic techniques, it has been demonstrated that cells are sensitive to nanoscale dimensions and could react to objects as small as 5 nm (Curtis and Wilkinson 2001). These studies led to significant interest towards developing nanostructured scaffolds for tissue regeneration. Since collagen fibrils with diameters in the range of 50–500 nm are major components of the natural ECM, a logical approach is to fabricate nanoscale polymeric fibers as scaffolds to support tissue regeneration. Nanofabrication techniques such as electrospinning, phase separation and self assembly have been developed to form unique nanofibrous structures from both natural and synthetic polymers (Laurencin and Nair 2008). Among these, electrospinning has developed into a promising, versatile and economical technique to produce nanostructured scaffolds with fiber diameter ranging from approximately 1–1000 nm (Nair et al. 2004; Zhang et al. 2005).

The process of electrospinning is briefly described as follows: An electric field is applied to a pendant droplet of polymer solution at the tip of a needle or capillary attached to a syringe or pipette. A syringe pump is commonly used to feed the polymer solution at a constant flow rate to the needle/capillary. The electrode is either inserted in the polymer solution or connected to the tip of the needle. When an electric potential is applied to the droplet, the droplet will be subjected to mutually opposed forces. Some forces, such as surface tension and viscoelasticity, tend to retain the hemispherical shape of the droplet whereas forces such as the applied electric field tend to deform the droplet to form a conically-shaped Taylor cone. Beyond a threshold voltage, the electric forces in the droplet predominate and, at that point, a narrow charged polymer jet ejects from the tip of the Taylor cone. At optimal polymer viscosities, the ejected jet travels in a nearly straight line towards the rounded collector due to the stabilization imparted by the longitudinal stress of the external electric field on the charge carried by the jet. However, at some point on the path, the jet reaches a point of instability due to the repulsive forces arising from the opposing charges and the jet undertakes bending, winding, spiraling and looping paths in three dimensions. This process leads to the continuous stretching of the polymer jet and significant reduction of the fiber diameter along with the rapid evaporation of the solvent. The result is the formation of ultrathin fibers deposited on a grounded collector surface (Reneker and Chun 1996; Shin et al. 2001).

The versatility of the electrospinning process is achieved by varying system, solution and processing parameters (such as strength of the electric field, polymer solution viscosity and conductivity, polymer properties, flow rate, distance between the needle tip and the collector), as well as, by modifying the basic electrospinning set up. Small modifications significantly vary the morphology and the diameter of the polymeric fibers. By now, studies have demonstrated the feasibility of developing nanofibers from a range of natural and synthetic materials having different physical, chemical and mechanical properties. Such variety is achieved by co-spinning different materials and by incorporating nanoparticles or fillers to create scaffolds with varying strength, surface chemistry, and degradation properties (Leach et al. 2011; Bonani et al. 2011; Kim and Yoo 2010).

In wound healing, the proliferation and differentiation of cells is directly dependent on adhesion to the ECM. The interaction of dermal fibroblasts, keratinocytes, and other cell types with fibronectin, collagen, hyaluronic acid polysaccharide (HA), and other components of the basement membrane is critical in re-epithelialization and repair mechanisms (Loughlin and Artlett 2009). Because of this, the development of biocompatible nanofibers that mimic the morphology of the ECM is essential for the long-term compatibility of biomaterials within the microenvironment, and necessary for preventing an adverse reaction to exogenous therapies from the host immune response. Electrospinning methods have proven effective in developing collagen nanofibrous matrices that do just that (Rho et al. 2006). Nanofibers with diameters comparable to basement membrane collagen fibrils, and cross-linked fibers demonstrate qualities of high porosity, water-retention, large surface area, and biocompatibility. It has been reported that nanofiber matrices with small interstices and high effective surface area can promote hemostasis without using a haemostatic agent (Zhang et al. 2005). Similar to hydrogel dressings, polymeric nanofibers from hydrophilic polymers exhibit effective water absorption, an important wound dressing property that enables efficient removal of wound exudates. However, due to the significant increase in surface area to volume ratio in the case of polymeric nanofibers, the extent of water absorption is much better than polymeric film dressings (Zhang et al. 2005). Moreover, the unique nano-micro porous structure of nanofiber matrices also presents a nutrient and gas permeable membrane which can significantly prevent wound desiccation and, at the same time, prevent bacterial infection (Khil et al. 2003). Another advantage of the finer fiber sizes of the nanofiber matrix is the excellent conformability of these dressings compared to dressings with larger fiber sizes. The increased conformability allows for the better fitting of the dressing to irregular 3D contours of wounds. Apart from these properties, an interesting study by Sanders et al. demonstrated the immunomodulatory function of nanofiber structures which will have significant implications in developing nanofiber based wound dressings for chronic wounds such as diabetic and pressure ulcers (Sanders et al. 2000). The study showed that the thickness of the fibrous capsule, as well as macrophage density, was higher on larger fibers (diameters greater than 6 µm) compared to smaller diameter fibers (diameters less than 6 µm) whereas vascularization was higher in the case of small fibers. It is hypothesized that small fibers produce reduced immune response compared to larger fibers possibly due to reduced cell-material contact surface area or due to a curvature threshold effect that in turn triggers cell signaling. Moreover, it has been suggested that by providing an ECM mimic road map, nanofibrous matrices with appropriate properties might lead to scar free tissue regeneration (Zhang et al. 2005). The versatility of the electrospinning process and the high surface area of polymeric nanofibers also allow for drug encapsulation within the fibers and surface functionalization of the matrices, which can confer useful properties to bioactive wound dressings as discussed in Section 3. To sum up, the ECM mimic length scale of electrospun nanofibers, along with the ability of nanofiber matrices to satisfy many of the unique properties required for optimal wound healing, makes them potential candidates for developing bioactive wound dressings.

The scope of this chapter is to give an overview of the unique advantages of polymeric nanofibers towards modulating a cell response and as a drug/protein vehicle to improve wound healing, mainly in chronic wounds such as diabetic and pressure ulcers.

EFFECT OF FIBER DIAMETER ON CELL RESPONSE

Several studies have confirmed that cells prefer to live in a complex nanostructured environment composed of pores, ridges and fibers that mimics the structure of the ECM compared to two-dimensional films or microfiber matrices. To evaluate the difference in cell behavior towards fibers of different diameter, Kwon et al., developed fibrous matrices from poly(L-lactide-co-caprolactone) (PLL-CL) with fiber diameters ranging from ~0.3 μm, ~1.2 μm and 7 μm using the process of electrospinning (Kwon et al. 2005). The adhesion and proliferation of human umbilical vein endothelial cells (HUVEC) on three different types of PLL-CL fiber matrices were then followed using scanning electron microscopy and cell counting. It has been found that HUVECs adhered and proliferated well on the small diameter fiber matrices (0.3 and 1.2 μm diameter), whereas the fiber matrices with higher fiber diameter (7.0 μm) showed reduced cell adhesion, restricted cell spreading and no signs of proliferation. The authors attributed this minimal degree of cell proliferation on the larger diameter microfiber matrix to the fact that considerable interfiber distance and low surface density of fibers hindered cell adhesion between neighboring fibers.

Similar results were observed by Kumbar et al. using human skin fibroblasts confirming the favorable effect of lower fiber diameters on cellular response (Kumbar et al. 2008). The study reported the fabrication of poly(lactic acid-co-glycolic acid) (PLAGA) matrices with fiber diameters ranging from 150 nm to 6000 nm via electrospinning. Human skin fibroblasts showed a well-spread morphology and significant growth on matrices with fiber diameters less than 1100 nm. Higher fiber diameter showed cells with round morphology and low cell proliferation. Moreover, collagen type III gene expression was significantly up-regulated in the case of cells grown on matrices with fiber diameter in the range of 350–1100 nm.

A study by Schindler et al., investigated the ability of nanofiber matrix to promote *in vivo*-like organization and morphogenesis of cells in culture (Schindler et al. 2005). The synthetic nanofiber matrices composed of fiber diameters of ~180 nm and a pore diameter of ~700 nm were prepared by electrospinning polyamide onto glass cover slips. The surface smoothness of the matrix was within 5 nm over a length of 1.5 μm which is similar to the three-dimensional organization of fibers in basement membranes. The organizational and structural changes of the intracellular components (actin and focal adhesion components) of NIH 3T3 fibroblasts cultured on nanofibrous matrices and glass substrate were measured as a function of adhesion. Fibroblasts plated on the glass substrate were found to be well spread with an elaborate checkerboard pattern of stress fibers. However, the cells cultured on the nanofiber matrix showed significant changes in morphology and shape. They were found to be more elongated and bipolar with thinner actin fibers arranged parallel to the long axis of the cell. Notable increases in the formation of actin-rich lamellipodia, membrane ruffles and cortical actin were also observed. Staining of vinculin in fibroblasts cultured on glass substrate showed a parallel streaked structure. However, the streaked staining for vinculin within cells on nanostructured matrix was limited to the edge of the lamellipodia with a more diffuse staining throughout the cell cytoplasm. Similar to actin distribution, such a pattern of vinculin labeling on nanofiber matrix correlates with cellular differentiation and morphogenesis *in vivo*. The cells were also stained for focal adhesion kinase (FAK) which functions as a central mechano-sensing transducer in cells. Cells cultured on glass demonstrated a streaky pattern of FAK labeling similar to a vinculin pattern. However, the localization of FAK was found to be more punctuated and less well-defined for cells on nanofiber matrix. Similarly, the distribution of fibronectin on the cell surface cultured on glass substrate revealed a classic linear pattern of fibrils whereas, the cells on the nanofiber matrix showed a thicker network of more randomly deposited apically localized fibrils. This indicates that the fibrils are permissive for the assembly of a matrix that can promote the formation of 3D-matrix adhesions. These studies clearly demonstrate the advantages of using nanofiber matrices compared to microstructured or two-dimensional films to promote cell function for enhanced wound healing.

NANOFIBER MATRICES AS DIABETIC WOUND DRESSINGS

Being a very mild process, electrospinning can be used to encapsulate sensitive biomolecules within fibers. This quality, coupled with the high surface area of the fibers, allows for suitable presentation of these molecules to impart biological activity. Successful reproduction of collagen fibers through nanotechnology is an important step in mimicking the natural host environment. The next logical step includes the incorporation of additional ECM components during the electrospinning process itself. In order to steer tissue repair and regeneration in an effective direction, nanostructures must both structurally mimic the natural extracellular environment and be biologically interactive. This is evident from a recent study using collagen nanofibers, wherein coating the collagen nanofibers with matrix components such as laminin, fibronectin or additional (unspun) type I collagen, significantly increased its bioactivity and significant keratinocyte adhesion, migration and proliferation (Loughlin and Artlett 2009).

Choosing the appropriate molecules to incorporate into engineered nanostructures plays as important a role in achieving the goals of medical therapies as the physical makeup of the materials themselves. For example, hyaluronic acid (HA) is a linear polysaccharide consisting of alternating disaccharide units of b-1,4-D-glucuronic acid and b-1,3-N-acetyl-D-glucosamine that is essential in cell migration and proliferation in fetal development and repair (Chen and Abatangelo 1999). Using a mixture of 1,1,1,3,3,3-hexafluoro-2-propanol (HFIP) and formic acid as a solvent for collagen and HA, researchers were able to successfully electrospin a uniquely functional collagen-HA nanofiber scaffold (Hsu et al. 2010). The implications of such a procedure are notable in the development of a system that improves re-epithelialization without the development of scar tissue. The absence of scar formation is important in the context of internal wounds, surgeries, burns and lacerations, and can also be of cosmetic significance. Studies show that collagen-HA nanofibers decrease the expression of CD44 molecules associated with hypertrophic scar fibroblasts, and further down-regulate the inhibition of matrix metalloproteinases (MMPs) (Hsu et al. 2010). These effects increase the development of scarless wound healing and contribute to the re-epithelialization and ECM remodeling stages of wound repair. Depending on the goal of treatment, molecules with different functions can be utilized. This particular model is important in displaying the dynamic properties possible in the formation of nanofiber matrices geared toward scarless tissue formation not otherwise seen in adult repair processes. However, the already-high concentration of MMPs specific to diabetic wounds requires the application of slightly different nanofiber morphologies.

Increased presence of MMPs provides a specific example of the pathological characteristics of diabetic wounds. MMPs are Zinc-dependent proteases that are normally responsible for the cleavage and degradation of almost all components of the ECM (Lobmann et al. 2002). Such proteins are necessary for the appropriate interaction needed for typical matrix formation in a damaged state. However, the proper balance of MMP matrix remodeling and reconstruction must occur for them to be beneficial. MMP over-activity in diabetic wounds results in inhibition of wound closure and persistence of ulcers. Although MMPs are implicated in pathology in the diabetic state, they can nevertheless be useful in facilitating treatment with nanofiber technology. For example, MMP activity can be harnessed to cleave components of nanofiber systems designed to aid in wound healing upon release. Researchers employed such a method by electrospinning poly(ε-caprolactone) (PCL), a biocompatible and biodegradable polymer, along with polyethylene glycol(amine)$_2$ (PGA), to create a nanofibrous meshwork with readily exposed primary amine groups, capable of binding additional peptides (Kim and Yoo 2010). Researchers then used an MMP-cleavable hepta-peptide (DGPLGVC) to conjugate poly(ethyleneimine) (LPEI), along with DNA, to the nanofiber meshwork (Kim and Yoo 2010). To clarify, upon application of the PCL-PGA nanomeshwork to diabetic wounds, abundant concentrations of MMPs in the microenvironment cleaved the linker hepta-peptides, situated between the nanofibers and the LPEI-DNA ends, to release the LPEI-DNA in a controlled manner for transfection of target cells. Such an innovative design can be employed to deliver epidermal genes, small molecular weight molecules, proteins, or antibiotics to tissues for diabetic wound therapy.

Nanofibrous scaffolds with the ability to bind molecules without the application of cleavable linker-peptides compose another modality useful in diabetic wound therapy. Specific molecules chemically conjugated to nanofibers remain stable on the fiber surface. Due to this fixation, bioactivities of molecules remain more viable over the lifespan of the nanofiber within the body. Various growth factors are excellent candidates for these types of therapies. Specifically, members of the epithelial growth factor (EGF) family have been extensively studied for their effects in epithelial cell proliferation (Alemdaroğlu et al. 2006; Hashimoto 2000). Keratinocytes, the predominant cell in human epidermal tissue, have several classifications. For purposes of wound closure, the most important type is the epidermal stem cell (ESC), capable of differentiation and a large degree of proliferation. EGF directly interacts with EGF-receptors on the surface of ESCs and stimulates intracellular signaling cascades that result in differentiation and rapid division of ESCs to replace dead or damaged epithelia (Hashimoto 2000). While EGF release from Keratinocytes occurs naturally, it is hypothesized that exogenous EGF, conjugated to nanofibrous matrices, will boost this process and counteract the attenuated healing state. Researchers investigating these mechanisms in diabetic wounds have indeed produced results of accelerated wound closure (Choi et al. 2008). Using the previously described PCL-PGA nanofiber morphology, EGF was conjugated to the exposed amine groups on the fibers and the 50–500 nm diameter EGF-nanofibers were used as a dressing for diabetic animal dorsal wounds. By chemically conjugating EGF directly to nanofibers, the EGF is shielded from degradation in the wound microenvironment. Stabilized EGF molecules stimulated the rapid proliferation of keratinocytes while the PCL-PGA nanomatrix provided a suitable architecture for keratinocyte adhesion and migration. Additionally, EGF helped keratinocytes maintain phenotypic expression of keratinocyte-specific genes and decreased the representation of scar-fibroblasts that would otherwise contribute to deformation of the replaced skin (Choi et al. 2008). Such research demonstrates the versatility of nano-delivery techniques and it highlights the value of nanotechnology for the purposes of wound repair.

Selection of substances suitable for interaction within pathological microenvironments, such as those found in diabetic states, must directly address the presence of imbalanced biological mechanisms. Biocompatible molecules conducive to decreasing uncontrolled inflammation in diabetic ulcers are ideal for use in conjunction with nanotechnology. Encapsulation of such molecules within nanofibers for the purposes of accelerated healing is of particular interest. For example, studies show that compounds from the medicinal fungi *Phellinus linteus* and *Agaricus subrufescens*, and the plant *Curcuma longa* display anti-tumor, immunomodulatory, anti-inflammatory, and anti-allergenic properties (Choi et al. 2006a, b; Eybl et al. 2006). An additional benefit of these compounds is their low level of cytotoxicity to living tissue. An isolated molecule from *Curcuma longa*, known as curcumin, despite having a slightly greater cytotoxicity in high concentrations, has received particular attention due to its anti-inflammatory potency. In early studies, the effect of curcumin on *in vitro* mast cell degranulation confirmed the mechanism of curcumin in reducing inflammation (Choi et al. 2010). By limiting the activity of the G-protein coupled signaling cascade accelerated by compound 48/80, curcumin inhibits Ca^{2+} uptake by mast cells, ultimately resulting in an increased level of intracellular cAMP, inhibition of superoxide anion generation, and attenuated inflammatory mediator release (Choi et al. 2010). Curcumin also reduced the death of human foreskin fibroblasts (HFF-1) in the presence of hydrogen peroxide, and reduced the amount of IL-6 released from macrophages *in vitro*, thus highlighting its antioxidant properties (Choi et al. 2010). Researchers further found that oral administration of curcumin in diabetic rats resulted in a slight decrease in glycosylated hemoglobin (HbA1c) levels and a notable decrease in reactive oxygen species (ROS) *in vivo* (Rungseesantivanon et al. 2010).

Opposed to oral administration, curcumin released directly at the wound site has the potential to produce a focused healing response. At sources of ulceration, the most immediate and common immune cell infiltrates include monocytes capable of differentiating into macrophages and dendritic cells critical in debris clearance, infection prevention and wound closure. As previously discussed, dysfunction of these cells contributes to general malfunction of the healing process and persistent inflammation. Applicably, curcumin has been shown to reduce the action of nuclear factor-κB (NF-κB) in human monocytic (THP-1) cells and decreased activity of histone acetyltransferase (HAT), thus limiting THP-1 inflammatory cytokine release (Yun et al. 2010). To carry out these effects locally, curcumin can be delivered via common gel systems. More specifically, Poly(ethylene glycol) (PEG) derived hydrogels have been used to apply curcumin to diabetic wounds, resulting in increased rates of healing (Sidhu et al. 1999). PEG hydrogels are nontoxic, biocompatible polymers prepared by radiation crosslinking. They have a high degree of porosity, and their ability to hold and release compounds varies according to concentration prior to the crosslinking process (Peppas et al. 1999). If curcumin was directly applied to the wound, it would be immediately vulnerable to breakdown within the body, and the potential of its action would be limited within a small timeframe. The benefit of hydrogels and similar delivery systems is that they allow release of compounds at a controlled rate, thus protecting the compounds from breakdown or denaturation while simultaneously extending the time period of activity. However, hydrogels lack the structural integrity associated with nanofiber matrices, and tend to have more of a fast-burst rate of release during which 90% of molecules diffuse out of the gel within the first few hours (Peppas et al. 1999). A critical improvement of delivery mechanisms is one that allows a slow release of bioactive molecules over a long period of time. This will decrease the need for frequent wound dressing changes. Due to the fact that the therapeutic system is still effective over the course of hours to weeks or longer, replacing it would be less necessary.

To achieve longer effects, curcumin can thus be loaded onto nanofibers prior to the electrospinning process. As previously discussed, PCL is a commonly used polymer that is also ideal for incorporation of molecules within its structure. PCL is non-toxic and biodegrades within the body at a slow rate conducive to long-term application strategies. We have developed bead free curcumin loaded (3% and 17% w/w) PCL nanofibers with an average diameter of 300–400 nm (Merrell et al. 2009). Figure 1 shows PCL nanofibers loaded with curcumin. Despite the fact that PCL is a synthetic polyester, unlike natural collagen, nanofibers of such a diameter are still

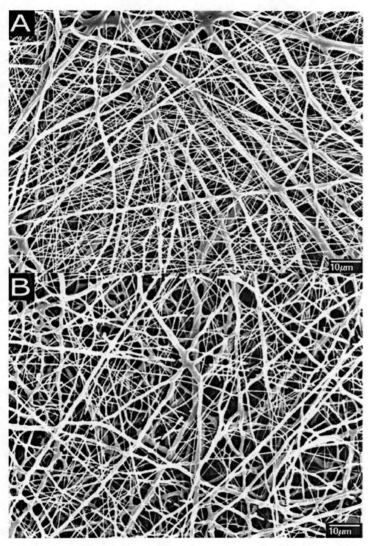

Fig. 1. Morphology of Curcumin Loaded PCL Nanofibers. (A) PCL nanofibers having 3% (w/w) of Curcumin; and (B) PCL nanofibers having 17% (w/w) of curcumin. (with permission from Wiley-Blackwell).

ideal for ECM biomimicry, and were found to release curcumin over the course of three days without a substantial burst-release. When loaded PCL nanofibers were applied to the dorsal wounds of diabetic mice, the curcumin actively reduced inflammation and accelerated wound closure over the course of 10 days (Fig. 2) (Merrell et al. 2009). The components of this unique delivery system consist of many of the previously described

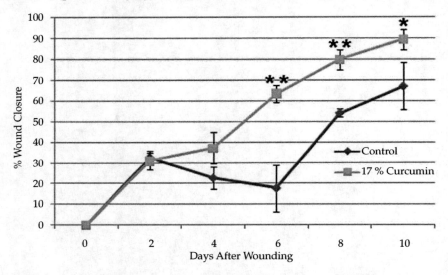

Fig. 2. Percent Wound Closure of Diabetic Wounds as a Function of Time. The graph reveals that wound closure was significantly higher in diabetic mice with 17% curcumin dressing after 6 days (*$p \leq 0.05$, **$p \leq 0.01$) compared with the closure rates of mice with a PCL-nanofiber control dressing. (with permission from Wiley-Blackwell).

specifications necessary for therapeutic systems for diabetic states. PCL acts as a biocompatible structure that induces the attachment, migration, and proliferation of cellular components involved in re-epithelialization. Simultaneously, curcumin acts at the wound site to reduce hyperglycemia-induced inflammation in order to enable commencement of the appropriate healing response. The duality of such a system is further enhanced by its ability to maintain stability over a period of several days while remaining a bioactive healing accelerant. Furthermore, due to their nontoxicity and biodegradability, the nanofibers do not have to be removed after wound closure, and can be enzymatically broken down during remodeling of the ECM.

Similar nanofiber systems are also beneficial in the controlled release of antibiotic, antifungal, and antimicrobial agents at localized wound sites. The vulnerability of diabetic individuals to infection is a common complication of unhealed ulceration and surgery. The electrospun biodegradable, biosynthetic, and biocompatible co-polymer poly(lactide-*co*-glycolide) (PLAGA) has proven useful in the loading and release of different antibiotics such as cephalosporins: cefoxitin and cefazolin (Kim et al. 2004). PLAGA conjugated with poly(ethylene glycol)-b-poly(lactide) (PEG-b-PLA) was efficient in encapsulating cefoxitin for a controlled release that successfully inhibited the growth of the bacterium *Staphylococcus aureus* (Kim et al. 2004). Biodegradable fiber matrices loaded with antibiotics, or any of the previously discussed molecules (curcumin for control of inflammation, or EGF for tissue regeneration) can potentially be placed on wounds and surgically modified tissues for infection prevention and limitation of post-operative complications.

The diverse delivery mechanisms discussed thus far involve several nanofiber morphologies and molecular activities. These modalities are not mutually exclusive, however, and can be combined and integrated for the production of unique results. For instance, nanofibers have the potential to encapsulate loaded molecules that exhibit the controlled release seen in curcumin-PCL models, and simultaneously conjugate other compounds for immobile stimulation noted in EGF-PCL-PGA stimulation of tissue regeneration (Choi et al. 2008; Merrell et al. 2009). In one particular model researchers used a dual-nozzle electrospinning process to incorporate bovine serum albumin (BSA) protein into the core of a co-spun nanofiber matrix consisting of PCL and PCL-PGA (Choi and Yo 2010). The dual-nozzle system is a modification that allows combination of the two immiscible components of the nanofiber. In this example, Choi and Yo mixed the BSA aqueous phase with poly(vinyl alcohol) (PVA) to increase viscosity and stabilization upon cospinning with a PCL/PCL-PGA polymer organic phase (Choi and Yo 2010). Following production of the nanofiber matrix with an encapsulated BSA core, researchers chemically conjugated BSA to the exposed primary amine groups of PGA. BSA protein encapsulated in the fiber core exhibited fast-release diffusion, while conjugated BSA exhibited a minimum release over the 4-day recording period (Choi and Yo 2010). By combining methods of encapsulation and conjugation, it is possible to create biphasic protein release from a nanofiber scaffold. Depending upon the focus of the therapy, molecules can simultaneously be released for an immediate response and remain conjugated for stable, long-term activity.

Application to Areas of Health and Disease

While the biphasic model is the most integrative, it remains just one example of the enormous potential of electrospun nanofiber applications. Implications of such technology within the diabetic state must simultaneously address the multifaceted obstacles characteristic of the hyperglycemic environment. Of these pathologies, the four highlighted to this point have been: 1. Uncontrolled chronic inflammation; 2. Insufficient ECM deposition; 3. A lack of proliferative re-epithelialization; and 4. Increased susceptibility to infection. Such factors are primarily responsible for the abnormal persistence of diabetic ulcers and, thus, must be directly addressed when considering treatment protocols. Electrospun nanofiber matrices provide researchers with a dynamic set of options from which to formulate the most effective therapy. Future applications have the capacity to integrate the aforementioned techniques to selectively solve diabetic wound pathologies. The use of anti-inflammatory molecules such as curcumin, loaded within nanofiber polymers, can first and foremost attenuate the inflammatory response and make way for the successive steps of healing. Through their biologically interactive properties, the nanofibers themselves act as appropriate scaffolding to bridge the expanse of the wound and recruit the appropriate cells necessary for ECM deposition. Those same fibers, through the chemical conjugation of EGF and other growth factors, can stimulate the proper proliferative pathways of keratinocytes and other cells responsible for re-epithelialization with a decreased incidence of scar tissue formation. Finally, innovative alterations of nanofiber composition and electrospinning processes provide the potential for a single dressing to additionally conjugate or encapsulate antibiotics or other drug therapies critical in the prevention of infection. Through such state-of-the-art modalities all four of the emphasized diabetic wound pathologies can be adequately addressed and remedied in one novel therapy.

Summary Points

- To improve upon current technologies for wound healing, new and future constructs, including electrospun nanofibers, must meet several important requirements.
- They should actively engage the microenvironment through mimicry of biostructural frameworks.
- An appropriate imitation of ECM components requires fiber and mesh diameters to be on the nanoscale.
- New modalities must also actively regulate cellular biological functions through the retention and release of small molecular weight molecules, such as anti-inflammatory mediators, cytokines, antibiotics, growth factors, and so on.
- Ability to conjugate biomolecules to exogenous therapeutic structural frameworks is equally as important as secreting them in a controlled manner for long-term delivery.

- Materials involved should be adequately biocompatible and biodegradable, and suitable for cell migration, differentiation, and proliferation.
- Electrospun nanofibers and matrices offer several promising options as wound dressings and delivery systems for individuals at risk for chronic wounds and those with biologically-compromised healing mechanisms.

Abbreviations

BSA	:	Bovine Serum Albumin
cAMP	:	Cyclic Adenosine Monophosphate
CD44	:	Receptor for Hyaluronic Acid
ECM	:	Extracellular Matrix
EGF	:	Epithelial Growth Factor
ESC	:	Epidermal Stem Cell
FAK	:	Focal Adhesion Kinase
HA	:	Hyaluronic Acid (Polysaccharide)
HAT	:	Histone Acetyltransferase
HbA1c	:	Glycosylated Hemoglobin
HFF-1	:	Human Foreskin Fibroblasts
HFIP	:	1,1,1,3,3,3-hexafluoro-2-propanol
HUVEC	:	Human Umbilical Vein Endothelial Cell
IL-6	:	Interleukin-6
IL-8 (CXCL8)	:	Interleukin-8
LPEI	:	Poly(ethyleneimine)
MCP-1	:	Monocyte Chemo-attractant Protein-1
MMP	:	Matrix Metalloproteinase
NF-κB	:	Nuclear Factor-κB
NIH 3T3	:	Mouse Embryonic Fibroblast Cell Line (3-day Transfer, inoculum 3×10^5 cells)
PCL	:	Poly(ε-caprolactone)
PEG	:	Poly(ethylene glycol)
PEG-b-PLA	:	Poly(ethylene glycol)-b-poly(lactide)
PGA	:	Polyethylene Glycol(amine)$_2$
PLAGA	:	Poly(lactic acid-co-glycolic acid)
PLL-CL	:	Poly(L-lactide-co-caprolactone)
PVA	:	Poly(vinyl alcohol)
ROS	:	Reactive Oxygen Species
THP-1	:	Human Monocytic Cell Line
TNFα	:	Tumor Necrosis Factor α

References

Alemdaroğlu, C., Z. Değim, N. Çelebi, F. Zor, S. Öztürk and D. Erdoğan. 2006. An investigation on burn wound healing in rats with chitosan gel formation containing epidermal growth factor. Burns. 32: 319–327.

Bonani, W., D. Maniglio, A. Motta, W. Tan and C. Migliaresi. 2011. Biohybrid nanofiber constructs with anisotropic biomechanical properties. J. Biomed. Mater. Res. B. Appl. Biomater. 96: 276–286.

Chen, W.Y.J., and G. Abatangelo. 1999. Functions of hyaluronan in wound repair. Wound Repair Regen. 7: 79–89.

Choi, J.S., K.W. Leong and H.S. Yooa. 2008. In vivo wound healing of diabetic ulcers using electrospun nanofibers immobilized with human epidermal growth factor (EGF). Biomaterials 29: 587–596.

Choi, J.S., and H.S. Yo. 2010. Nano-inspired fibrous matrix with bi-phasic release of proteins. J. Nanosci. Nanotechno. 10: 3038–3045.

Choi, Y.H., G.H. Yan, O.H. Chai, J.M. Lim, S.Y. Sung, X. Zhang, J.-H. Kim, S.H. Choi, M.S. Lee, E.-H. Han, H.T. Kim and T.H. Song. 2006a. Inhibition of anaphylaxis-like reaction and mast cell activation by water extract from the fruiting body of *Phellinus linteus*. Biol. Pharm. Bull. 29: 1360–1365.

Choi, Y.H., G.H. Yan, O.H. Chai, Y.H. Choi, X. Zhang, J.M. Lim, J-H. Kim, M.S. Lee, E-H. Han, H.T. Kim and C.H. Song. 2006b. Inhibitory effects of *Agaricus blazei* on mast cell-mediated anaphylaxis-like reactions. Biol. Pharm. Bull. 29: 1366–1371.

Choi, Y.H., G.H. Yan, O.H. Chai and C.H. Song. 2010. Inhibitory effects of curcumin on passive cutaneous anaphylactoid response and compound 48/80–induced mast cell activation. Anat. Cell Biol. 43: 36–43.

Curtis, A., and C. Wilkinson. 2001. Nanotechniques and approaches in biotechnology. Trends Biotechnol. 19: 97–101.

Duraisamy, Y., M. Slevin, N. Smith, J. Bailey, J. Zweit, C. Smith, N. Ahmed and J. Gaffney. 2001. Effect of glycation on basic fibroblast growth factor induced angiogenesis and activation of associated signal transduction pathways in vascular endothelial cells: possible relevance to wound healing in diabetes. Angiogenesis 4: 277–288.

Eybl, V., D. Kotyzova and J. Koutensky. 2006. Comparative study of natural antioxidants—curcumin, resveratrol and melatonin—in cadmium-induced oxidative damage in mice. Toxicology 225: 150–156.

Hashimoto, K. 2000. Regulation of keratinocyte function by growth factors. J. Dermatol. Sci. 24: 46–50.

Hsu, F.-Y., Y.-S. Hung, H.-M. Liou and C.-H. Shen. 2010. Electrospun hyaluronate–collagen nanofibrous matrix and the effects of varying the concentration of hyaluronate on the characteristics of foreskin fibroblast cells. Acta. Biomater. 6: 2140–2147.

Khanna, S., S. Biswas, Y. Shang, E. Collard, A. Azad, C. Kauh, V. Bhasker, G.M. Gordillo, C.K. Sen and S. Roy. 2010. Macrophage dysfunction impairs resolution of inflammation in the wounds of diabetic mice. Public Library of Science One. 5(3): e9539.

Khil, M.S., D.I. Cha, H.Y. Kim, I.S. Kim and N. Bhattarai. 2003. Electrospun nanofibrous polyurethane membrane as wound dressing. J. Biomed. Mater. Res. B. Appl. Biomater. 67: 675–679.

Kim, K., J.K. Luu, C. Chang, D. Fang, B.S. Hsiao, B. Chu and M. Hadjiargyrou. 2004. Incorporation and controlled release of a hydrophilic antibiotic using poly(lactide-co-glycolide)-based electrospun nanofibrous scaffolds. J. Control Release 98: 47–56.

Kim, H.S., and H.S. Yoo. 2010. MMPs-responsive release of DNA from electrospun nanofibrous matrix for local gene therapy: In vitro and in vivo evaluation. J. Control. Release 145: 264–271.

Kumbar, S.G., S.P. Nukavarapu, J. Roshan, L.S. Nair and C.T. Laurencin. 2008. Electrospun poly(lactic acid-co-glycolic acid) scaffolds for skin tissue engineering. Biomaterials 29: 4100–4107.

Kwon, K., S. Kodoaki and T. Matsuda. 2005. Electrospun nano to microfiber fabrics made of biodegradable copolyesters : structural characteristics, mechanical properties and cell adhesion potential. Biomaterials 26: 3929–3939.

Landis, R.C., B.J. Evans, N. Chaturvedi and D.O. Haskard. 2010. Persistence of TNFα in diabetic wounds. Diabetologia 53:1537–1538.

Laurencin, C.T., L.S. Nair. [Eds.] 2008. Nanotechnology and Tissue Engineering: The Scaffold. CRC press, Boca Raton, Florida, USA.

Lay-Flurrie, K. 2004. The properties of hydrogel dressings and their impact on wound healing. Prof. Nurse. 19: 269–73.

Leach, M.K., Z.Q. Feng, S.J. Tuck and J.M. Corey. 2011. Electrospinning fundamentals: optimizing solution and apparatus parameters. J. Vis. Exp. 47: e2494.

Lobmann, R., A. Ambrosch, G. Schultz, K. Waldmann, S. Schiweck and H. Lehnert. 2002. Expression of matrix-metalloproteinases and their inhibitors in the wounds of diabetic and non-diabetic patients. Diabetologia 45: 1011–1016.

Loots, M.A., E.N. Lamme, J. Zeegelaar, J.R. Mekkes, J.D. Bos and E. Middelkoop. 1998. Differences in cellular infiltrate and extracellular matrix of chronic diabetic and venous ulcers versus acute wounds. J. Invest. Dermatol. 111: 850–857.

Loughlin, D.T., and C.M. Artlett. 2009. 3-deoxyglucosone-collagen alters human dermal fibroblast migration and adhesion: implications for impaired wound healing in patients with diabetes. Wound Repair Regen. 17: 739–749.

Merrell, J.G., S.W. McLaughlin, L. Tie, C.T. Laurencin, A.F. Chen and L.S. Nair. 2009. Curcumin-loaded poly(ε-caprolactone) nanofibers: diabetic wound dressing with anti-oxidant and anti-inflammatory properties. Clin. Ex. Pharmacol. P. 36: 1149–1156.

Nair, L.S., S. Bhattacharyya and C.T. Laurencin. 2004. Development of novel tissue engineering scaffolds via electrospinning. Expert Opin. Biol. Ther. 4: 1–10.

Nair, L.S., S. Bhattacharya and C.T. Laurencin. Nanotechnology and Tissue Engineering: The Scaffold based Approach. pp. 1–56. In: C. Kumar. [Ed.] 2006. Tissue, Cell and Organ Engineering, Nanotechnologies for the Life Sciences series. Wiley-VCH Verlag GmBH & Co., Weinheim, Germany.

Ovington, L.G., B. Pierce. Wound dressings: form, function, feasibility and facts. pp. 311–319. In: D.L. Krasner, G.T. Rodeheaver, R.G. Sibbald. [eds.] 2001.

Chronic Wound Care: a Clinical Sourcebook for Healthcare Professionals 3rd ed. Wayne, Pennsylvania, USA.

Peppas, N.A., K.B. Keys, M. Torres-Lugo and A.M. Lowman. 1999. Poly(ethylene glycol)-containing hydrogels in drug delivery. J. Control Release 62: 81–87.

Pradhan, L., C. Nabzdyk, N.D. Andersen, F.W. LoGerfo and A. Veves. 2009. Inflammation and neuropeptides: the connection in diabetic wound healing. Expert Rev. Mol. Med. 11: e2.

Pradhan, L., X. Cai, S. Wu, N.D. Andersen, M. Martin, J. Malek, P. Guthrie, A. Veves and F.W. LoGerfo. 2010. Gene Expression of pro-inflammatory cytokines and neuropeptides in diabetic wound healing. J. Surg. Res. (in press).

Reneker, D.H., and I. Chun. 1996. Nanometric diameter fibers of polymer produced by electrospinning. Nanotechnology. 7: 216–223.

Rho, K.S., L. Jeong, G. Lee, B.-M. Seo, Y.J. Park, S.-D. Hong, S. Roh, J.J. Cho, W.H. Park and B-M. Min. 2006. Electrospinning of collagen nanofibers: effects on the behavior of normal human keratinocytes and early-stage wound healing. Biomaterials 27: 1452–1461.

Rungseesantivanon, S., N. Thenchaisri, P. Ruangvejvorachai and S. Patumraj. 2010. Curcumin supplementation could improve diabetes-induced endothelial dysfunction associated with decreased vascular superoxide production and PKC inhibition. BMC Complem. Altern. M. 10: 57.

Sanders, J.E., C.E. Stiles and C.L. Hayes. 2000. Tissue response to single-polymer fibers of varying diameters: evaluation of fibrous encapsulation and macrophage density. J. Biomed. Mater Res. 52: 231–237.

Schindler, M., I. Ahmed, J. Kamal, A. Nur-E-Kamal, T.H. Grafe, H.Y. Chung and S. Meiners. 2005. A synthetic nanofibrillar matrix to promote in vivo like organization and morphogenesis for cell in culture. Biomaterials 26: 5624–5631.

Shin, M.Y., M.M. Hohmann, M. Brenner and G.C. Ruteldge. 2001. Experimental characterization of electrospinning: The electrically forced jet and instabilities. Polymer 42: 9955–9967.

Sidhu, G.S., H. Mani, J.P. Gaddipati, A.K. Singh, P. Seth, K.K. Banaudha, G.K. Patnaik and R.K. Maheshwari. 1999. Curcumin enhances wound healing in streptozotocin induced diabetic rats and genetically diabetic mice. Wound Repair Regen. 7: 362–374.

Yun, J.M., I. Jialal and S. Devaraj. 2010. Epigenetic regulation of high glucose-induced proinflammatory cytokine production in monocytes by curcumin. J. Nutr. Biochem. (in press).

Zhang, Y., C.T. Lim, S. Ramakrishna and Z.-M. Huang. 2005. Recent development of polymeric nanofibers for biomedical and biotechnological applications. J Mater. Sci: Mater in Medicine 16: 933–946.

18

Poly-N-acetyl Glucosamine Nanofibers Derived from a Marine Diatom: Applications in Diabetic Wound Healing and Tissue Regeneration

John N. Voumakis,[3,a,5] Thomas Fischer,[2] Haley Buff Lindner,[1,a] Marina Demcheva,[3,b] Arun Seth[4] and Robin C. Muise-Helmericks[1,b]

INTRODUCTION

Natural products derived from marine organisms are useful for the development of drugs used for human disease. Marine organisms have proved to be a rich source of compounds that are either in clinical or

---

[1]Department of Regenerative Medicine and Cell Biology, Medical University of South Carolina, Charleston, SC 29425, USA.
[a]E-mail: hab2@musc.edu
[b]E-mail: musehelm@musc.edu
[2]Francis Owen Blood Research Lab, Department of Pathology and Laboratory Medicine, University of North Carolina at Chapel Hill, 125 University Lake Rd, Chapel Hill, NC 27516, USA.
E-mail: thomas_fischer@med.unc.edu
[3]Marine Polymer Technologies, Inc., Danvers, MA 01923.
[a]E-mail: johnv@webmpt.com
[b]E-mail: marina_d@webmpt.com
[4]Sunnybrook Research Institute, University of Toronto, Ontario, Canada, M5G 1G6.
E-mail: arun.seth@sri.utoronto.ca
[5]Department of Graduate Studies, Medical University of South Carolina, Charleston, SC 29425, USA.

List of abbreviations after the text.

preclinical studies. In the 1950s, arabino- and ribo-pentolsyl nucleosides derived from marine sponges were discovered, eventually leading to the development of Ara-A and Ara-C, anticancer therapeutics that have been in the clinic for many years (Chin, Yoon et al. 2009; Molinski, Dalisay et al. 2009). Numerous drugs have been isolated from marine organisms for use in cancer therapy by inhibition of cell cycle progression, growth factor secretion, and intracellular signaling. Many marine sulfated polysaccharides have anticoagulant and antithrombotic activities in part due to their activation of thrombin inhibitors. Some of these algal polysaccharides inhibit viral replication and/or attachment to cell surfaces while some function as anti-inflammatory agents. Injection of fucoidan, a sulfated polysaccharide, inhibits migration of leukocytes to sites of inflammation (Sun, Mao et al.; Zhang, Fujii et al. 2008). In this chapter we will discuss the purification, structural and chemical characteristics of a poly-N-acetyl-glucosamine nanofiber derived from a marine diatom (microalga) which is showing promise for the treatment of chronic cutaneous wounds commonly developed in patients with Type II diabetes.

ISOLATION AND CHARACTERIZATION OF pGlcNAc NANOFIBERS ISOLATED FROM A MARINE DIATOM

There are more than 200 known genera of diatoms and as many as 200,000 species, living in oceans, freshwater, and in the soil. The majority of species are found in the oceans where they are major contributors to oxygen generation via photosynthesis, and are the basis of the food chain for numerous marine and fresh water species (Hasle and Syvertsen 1997). Some diatoms produce a nanofiber component consisting of a high molecular weight polysaccharide. The nanofibers include many individual polymers aligned in parallel and tightly hydrogen-bonded resulting in a unique three dimensional structure. These fibers have been visualized and characterized by electron microscopy. Figure 1A shows a high resolution (250,000X) electron micrograph of a dried membrane "patch" of these nanofibers. The individual pGlcNAc nanofibers have widths within the 0.1µ range. It is this "patch" that is used as described below in cutaneous wound healing. Figures 1B and 1C show the nanofibers and sub-fibers contained within and dimensions of both the nanofibers and their associated sub-fiber structures. Chemical analyses have shown that these nanofibers consist of long chains of N-acetyl glucosamine in a β1–4 linkage (Fig. 2). As measured using intrinsic viscosity, the nanofibers consist of high molecular weight fully acetylated pGlcNAc polymers averaging molecular weight of 2.8×10^6 Daltons and containing 80 to 120 pGlcNAc polymers per nanofiber. Figure 3 shows an infrared spectrum (FTIR) of the

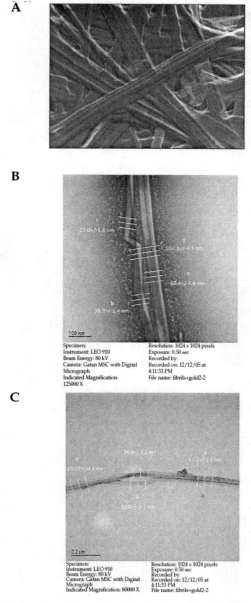

Fig. 1. pGlcNAc polymers exist as 0.1μ nanofibers with sub-fiber structures. Scanning electronmicroscopy showing in (A) a dried membrane of concentrated, purified nanofibers, (B and C) higher magnifications of nanofibers and their associated sub-fibers. In each micrograph, nm measurements are indicated.

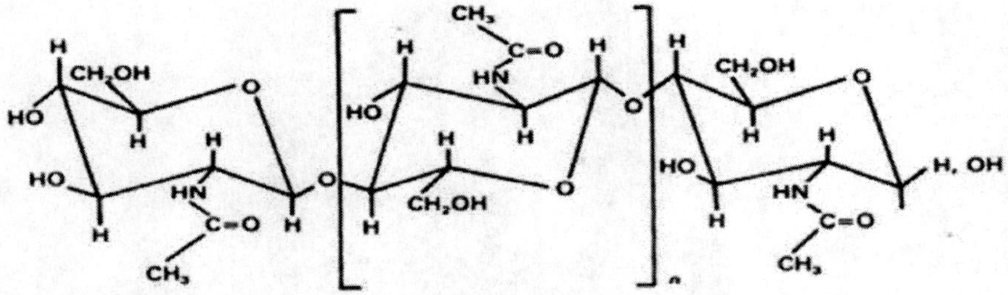

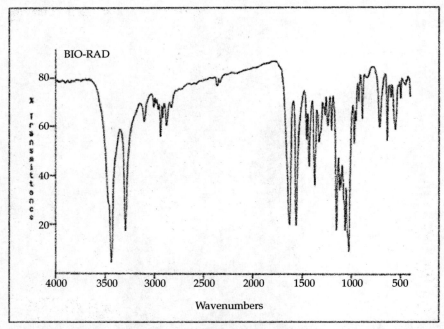

Fig. 2. Chemical structure of the pGlcNAc nanofiber showing the β1-4 linkage.

Fig. 3. FTIR spectrum of the pGlcNAc nanofibers. An infra-red spectrum of the pGlcNAc nanofiber material is shown. Notice the sharpness of the peaks indicating a regular three dimensional crystalline structure.

pGlcNAc nanofibers indicating their unique covalent chemical structure while X-ray scattering studies show they are organized in a unique beta quaternary structure. This unique 3D structure is the basis for their many biological activities.

pGlcNAc NANOFIBERS HAVE BIOLOGICAL ACTIVITY

Polymers containing N-acetyl glucosamine such as hyaluronic acid are the focus of numerous studies for use in anti-coagulant therapies and treatment of joint pain. The identification of the unique fiber structure found in the marine diatom pGlcNAc nanofibers has lead to their use as a hemostatic agent and to the initial understanding of their biological properties. These pure nanofiber preparations have been tested for their biocompatibility and show no toxic, allergic or deleterious effects and are therefore considered completely biocompatible.

Hemostasis

It has been shown that materials consisting entirely of pGlcNAc nanofibers are useful for the control of bleeding. When a vascular injury occurs, endothelial cells lining the blood vessels interact with circulating platelets. Adhesion to the basal lamina of the damaged vessel exposes the platelets to von Willebrand factor and collagen resulting in the adherence of platelets to the basal lamina and for the interaction between collagen and $\alpha_2\beta_3$ integrin thus stimulating the $\alpha_{IIb}\beta_3$ integrin and the secretion of ADP and thromboxane. Other pathways can also activate $\alpha_{IIb}\beta_3$ which causes tight binding to fibrinogen. Dimeric fibrinogen links platelets into aggregates, resulting in fibrin polymerization and the formation of a primary plug.

Platelets become activated by contact to foreign materials such as metals, glass and plastic. Accentuation of this property is important for the development of new hemostatic devices. Numerous studies have shown that pGlcNAc nanofibers reduce bleeding in several experimental systems (Pusateri, Modrow et al. 2003; Jensen, Machicado et al. 2004; Schwaitzberg, Chan et al. 2004) including liver injury models in pigs, canine models of esophageal variceal hemorrhage, and in several human clinical trials (Cole, Connolly et al. 1999; Hirsch 2003; Weiner, Fischer et al. 2003; Palmer, Gantt et al. 2004). Activation of platelets is via specific interactions between the nanofibers and platelet cell surface receptor proteins. Contact with the nanofibers result in marked platelet shape change and pseudopodia extensions characteristic of a full and irreversible platelet activation (Fig. 4), and Ca2+ immobilization and activation of integrin αIIbβ3 and fibrin polymerization (Thatte, Zagarins et al. 2004). It has been found that pGlcNAc nanofibers specifically bind to integrin β3 and activate platelets via an outside-in integrin signaling cascade (Fischer, Thatte et al. 2005). It has been recently demonstrated that red blood cells also undergo a dramatic shape change from discoid to stomatocytic upon receptor-specific interactions with the pGlcNAc nanofibers. These changes

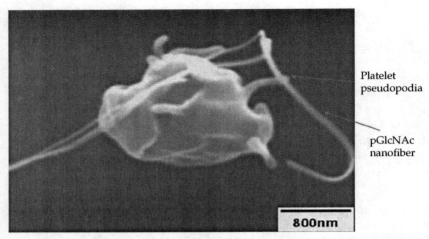

Fig. 4. Platelet activation by pGlcNAc nanofibers results in shape change and pseudopodia formation. Electron microscopy shows the interaction between a single platelet and the pGlcNAc nanofiber. Positions of the pseudopodia and the nanofiber are indicated by the arrows.

also result in activation of the clotting cascade in a manner similar to that occurring on platelets (Fischer, Valeri et al. 2008; Smith, Vournakis et al. 2008).

Given the marked effect of the pGlcNAc nanofibers on hemostasis, a number of patch products consisting of the nanofibers are FDA cleared and used in the clinical arena to control bleeding during cardiac catheterization (Hirsch, Reddy et al. 2003; Palmer, Gantt et al. 2004) and are marketed as the Syvek Patch family of products. In addition, a newer larger product, the mRDH bandage, is used under conditions of severe bleeding in surgery and trauma ((Salerno, Gaughan et al. 2009), J.Trauma, in press). The use of the mRDH for controlling hemorrhaging due to the severe injury effects of explosive devices (IEDs) in war theaters, and post operatively in obese, diabetic patients, show rapid hemostasis (J.Trauma, in press).

pGlcNAc NANOFIBERS IN DIABETIC WOUND HEALING

One of the well-described effects of platelet activation is the secretion of concentrated growth factors which promote wound healing (Smith and Roukis 2009; Smyth, McEver et al. 2009). Wound healing involves a complex interaction between a series of cell types, cytokines and growth factors. Each component participates in a contiguous process involving hemostasis, inflammation, cellular proliferation and remodeling. Many chronic wounds, such as those that occur in diabetic patients, remain resistant to treatments and either do not heal or heal very slowly. Efforts have been made to not only increase wound healing in chronic wounds but to also increase the rate of wound closure of acute wounds. Many of these efforts, including the application of recombinant growth factors or creation of dermal substitutes have been somewhat effective; however, chronic wound healing remains a matter of high clinical significance. Given the marked effect of pGlcNAc nanofibers on hemostasis, pure preparations of these fibers were formed into membranes and tested for their ability to increase wound closure in diabetic mouse models.

Leptin and Leptin Receptor Diabetic Mouse Models

There are two leptin signaling based mouse models of diabetes. One model is an abnormal splice variant that prevents expression of a full length leptin receptor (db/db) (Chen, Charlat et al. 1996). Leptin, a fat cell specific hormone, binds this receptor and regulates adipose tissue accumulation. In the absence of appropriate signaling, these mice and their leptin deficient counterparts (ob/ob) become obese and insulin resistant and are therefore considered to be mouse models for Type 2 diabetes (Hattangady and Rajadhyaksha 2009). These mice also show significant wound healing deficiencies as compared to wild type counterparts making these mice useful for wound healing studies (Tomlinson and Ferguson 2003; Odorisio, Cianfarani et al. 2006). These models allow for the quantitative measurement of a series of wound healing criteria including formation of granulation tissue, angiogenesis, wound closure and collagen synthesis and also allow for the testing of new therapies.

Recent findings show that treatment of cutaneous wounds with pGlcNAc fiber-derived membranes results in an increased kinetics of wound healing. Using the db/db mouse model, treatment of full thickness skin wounds with a membrane consisting entirely of pGlcNAc nanofibers resulted in a 90% wound closure rate that was 9 days faster than the untreated db/db mouse controls. Treated wounds showed an increase proliferation within granulation tissue as well as increased angiogenesis, as measured by PECAM staining. These findings suggest that pGlcNAc nanofibers, in addition to the effects on hemostasis, also activate a wound healing genetic program (Pietramaggiori, Yang et al. 2008).

To create a shorter fully biodegradable nanofiber, the native nanofibers (NAG) were shortened by irradiation (sNAG). Using the db/db mouse system, these thin sNAG membranes placed directly into the wound bed profoundly accelerated wound closure mainly by re-epithelialization and increased keratinocyte migration, granulation tissue formation, cell proliferation, and vascularization compared with control wound treatments. Expression of markers of angiogenesis (Vournakis, Eldridge et al. 2008; Scherer, Pietramaggiori et al. 2009) (VEGF), cell migration (uPAR) and ECM remodeling (MMP3, MMP9) were also up-regulated in sNAG treated wounds compared with controls. Erba et al. showed that a combinatorial treatment of VAC (vacuum assisted closure) in combination with sNAG application is synergistic for wound closure and leads to increased wound repair as compared to VAC alone (J.Trauma, *in press*). These findings indicate that pGlcNAc nanofibers can be used to not only provide hemostasis but also to increase wound healing kinetics, thus providing combinatorial therapy using a FDA approved single agent.

MECHANISMS OF pGlcNAc NANOFIBER FUNCTION IN WOUND HEALING

pGlcNAc Nanofiber Stimulation of Angiogenesis

One effect of pGlcNAc nanofiber treatment is a marked increase in angiogenesis (Shiojima and Walsh 2002). Increased angiogenesis is a hallmark of cutaneous wound healing and is necessary to support new tissue formation (Singer and Clark 1999) and has long been known to be deficient in diabetic wound healing (Bohlen and Niggl 1979). VEGF production is strongly up-regulated in wound healing; secreted by activated macrophage and keratinocytes working in consort to stimulate new capillary production within the wounded area. Impairment of new vessel formation results in decreased wound healing abilities (Tonnesen, Feng et al. 2000; Galiano, Tepper et al. 2004; Hong, Lange-Asschenfeldt et al. 2004). Concentrated efforts have been made to increase vascularization for tissue regeneration and repair of chronic, non-healing ischemic wounds. New therapies using recombinant growth factors or vascular progenitor cells to foster the formation of new blood vessels have been proposed, some of which are presently in phase II/III trials (Yla-Herttuala, Markkanen et al. 2004; Katsube, Bishop et al. 2005).

We tested whether the increase in angiogenesis resulting from treatment of cutaneous wounds with pGlcNAc fibers was due to a direct effect on endothelial cells per se and whether the effect of pGlcNAc was integrin and/or VEGF dependent. We found that pGlcNAc treatment, in the absence of added growth factor or serum, induces endothelial cells motility and increased *in vitro* angiogenesis as measured by cord formation in Matrigel assays. pGlcNAc-induced cell motility is found to be integrin mediated and results in the activation of MAPK and the increased expression of the transcription factor Ets1, which is required for pGlcNAc-induced cell motility. Activation of the Ets1 transcription factor has emerged as an important downstream regulator of angiogenic cell movement. Endothelial cell growth factors such as vascular endothelial growth factor (VEGF) up-regulate the activity of Ets1 in primary endothelial cells (Lavenburg, Ivey et al. 2003) resulting in the activation of downstream target genes, such as metalloproteinases and vimentin (Sementchenko and Watson 2000) that are important for matrix degradation and cell migration.

We also show that pGlcNAc induction of Ets1 expression is dependent on integrin activation. Indeed, antibody blockade of α5β1 integrin activation results in a decreased induction of Ets1 by pGlcNAc. Integrins play important roles in both the migration of cells in the wound bed and in the regulation of basement membrane and extracellular matrix (ECM) interactions. Integrin ligation (or association with the cognate matrix protein) activates a series of signaling pathways, including a Rac-dependent pathway generally thought to result in migration, MAP kinase (proliferation), PI3K/Akt (survival) and IKK/NFκB (survival). Stimulation of cells with pGlcNAc nanofibers must then result in the activation of integrin dependent signaling and depending on cellular context could therefore activate multiple integrins and multiple pathways. Stimulation of integrin dependent signaling in a wound bed then may result in activation of a series of complex signaling pathways that culminate in a stimulation of wound healing.

Impact of pGlcNAc Nanofibers on Innate Immunity-antimicrobial activity

In addition to its effects on Ets-induced cell motility, pGlcNAc treatment results in an increased expression of IL-1 and VEGF (Vournakis, Eldridge et al. 2008). It is well known that the wound healing process is regulated by a coordinated secretion of chemokines and growth factors. VEGF secretion allows for the recruitment of new vessels both via activation of resident endothelial cells and the recruitment of endothelial progenitors (Tonnesen, Feng et al. 2000; Galiano, Tepper et al. 2004). IL-1, secreted by macrophages and other cells activates keratinocytes and recruits other lymphocytes. Interestingly, pGlcNAc treatment of endothelial cells does not result in an up-regulation of IL-8, another important cytokine involved in wound healing nor does it affect fibroblast growth factors 1 or 2, lending to the idea of specificity in the stimulation of pGlcNAc induced cytokine, growth factor secretion.

To test for pGlcNAc nanofiber dependent changes in gene expression, a series of different microarray analyses have been performed using primary human endothelial cells, keratinocytes and fibroblasts. These analyses have indicated the regulation of a number of genes involved in transcription cell cycle regulation and cytokine expression. Interestingly, pGlcNAc nanofiber treatment of each cell type results in the induction of a gene cluster that functions within innate immunity. A list of some of these genes is shown in Table 1, below. Interestingly, each of these genes functions in the detection and clearance of bacterial infections.

Table 1. Genes up-regulated in response to sNAG nanofiber stimulation.

IL-1—Pro-inflammatory cytokine involved in immune defense
CEACAM3—Cell adhesion molecule which directs phagocytosis of several bacterial species
SPAG11—β-defensin-3 like molecule that exhibits antimicrobial properties
Defensins—Several defensins that exhibit antimicrobial activity
TLRs—Several Toll-like Receptors: important for stimulation of cellular responses toward infection

Defensins and Defensin Like Molecules

Defensins are small (3-4 kDa), cysteine-rich cationic peptides found in mammals, insects, and plants that are classified into different families (α, β, and θ) based on their pattern of disulfide bonding. α-defensins are neutrophil specific, comprising approximately 5–7% of the total cellular neutrophil protein (Ganz 1994; Ganz and Lehrer 1994). β-defensins are constitutively expressed by epithelial cells of the tracheobronchial lining, skin, and kidney where they can be up-regulated in response to infectious or inflammatory stimuli (Ganz 1994; Ganz and Lehrer 1994). Although the exact mechanism is not known, these small peptides possess antimicrobial properties that are active against many gram positive and negative bacteria, fungi, and viruses. As shown in Table 1, pGlcNAc nanofiber stimulation of both primary endothelial cells and keratinocytes results in increased expression of a series of defensin peptides and a defensin related peptide, SPAG11 (Fei, Chen et al. 2007). Our published results show that sNAG treatment is antibacterial via defensin up regulation in a *Staphylococcus aureus* infected cutaneous wound healing model (Lindner, Zhang et al.).

As shown in Fig. 5A, sNAG stimulates at least a 3-fold increased expression of defensins. In addition, using a *Staphylococcus aureus* infected wound model, we show that defensin peptide expedites bacterial clearance as shown by the tissue gram staining in Fig. 5B. The dark areas of staining indicate bacterial staining which is much diminished in the defensin treated infected wounds. Given that TLRs and defensins are part of the innate immune system, activation of these pathways will preclude the generation of resistant organisms as well as allow for the antibiotic-independent clearance of bacterial infection. Use of pGlcNAc nanofibers in the hospital setting would defray much of the cost and markedly reduce the production of antibiotic resistant species.

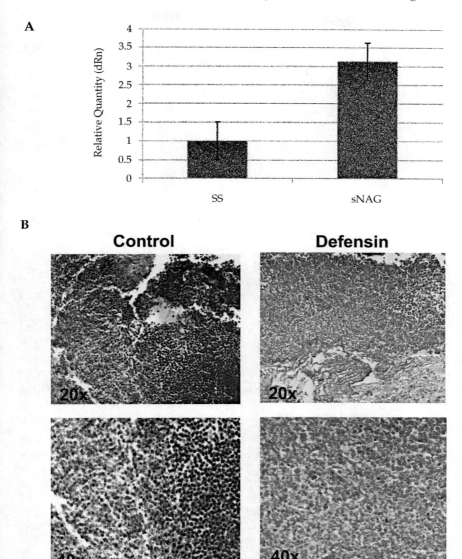

Fig. 5. pGlcNAc nanofiber treatment promotes defensin expression and antibacterial responses. (A) Real Time PCR of sNAG treated cells showing increased defensin expression. (B) *Staph aureus* infected cutaneous wounds were untreated or treated with defensin peptide and assessed for *S. aureus* bacterial staining 3 days post wound/infection.

Toll-like Receptors

Toll-like receptors (TLRs) are transmembrane signaling receptors that recognize bacterial components as "non-self" and trigger a series of signaling cascades that result in the microbial killing or activation of immune cells (Akira 2003; Underhill 2004; Takeda and Akira 2007; Randhawa and Hawn 2008). TLR receptors are expressed on the surface of macrophages and are required for the initiation of the innate immune response and for the generation of the adaptive immune response. TLRs are a family of ten different receptors that recognize ligands such as LPS and bacterial lipoproteins and flagellin (Heine, Kirschning et al. 1999; Lien, Sellati et al. 1999; McCurdy, Lin et al. 2001; Henneke, Takeuchi et al. 2002). Different TLRs are thought to provide specificity to the immune response. For example, TLR4 deficient mice do not respond to LPS and are more susceptible to infection by gram negative bacteria whereas mice deficient in TLR2 are more susceptible to infection by gram positive bacteria (Heine, Kirschning et al. 1999; Lien, Sellati et al. 1999; McCurdy, Lin et al. 2001; Henneke, Takeuchi et al. 2002). Signaling pathways initiated downstream of TLRs are complex, but are known to include the NFκB and MAP kinase pathways which lead to cytokine production and interferon inducible genes.

Using a custom gene chip containing 14 toll-like receptor genes and several housekeeping genes as controls we have shown that 4 TLRs, a downstream signaling intermediate (TRAF6) and an IL1-like receptor SIGIRR which modulates TLR activity are up-regulated by sNAG treatment of primary human endothelial cells following a 5 hour stimulation. These results are shown in Table 2 and include the known activators of these receptors and the fold increase upon sNAG stimulation. Gene array results have been confirmed using qPCR (data not shown). These are "pattern recognition receptors" which recognize molecules that are shared by pathogens such as bacteria, thus recognizing bacteria and allowing for antimicrobial responses (Chao 2009). Given that these receptors recognize patterns and that pGlcNAc nanofibers are chemically, although not structurally, similar to bacterial cell walls, treatment with these nanofibers may actually be mimicking a bacterial infection, thus activating an innate immune response.

Table 2. pGlcNAc nanofibers stimulate Toll-like receptor expression.

TLR1—Triacyl lipopeptides from bacteria and mycobacteria; 7.6 fold
TLR4—LPS, viral proteins, Hsp60 (Chlamydia); 5.1 fold
TLR7—synthetic compounds; 3.3 fold
TLR8—synthetic compounds; 2.1 fold
TRAF6—Downstream signaling modulator; 6.2 fold
SIGRR - IL-1 receptor related TLR modulator; 5.9 fold

pGlcNAc Nanofibers Promote Cellular Recruitment

In addition to reduced angiogenesis, diabetic wound healing also presents with decreases in cellular recruitment and proliferation. pGlcNAc nanofibers stimulate the expression such as IL-1 and VEGF that allow for cellular recruitment. Defensins exhibit biological activities beyond the inhibition of microbial cells, including their contribution to the adaptive immune response by exhibiting chemotactic activity on dendritic (Hubert, Herman et al. 2007) and T cells, monocytes, and macrophages (Liu, Verin et al. 2001) and keratinocytes (Niyonsaba, Ushio et al. 2007). Interestingly, we show that pGlcNAc nanofiber treatment induces both α and β-defensin expression in endothelial cells and β-defensins in keratinocytes, (Lindner, Zhang et al.). Gene array analysis of primary human keratinocytes, endothelial cells and fibroblasts has also shown the pGlcNAc-dependent regulation of other chemotactic factors such as CX3CR1, the receptor for fractalkine, which is involved in the adhesion and migration of leukocytes such as monocytes (Fong, Robinson et al. 1998; Goda, Imai et al. 2000; Foussat, Bouchet-Delbos et al. 2001; Shulby, Dolloff et al. 2004) is up-regulated 2 to 3-fold. Indeed, Erba et al. (J.Trauma, in press) shows pGlcNAc nanofibers modulate the expression of EGF and TGFβ, growth factors also involved in the wound healing process. These findings suggest that pGlcNAc nanofiber treatment may activate endothelial cells and other cell types within a healing wound to secrete factors required for cellular recruitment.

To test the hypothesis that pGlcNAc-stimulated endothelial cells can cause increased cellular recruitment, modified transwell assays were performed. Endothelial cells stimulated with pGlcNAc nanofibers were used as a chemoattractant in transwell assays. As shown in Fig. 6, pGlcNAc treatment causes a 2 to 3-fold increase in fibroblast migration toward fibronectin. These findings support the hypothesis that pGlcNAc nanofibers promote endothelial-dependent recruitment of fibroblasts. One effect of pGlcNAc nanofiber treatment of cutaneous wounds, therefore, may be to stimulate the production of chemotactic factors involved in cell movement and recruitment thus resulting in increased wound repair.

Applications to Areas of Health and Disease

Published observations and data discussed in this chapter support the use of pGlcNAc nanofibers as a hemostatic agent to control bleeding. In addition, pGlcNAc nanofibers increase angiogenesis, which is important for cutaneous wound healing and is necessary for tissue regeneration (Singer and Clark 1999). pGlcNAc stimulation of endothelial cells also results in growth factor secretion and in the expression of genes that

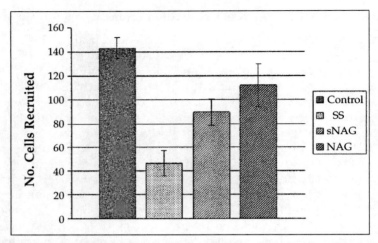

Fig. 6. pGlcNAc nanofiber treatment of primary endothelial cells results in increased cell recruitment. NIH3T3 cells were seeded on top of fibronectin coated transwells and allowed to migrate toward serum starved EC cells (SS) treated with pGlcNAc nanofibers (sNAG, 50μg/ml) or with 10% FCS containing media (control) for 12 hours.

function in innate immunity and bacterial clearance. These findings suggest that pGlcNAc nanofibers will be useful for both acute and chronic wounds and as an antibacterial therapy.

pGlcNAc in Diabetic Wound Healing—clinical study

A clinical trial of patients with venous leg ulcers (VLUs) treated with pGlcNAc nanofiber has demonstrated significant efficacy compared to standard of care controls. In over 80 patients, treatment one time per two weeks showed a greater than 86% wound closure rate in 20 weeks as compared to 45% in a control group. This study demonstrated that pGlcNAc nanofiber treatment is effective in treatment of VLUs and that its application was not associated with significant adverse effects (Kelechi et al. JAAD in press).

Conclusions

Our current understanding of the physiological consequences of pGlcNAc nanofiber treatment include hemostasis, increased cell motility, and increased wound healing due, at least in part, to increased cellular proliferation, viability, recruitment and differentiation and is an important regulator of an innate immune response to infection. In conclusion, we suggest that the pGlcNAc nanofiber materials presently used for hemostasis will be useful for wound healing and as an antibacterial agent.

Conceivably, these nanofibers would promote healing and specific tissue regeneration within a particular microenvironment, suggesting the future development of pGlcNAc nanofiber materials for a more comprehensive therapeutic for tissue regeneration.

Definitions of Explanations of Key Terms/genes/pathways

Integrins: Integrins are transmembrane receptors that are important for a cell's attachment and sensing of its external environment. These receptors are known to play a role in cell shape, cell motility, and cell cycle. Integrins are unique in that they can perform both the typical outside-in signaling as well as inside-out signaling, allowing for quick responses to environmental or cellular changes.

VEGF: Vascular endothelial growth factor is an important signaling protein that is produced and secreted by cells when there is a requirement for new blood vessel formation. The normal role of VEGF includes the formation of blood vessels during development and tissue repair.

MAPK: A group of kinases that respond to extracellular stimuli and regulate cellular activities such as cell migration, proliferation, cell division, gene expression, and cell survival.

Innate Immunity: Refers to the basic, first line of defense against infection that a species possesses. The cells of the innate immune system recognize and respond to pathogens quickly and in a non-specific way, unlike the adaptive immune system.

Summary Points

- Natural products that are isolated from marine organisms can be useful for the development of drugs or products for human disease.
- pGlcNAc nanofibers, isolated from *Thallasosira fluviatalis*, consist of high molecular weight polysaccharides that are organized in a unique 3D structure giving these fibers important biological properties.
- pGlcNAc nanofibers are useful as a hemostatic therapy to control bleeding.
- Treatment of cutaneous wounds with pGlcNAc nanofibers results in increased wound healing kinetics, partly due to increased angiogenesis, increased cell motility, and increased cell proliferation.
- pGlcNAc nanofibers treatment of endothelial cells stimulates factors that are important for innate immune function and bacterial clearance.
- Many diabetic patients suffer from chronic wounds which may benefit from pGlcNAc therapy.

Abbreviations

Ara-A	:	arabinosyl adenine
Ara-C	:	arabinosyl cytosine
pGlcNAc	:	poly-N-acetyl glucosamine
FTIR	:	fourier transform infrared spectroscopy
FDA	:	Food and Drug Administration
ADP	:	adenosine diphosphate
IED	:	Improvised explosive device
VEGF	:	Vascular endothelial growth factor
PECAM	:	Platelet endothelial cell adhesion molecule
NAG	:	N-acetyl glucosamine
sNAG	:	short N-acetyl glucosamine
uPAR	:	Urokinase-type plasminogen activator receptor
ECM	:	extracellular matrix
MMP	:	Matrix metalloproteinases
VAC	:	vacuum assisted closure
MAPK	:	Mitogen-activated protein kinase
FGF	:	Fibroblast growth factor
PI3K	:	Phosphatidylinositol 3-kinase
NFκB	:	nuclear factor kappa-light-chain-enhancer of activated B cells
MTT	:	(3-(4,5-Dimethylthiazol-2-yl)-2,5-diphenyltetrazolium bromide
IL-8	:	IL-1, Interleukin 1 Interleukin 8
SPAG11	:	Sperm-associated antigen 11
CEACAM3:		carcinoembryonic antigen-related cell adhesion molecule 3
TLR	:	Toll-like receptor
LPS	:	lipopolysaccharide
qPCR	:	quantitative polymerase chain reaction
TRAF	:	TNF receptor associated factor
TNFα	:	Tumor necrosis factor-alpha
CCR2	:	chemokine receptor 2
CCR6	:	chemokine receptor 6
EGF	:	Epidermal growth factor
TGFβ	:	Transforming growth factor beta
EC	:	endothelial cell
VLU	:	venous leg ulcers

References

Akira, S. 2003. "Mammalian Toll-like receptors." Curr. Opin. Immunol. 15(1): 5–11.
Bohlen, H.G. and B.A. Niggl. 1979. "Adult microvascular disturbances as a result of juvenile onset diabetes in Db/Db mice." Blood Vessels 16(5): 269–76.
Chao, W. 2009. "Toll-like receptor signaling: a critical modulator of cell survival and ischemic injury in the heart." Am. J. Physiol. Heart Circ. Physiol. 296(1): H1-12.
Chen, H., O. Charlat, et al. 1996. "Evidence that the diabetes gene encodes the leptin receptor: identification of a mutation in the leptin receptor gene in db/db mice." Cell 84(3): 491–5.
Chin, Y.W., K.D. Yoon, et al. 2009. "Cytotoxic anticancer candidates from terrestrial plants." Anticancer Agents Med. Chem. 9(8): 913–42.
Cole, D.J., R.J. Connolly, et al. 1999. "A pilot study evaluating the efficacy of a fully acetylated poly-N-acetyl glucosamine membrane formulation as a topical hemostatic agent." Surgery 126(3): 510–7.
Erba, P., et al., Poly-N-acetyl glucosamine fibers are synergistic with vacuum-assisted closure in augmenting the healing response of diabetic mice. J Trauma. 71(2 Suppl 1): p. S187–93.
Fei, Z., Z. Chen, et al. 2007. "Conditional RNA interference achieved by Oct-1 POU/rtTA fusion protein activator and a modified TRE-mouse U6 promoter." Biochem. Biophys. Res. Commun 354(4): 906–12.
Fischer, T.H., H.S. Thatte, et al. 2005. "Synergistic platelet integrin signaling and factor XII activation in poly-N-acetyl glucosamine fiber-mediated hemostasis." Biomaterials 26(27): 5433–43.
Fischer, T.H., C.R. Valeri, et al. 2008. "Non-classical processes in surface hemostasis: mechanisms for the poly-N-acetyl glucosamine-induced alteration of red blood cell morphology and surface prothrombogenicity." Biomed. Mater. 3(1): 015009.
Fong, A.M., L.A. Robinson, et al. 1998. "Fractalkine and CX3CR1 mediate a novel mechanism of leukocyte capture, firm adhesion, and activation under physiologic flow." J. Exp. Med. 188(8): 1413–9.
Foussat, A., L. Bouchet-Delbos, et al. 2001. "Deregulation of the expression of the fractalkine/fractalkine receptor complex in HIV-1-infected patients." Blood 98(6): 1678–86.
Galiano, R.D., O.M. Tepper, et al. 2004. "Topical vascular endothelial growth factor accelerates diabetic wound healing through increased angiogenesis and by mobilizing and recruiting bone marrow-derived cells." Am. J. Pathol. 164(6): 1935–47.
Ganz, T. 1994. "Biosynthesis of defensins and other antimicrobial peptides." Ciba Found Symp. 186: 62–71; discussion 71–6.
Ganz, T., and R.I. Lehrer. 1994. "Defensins." Curr. Opin. Immunol. 6(4): 584–9.
Goda, S., T. Imai, et al. 2000. "CX3C-chemokine, fractalkine-enhanced adhesion of THP-1 cells to endothelial cells through integrin-dependent and -independent mechanisms." J. Immunol. 164(8): 4313–20.
Hasle, G.R., and E.E. Syvertsen. 1997. Identifying Marine Diatoms and Dinoflagellates. Marine Diatoms Academic Press. In: C.R. Tomas pp. 5–385.

Hattangady, N.G. and M.S. Rajadhyaksha. 2009. "A brief review of *in vitro* models of diabetic neuropathy." Int. J. Diabetes Dev. Ctries 29(4): 143–9.

Heine, H., C.J. Kirschning, et al. 1999. "Cutting edge: cells that carry A null allele for toll-like receptor 2 are capable of responding to endotoxin." J. Immunol. 162(12): 6971–5.

Henneke, P., O. Takeuchi, et al. 2002. "Cellular activation, phagocytosis, and bactericidal activity against group B streptococcus involve parallel myeloid differentiation factor 88-dependent and independent signaling pathways." J. Immunol. 169(7): 3970–7.

Hirsch, J.A. 2003. "Harvard symposium on the clinical efficacy and hemostatic mechanism of action of poly-N-acetyl glucosamine." J. Invasive Cardiol. 15(9): 1–4.

Hirsch, J.A., S.A. Reddy, et al. 2003. "Non-invasive hemostatic closure devices: "patches and pads"." Tech. Vasc. Interv. Radiol. 6(2): 92–5.

Hong, Y.K., B. Lange-Asschenfeldt, et al. 2004. "VEGF-A promotes tissue repair-associated lymphatic vessel formation via VEGFR-2 and the alpha1beta1 and alpha2beta1 integrins." Faseb. J. 18(10): 1111–3.

Hubert, P., L. Herman, et al. 2007. "Defensins induce the recruitment of dendritic cells in cervical human papillomavirus-associated (pre)neoplastic lesions formed *in vitro* and transplanted *in vivo*." Faseb J.

Jensen, D.M., G.A. Machicado, et al. 2004. "Randomized double-blind studies of polysaccharide gel compared with glue and other agents for hemostasis of large veins and bleeding canine esophageal or gastric varices." J. Trauma 57 (1 Suppl): S33–7.

Katsube, K., A.T. Bishop, et al. 2005. "Vascular endothelial growth factor (VEGF) gene transfer enhances surgical revascularization of necrotic bone." J. Orthop. Res. 23(2): 469–74.

Kelechi, T.J., et al., A randomized, investigator-blinded, controlled pilot study to evaluate the safety and efficacy of a poly-N-acetyl glucosamine-derived membrane material in patients with venous leg ulcers. J Am Acad Dermatol.

Lavenburg, K.R., J. Ivey, et al. 2003. "Coordinated functions of Akt/PKB and ETS1 in tubule formation." Faseb. J. 17: 2278–80.

Lien, E., T.J. Sellati, et al. 1999. "Toll-like receptor 2 functions as a pattern recognition receptor for diverse bacterial products." J. Biol. Chem. 274(47): 33419–25.

Lindner, H.B., A. Zhang, et al. "Anti-bacterial effects of poly-N-acetyl-glucosamine nanofibers in cutaneous wound healing: requirement for akt1." PLoS One 6(4): e18996.

Liu, F., A.D. Verin, et al. 2001. "Differential regulation of sphingosine-1-phosphate- and VEGF-induced endothelial cell chemotaxis. Involvement of G(ialpha2)-linked Rho kinase activity." Am. J. Respir. Cell Mol. Biol. 24(6): 711–9.

McCurdy, J.D., T.J. Lin, et al. 2001. "Toll-like receptor 4-mediated activation of murine mast cells." J. Leukoc. Biol. 70(6): 977–84.

Molinski, T.F., D.S. Dalisay, et al. 2009. "Drug development from marine natural products." Nat. Rev. Drug Discov. 8(1): 69–85.

Niyonsaba, F., H. Ushio, et al. 2007. "Antimicrobial peptides human beta-defensins stimulate epidermal keratinocyte migration, proliferation and production

of proinflammatory cytokines and chemokines." J. Invest. Dermatol. 127(3): 594–604.
Odorisio, T., F. Cianfarani, et al. 2006. "The placenta growth factor in skin angiogenesis." J. Dermatol. Sci. 41(1): 11–9.
Palmer, B.L., D.S. Gantt, et al. 2004. "Effectiveness and safety of manual hemostasis facilitated by the SyvekPatch with one hour of bedrest after coronary angiography using six-French catheters." Am. J. Cardiol 93(1): 96–7.
Pietramaggiori, G., H.J. Yang, et al. 2008. "Effects of poly-N-acetyl glucosamine (pGlcNAc) patch on wound healing in db/db mouse." J. Trauma. 64(3): 803–8.
Pusateri, A.E., H.E. Modrow, et al. 2003. "Advanced hemostatic dressing development program: animal model selection criteria and results of a study of nine hemostatic dressings in a model of severe large venous hemorrhage and hepatic injury in Swine." J. Trauma. 55(3): 518–26.
Randhawa, A.K., and T.R. Hawn. 2008. "Toll-like receptors: their roles in bacterial recognition and respiratory infections." Expert Rev. Anti. Infect. Ther. 6(4): 479–95.
Salerno, T.A., C. Gaughan, et al. 2009. "Control of troublesome bleeding during repair of acute type A dissection with use of modified rapid deployment hemostat (MRDH)." J. Card Surg. 24(6): 722–4.
Scherer, S.S., G. Pietramaggiori, et al. 2009. "Poly-N-acetyl glucosamine nanofibers: a new bioactive material to enhance diabetic wound healing by cell migration and angiogenesis." Ann. Surg. 250(2): 322–30.
Schwaitzberg, S.D., M.W. Chan, et al. 2004. "Comparison of poly-N-acetyl glucosamine with commercially available topical hemostats for achieving hemostasis in coagulopathic models of splenic hemorrhage." J. Trauma. 57 (1 Suppl): S29–32.
Sementchenko, V.I., and D.K. Watson. 2000. "Ets target genes: past, present and future." Oncogene 19(55): 6533–48.
Shiojima, I., and K. Walsh. 2002. "Role of Akt signaling in vascular homeostasis and angiogenesis." Circ. Res. 90(12): 1243–50.
Shulby, S.A., N.G. Dolloff, et al. 2004. "CX3CR1-fractalkine expression regulates cellular mechanisms involved in adhesion, migration, and survival of human prostate cancer cells." Cancer Res. 64(14): 4693–8.
Singer, A.J., and R.A. Clark. 1999. "Cutaneous wound healing." N. Engl. J. Med. 341(10): 738–46.
Smith, C.J., J.N. Vournakis, et al. 2008. "Differential effect of materials for surface hemostasis on red blood cell morphology." Microsc. Res. Tech. 71(10): 721–9.
Smith, S.E. and T.S. Roukis. 2009. "Bone and wound healing augmentation with platelet-rich plasma." Clin. Podiatr Med. Surg. 26(4): 559–88.
Smyth, S.S., R.P. McEver, et al. 2009. "Platelet functions beyond hemostasis." J. Thromb. Haemost. 7(11): 1759–66.
Sun, H.H., W.J. Mao, et al. "Structural Characterization of Extracellular Polysaccharides Produced by the Marine Fungus Epicoccum nigrum JJY-40 and Their Antioxidant Activities." Mar. Biotechnol. (NY).
Takeda, K., and S. Akira. 2007. "Toll-like receptors." Curr. Protoc. Immunol. Chapter 14: Unit 14 12.

Thatte, H.S., S. Zagarins, et al. 2004. "Mechanisms of poly-N-acetyl glucosamine polymer-mediated hemostasis: platelet interactions." J. Trauma. 57(1 Suppl): S13–21.

Tomlinson, A., and M.W. Ferguson. 2003. "Wound healing: a model of dermal wound repair." Methods Mol. Biol. 225: 249–60.

Tonnesen, M.G., X. Feng, et al. 2000. "Angiogenesis in wound healing." J. Investig. Dermatol. Symp. Proc. 5(1): 40–6.

Underhill, D.M. 2004. "Toll-like receptors and microbes take aim at each other." Curr. Opin. Immunol. 16(4): 483–7.

Vournakis, J.N., J. Eldridge, et al. 2008. "Poly-N-acetyl glucosamine nanofibers regulate endothelial cell movement and angiogenesis: dependency on integrin activation of Ets1." J. Vasc. Res. 45(3): 222–32.

Weiner, B., T. Fischer, et al. 2003. "Hemostasis in the era of the chronic anticoagulated patient." J. Invasive Cardiol. 15(11): 669–73; quiz 674.

Yla-Herttuala, S., J.E. Markkanen, et al. 2004. "Gene therapy for ischemic cardiovascular diseases: some lessons learned from the first clinical trials." Trends Cardiovasc. Med. 14(8): 295–300.

Zhang, D., I. Fujii, et al. 2008. "The stimulatory activities of polysaccharide compounds derived from algae extracts on insulin secretion *in vitro*." Biol. Pharm. Bull 31(5): 921–4.

19

Nanotechnology Footsocks for Diabetic Foot

Alberto Piaggesi,[1,a] *Elisabetta Iacopi,*[1,b]
Elisa Banchellini[1,c] *and Laura Ambrosini Nobili*[1,d]

ABSTRACT

The complications of diabetes mellitus at the lower limbs, known also as "Diabetic Foot" (DF), are actually a complex syndrome with different but equally relevant components that cooperate to determine a severe and evolutive pathology which is, still now, the most frequent cause of amputation of lower limbs on planet Earth. In most cases the initial cause of such a dramatic condition depends from skin lesions that could be hopefully prevented with timely application of effective preventative measures. The recent application of nanotechnology to biomedical engineering made possible to realize moisturizer-charged nanocapsules which, incorporated in socks made of special fabric (Difoprev System

---

[1]Diabetic Foot Section, Department of Endocrinology and Metabolism, Azienda Ospedaliera Universitaria Pisana, Via Paradisa, 2–56124 Pisa Italy.
[a]E-mail: piaggesi@immr.med.unipi.it
[b]E-mail: elisabetta2009@alice.it
[c]E-mail: elisabanchellini@virgilio.it
[d]E-mail: laurambrosini@yahoo.it
List of abbreviations after the text.

LVM Bologna Italia ®), constituted a significant improvement in the effectiveness of preventative strategies by both maximizing the intensity and application time of moisturizing agents to dry an hypotrophic skin while on the other side overcoming the compliance of the patients regarding to treatment. In this chapter we will discuss both the technical and clinical aspects of the applications of such nanotechnology in the management of both neuropathic and ischemic DF. We will describe the system and its technical details and will introduce the clinical studies that were carried out in our Department on both neuropathic and ischemic DF providing data supporting the evidence of the safety and effectiveness of the application of nanotechnologies in the prevention of lower limb amputations in diabetic patients.

INTRODUCTION

Foot ulceration is the most important and prevalent complication of diabetes mellitus and represents the most frequent cause of lower limb amputations in these patients, determining an excess of risk which is twenty times higher compared to general population (Diamant et al. 2007; Prompers et al. 2007).

The costs of the management of this pathology, frequently complicated by infections that require admission of the patients and use of parenteral antibiotic therapy regimens, are probably the most relevant among those related to diabetes and its complications (Boulton et al. 2005; Reiber and Raugi 2005).

The pathogenesis of foot ulceration is well understood and recognizes in peripheral neuropathy the main predisposing factor, which, under the trigger effect of trauma, lead over the years to the development of the ulcer (Urbancic-Rovan 2005; Pecoraro 1991).

All of the component of diabetic neuropathy are involved in the pathogenesis of ulceration: sensory neuropathy reduces the ability to perceive external sensation as dangerous, motor neuropathy induces deformities in the foot that predispose to traumas, while autonomic neuropathy acts reducing the ability of sweating in the feet, because of the sympathetic denervation of the sweat glands, which consequently undergo to atrophy, thus determining a severe anhydrosis of the foot, which accompanies with dry skin, fissuring and which often represents the first lesion eventually followed by a ulcer (Low et al. 2006; Low et al. 2004).

Anhydrosis, accompanies with hyperkeratosis, due to localized hyper-pressure, and is the most typical sign of peripheral neuropathy;

both of them are considered pre-ulcerative conditions (Yosipovitch et al. 1998; Sakai et al. 2005).

The possibility of preventing the development and the evolution of the anhydrosis relies in the local application of moisturizers in the foot and the leg. This has proven to be an effective measure to prevent the ulcers, but it depends from the effectiveness of the moisturising agent and by the compliance of the patient, which is essential to guarantee the success of this approach (Johnston et al. 2006).

Recently a biomedical application of the nanotechnologies made possible to design and realize a tool that allows both to maximize the moisturizing action and to by-pass the patients' compliance.

THE DIFOPREV® SYSTEM

The system is based on a technology derived from the *nanostructured lipid carriers* (NLC), which has been demonstrated to be highly effective as a vehicle for both cosmetic and dermatological preparations. Their high loading capacity, their physical and chemical long-term stability and the triggered release are features which are all important for exerting their action (Muller et al. 2007).

It consists of a sock made of a special synthetic fibre, based on polyamide, which both has the property of being smooth and not irritant for the skin, and to be charged with the microcapsules containing the active moisturiser because of negative/positive surface charges interaction.

The sock is re-charged with the nanocapsules at each washing, since they are contained in single-dose dispensers with the washing detergent; the dispensers are manufactured as blisters, and in each box there are washing units for one-month usage.

The moisturising agent is a nanoemulsion of liposomes, made by a core of lipids (phosphatidylcholine) coated by the glycoprotein produced by the bacterium *Pseudoalteromonas Antarctica*, living in low-temperature environments in which acts like an antifreeze agent (Sung and Joung 2007; Nevot et al. 2006).

This protein, alongside with others found in microorganisms, is able to coat liposomes, protecting them by surfactant agents, so that they increase their solubility and disposability (De la Maza et al. 1999).

The liposomes, positively charged, can exerts a range of activities in the skin, interacting with the extracellular lipid junctions in the *stratum corneum* (SC), and isolating the deeper layer from the outer ones; a positive activity on fibroblasts functions has been also demonstrated *in vitro* (De la Maza et al. 1998).

Once worn the socks, the microcapsules are constantly released to the patient for all the time the sock is in contact with the skin - one charge lasts for 36 hours of application—thus eliminating the necessity for the patient to apply the moisturizer and maximising the time of application of the active agent.

In Fig. 1 is displayed a microphotography of a sock fibre charged with a microcapsule charged with the moisturizing agent.

To test the safety and effectiveness of the application of a system constituted by the micro-encapsulated (= active) socks in treating the pre-ulcerative conditions of both the neuropathic and ischemic foot which are risk factors for chronic ulceration in diabetic patients, we carried out two different study, one which has been published, while the other has been submitted (Banchellini et al. 2008).

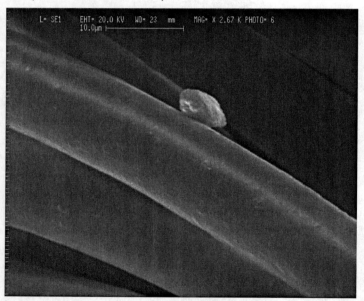

Fig. 1. A charged fibre. This microphotography shows socks fibres charged with moisturizing microcapsules (With permission from LVM Technologies, Bologna, Italy).

The Neuropathic Study

In the neuropathic Study, all the diabetic outpatients attending our foot clinic during the months of June and July of 2005 were screened according to the following inclusion criteria: they should have peripheral neuropathy, according to the definition of the American Diabetes Association (Boulton et al. 2005), and they should have bilateral anhydrosis of the foot, identified with clinical examination and confirmed with the Neuropad® test according to the methods previously described by Bilen et al. (Bilen et al. 2007).

Exclusion criteria were the presence of active ulceration in the foot, peripheral arterial disease, defined as an Ankle Brachial Pressure Index (ABPI) <0.9, therapy with beta-blockers, serum creatinine > 2 mg/dl and any local or systemic condition potentially interfering with skin structure and function.

After obtaining the informed consent patients were randomised into two groups: Group A was treated with the application of the active socks' according with the indication of manufacturers for six weeks, while group B did wear only the socks, without the microcapsules for the same amount of time. Patients and relatives were carefully instructed on the procedures for correctly use the active socks system and were verified in their ability in performing all the operations for re-charging the socks with the hydrating agent; to avoid the bias due to re-charging operations, patients in Group B were taught to do a shame re-charging. Patients of both groups were instructed in washing and re-charging the socks every other day, according to the instruction of manufacturers; two pairs of socks were given to each of them to guarantee the continuity of the study. Patients were evaluated at baseline and after six-week with a quantitative hydration score ranging from 1 to 5 (Table 1), skin moisture (Scalar Moisture Checker®, STR, Scotts Valley, CA), Trans-epidermal water loss (TEWL) (Vapometer®, Delfin Technologies Ltd., Kuopio, F), skin temperature with infrared thermometry (La Crosse Technology, Cinisello Balsamo, I) and skin hardness with a Durometer (Mod 3001/A®, AFFRI, Induno Olona, I)' (Hoedlftke 2001; Piaggesi et al. 1999; Pi-Chang et al. 2006; Levin and Maibach 2005) at baseline and after six weeks of continuous treatment.

The coefficient of variation (SD/Mean) and the test-retest repeatability ratio, both measured in normal skin before the study, were respectively <15% and comprised between 0.77 and 0.84 for all the instruments. In order to improve their reproducibility all the measurements were done by the same podologist (EB) blind to the treatment of the patients, always in the same point in the most severely affected foot: the summit of the plantar arch in a non-weight bearing position.

The feet were all digitally photographed before and after the six-weeks treatment. Patients were instructed to do their regular activities and to record in a logbook any eventual adverse event, even if not apparently related to the treatment, and a telephone number was provided in case of urgencies. Patients were also requested to keep all the remnants from the used boxes of the refills, so that a calculation about the effective usage of the system could be made retrospectively.

Table 1. The skin hydration score.

Score	Definition	Characteristics
1	Anhydrosis	Absent sweating, fissuration, desquamation, inelastic skin, absent plicability
2	Hypohidrosis	Sweating present only after stimulation, reduction or absence of dermatogliphics, low plicability
3	Normohydration	Spontaneous sweating, normal plication and glyphics, skin normoelastic
4	Hyperhidrosis	Excessive sweating, after stimulation, hypertrophic dermatogliphics, sporadic maceration in the interdigital areas
5	Maceration	Spontaneous excessive sweating, diffuse maceration, wet skin, eventual eczema

This is a quantitative hydration score ranging skin moisture from 1 to 5 (Scalar Moisture Checker®, STR, Scotts Valley, CA) used to evaluate patients at baseline and after six-week treatment.
Unpublished material of the author.

Baseline clinical and instrumental characteristics of patients were compared with those of a group of sex and age-matched healthy volunteers.

The Ischemic Study

The same study design was applied to a group of mildly [0.7<Ankle-Brachial Pressure Index (ABPI)<0.9] ischemic diabetic outpatients free from active lesions, for the ischemic component of diabetic foot pathology; 40 consecutive patients were randomly assigned to a treatment group with the complete Difoprev system or alternatively received only the uncharged socks and were compared to 12 normal controls. The length of the study and the parameters tested were the same as in the neuropathy study.

Data, expressed as median (95% confidence interval—CI), were analysed by means of a commercial software (Statview, SAS institute, Gary, Ill) running on a personal computer. The statistical test performed were chi-square test for the dichotomous variables and Mann-Whitney test for non-parametric data.

New Findings after the Clinical Studies with Difoprev

In the Neuropathic study, thirty-six (36) patients were screened between January and March of 2005 according to the inclusion and exclusion criteria, but only 30 accepted to participate in the study and were actually enrolled and randomized in the two groups (15 patients for group). They were very similar with respect to demographic characteristics (age Group A 59.6±13.8 yrs, Group B 61.4±15.5 yrs, Control Group 60.5±11.4 yrs). Between the two Groups of randomization there were not significant differences in duration of diabetes (16.1±9.0 yrs in Group A and 15.7±6.9 yrs in Group B) and level of glycometabolic control (HbA1c 8.6±1.3% in Group A and 8.9±1.7% in Group B).

No differences emerged between the two groups at baseline both for clinical features and instrumental evaluation of the severity of the dyshidrosis, while they were significantly different from those measured in healthy controls (Table 2).

Table 2. Baseline clinical features of patients studied.

	Group A	Group B	Controls
Skin Hydration Score	1.1 (0.4)	1.2 (0.8)	3.5 (0.6)*
Skin Hardness (UI)	52.4 (7.8)	47.7 (7.1)	35.4 (8.8)*
TEWL (g/h/m^2)	118.9 (45.7)	127.6 (49.8)	29.9 (16.4)*
Skin Moisture (%)	26.9 (14.8)	24.2 (13.6)	48.3 (12.2)*
Skin Temperature (°C)	34.1 (2.0)	33.8 (0.9)	34.1 (0.6)

Both clinical features and instrumental evaluation of the severity of the dyshidrosis were similar between two study groups, while they were significantly different from those measured in healthy controls.
Unpublished material of the author.
*p<0.05 vs. Group A and Group B.

In Fig. 2 (a–b) a case is reported among those treated with the complete application of the active socks system; it is evident the change in the hydration of the skin, as well as the reduction of hyperkeratosis and of wrinkles.

All the patients enrolled in both groups completed the study and all of them took back the empty capsules used for re-charging procedures: the calculations made indicated that the re-charging procedures were done every 2.0 (0.5) days, with no differences between the two groups.

No adverse events in both group A and Group B were reported.

After six weeks of treatment all the parameters measured in Group A patients showed a significant (p<0.05) improvement, while those in group B did not changed significantly. The Results are reported in Table 3. Only skin temperature did not change significantly in group A before and after the study. In Fig. 3 the percentage difference from baseline is reported for each parameter for both groups.

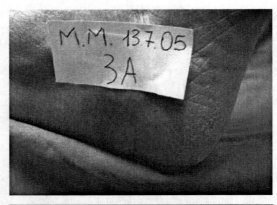
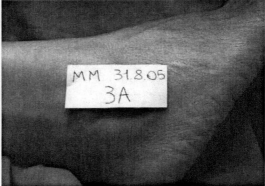

Fig. 2. A case report of a patient treated with the Difoprev® system. a) the patient's skin was dehydrated and inelastic, and a score of 1 was attributed. The same patient after 6 weeks of treatment; b) the skin is normally hydrated and the score attributed was 3 (compares the text in the *Patients and Methods* section).
Unpublished material of the author.

Color image of this figure appears in the color plate section at the end of the book.

The results in the ischemic study were superimposable to those in the neuropathy study, with the important difference that there was not a reduction in skin hardness like in the neuropathic group, in relation to the fact that at baseline the hardness of the skin of ischemic patients was reduced compared to controls, and not increased as in neuropathic patients.

Methodological Consideration

Our study demonstrates how the use of socks carrying a microencapsulated hydrating agent for six weeks is able to restore the normal hydration of the skin due to diabetic autonomic neuropathy, and to reduce the changes

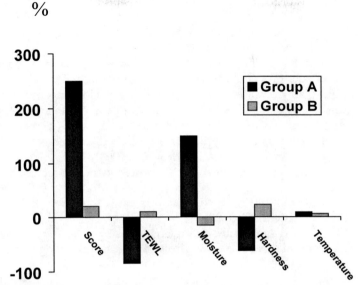

Fig. 3. The rate of change from baseline in all the parameters evaluated at the end of the study. We observed an increase of 250% of skin hydration score, and of 150% in skin moisture, paralleled by a reduction of 85% of TEWL and of 62% of skin hardness. All the differences observed in group A patients, except for skin temperature were significant, while no significant modification was observed in Group B patients.
Unpublished material of the author.

Table 3. Results of the clinical and instrumental evaluation in both groups.

	Group A Baseline	Group A + 6 weeks	Group B Baseline	Group B + 6 weeks
Skin Hydration Score	1.1 (0.4)	2.9 (0.8)*	1.2 (0.8)	1.4 (0.9)
Skin Hardness (UI)	52.4 (7.8)	26.4 (4.3)°	47.7 (7.1)	49.9 (6.9)
TEWL (g/h/m^2)	118.9 (45.7)	44.1 (18.3)*	127.6 (49.8)	141.9 (57.2)
Skin Moisture (%)	26.9 (14.8)	49.8 (19.8)*	24.2 (13.6)	27.4 (15.7)
Skin Temperature (°C)	34.1 (2.0)	34.7 (1.7)	33.8 (0.9)	34.0 (2.1)

After six weeks of treatment all the parameters measured in Group A patients showed a significant (p<0.05) improvement, while those in Group B did not changed significantly.
Unpublished material of the author.
* p<0.01 and °p<0.05 vs. baseline.

induced by hypotrophy because of chronic ischemia, thus reducing the risks for the development of ulceration.

This is an important issue, since the increasing prevalence of diabetes throughout the world, now estimated around 300 million people, emphasizes the necessity of managing this condition in almost 100 million people in the next few years (Boulton et al. 2005).

Prevention is crucial to avoid the progression of the disease from the early stages in which only predisposing conditions are present, to the ulcerative stages which are much more difficult to manage, with a lesser rate of success (Jeffcoate and Harding 2005).

The possibility of improving the hydration of the skin is important not only because it prevents the skin breaks, eventually complicated with infection, but also from a biomechanical point of view; normo-hydrated skin is more elastic and resistant, especially to the shear stress that the round-shaped margins of the foot apply during the gait. In this way the possibility of a skin break is significantly reduced as well as the related risk of developing a frank ulceration (Boulton 2004).

In our study the application of microencapsulated socks has demonstrated to be effective in arresting dehydration of the skin in neuropathic patients and to restore the normal trophic conditions of the foot (Verdie-Sevrain and Bontè 2005).

Both the clinical evaluation quantified in the *Skin Hydration Score* and all the instrumental measurements showed significant changes, all in the direction of a rescue of hydration of the foot.

The multi-instrumental approach that we choose to evaluate the activity of the microencapsulation technology allows us to measure it by different points of view: the skin moisture checker determines the direct hydration of the skin, while TEWL measures the barrier function of the skin toward evaporation and skin hardness is a function of the water content of the skin (Hoedlftke 2001; Levin and Maibach 2005).

These variables are all relevant to the reduction of ulcerative risk in the diabetic foot, because the actually measure the capability of the skin to resist to mechanical stresses, in particular to shear stress (Piaggesi et al. 1999).

All the changes observed in these parameters reflect a significant increase in the hydration of the skin in the group of patients treated with the active socks, compared to the group treated with the inactive socks.

Since in the patients of both groups the sudomotor activity was severely impaired as a consequence of autonomic neuropathy, we interpret these changes as a result of the hydrating activity of the lipids, vehiculated by the nano-particles present in the activated system.

The way by which the glycoprotein restore the normal hydration of the skin is probably related to its interaction with the *stratum corneum* (SC) of the epidermis that in the diabetic patients has been found to be less hydrated and similar to that of senile patients (Sakai et al. 2005).

In our patients the barrier function of SC was also restored with this approach, as demonstrated by a highly significant reduction of TEWL in group A patients. This is not surprising since nano-emulsions containing lipids have shown to interact with the inner layers of the epidermis, modifying their structure (Yilmaz and Borchert 2006).

Positively charged nano-emulsions, like the one we used, are particularly effective, because of the electric attraction between the particles and the negatively charged structures of the skin (Rojanasakul et al. 1992).

Spreadability is also a key-factor in enhancing the effectiveness of the emulsion; according to the Young-Dupre equation the higher the extent of spreading the higher is the attraction or adhesion between the emulsion and the skin (Yilmaz and Borchert 2006).

Because of the small dimension of the particles and of the wide surface of application, the interaction between the skin and the glycoprotein-coated lipids is maximized.

Another positive consideration should be made regarding the strategy of application of the nano-particles: the possibility of incorporating them in a sock, worn by the patients, with no further action required for assuring the release of the active substance, both ensure a long time of effective contact between the skin and the particles and cut-off the compliance of the patient.

This is of paramount relevance as far as a chronic therapy is concerned, especially in the management of the diabetic foot, an area in which the contemporary presence of neuropathy and chronicity strongly reduces the compliance of the patients (Wu and Armstrong et al. 2006; Johnston et al. 2006).

We can conclude that the Difoprev® system is safe and effective for the re-hydration of the neuropathic and ischemic skin in diabetic patients. Its use, embedded in a comprehensive management strategy, may be of value for the prevention of ulceration of the diabetic foot.

Applications to Areas of Health and Disease

The possibility of using nanotechnologies in the management of diabetic foot opens a wild range of opportunities both in the field of prevention, treatment and rehabilitation of all the components of this pathology: neuropathy, ischaemia and infection.

Although the present chapter describes only the clinical evidence in the prevention in both neuropathy and ischaemia in this patients, further studies will eventually focus on other aspects of the applications of nanotechnology to DF and explore many other options.

In fact the flexibility of the system allows to imagine many different declinations of the system according with different molecules used to charge the nanocapsules. In these cases we may use Difoprev charged with local antiseptics to control infection or charged with vasoactive drugs to stimulate neoangiogenesis or, again, with analgesic drugs to control neuropathic or ischemic pain. The potentiality of the Difoprev system is such that we may only speculate on developments but by now we can assume that its use is fully justified as a preventative tool in the neuroischemic patients both in primary prevention and in secondary prevention. The limit of this approach is only related to the actual possibility to incorporate the molecules in nanocapsules. This depending on the chemical physical structure of both vehicles (= Nanocapsules) and active principles (=Molecules). For the research is required to fill the gap between potential application and availability of solutions, but the course is clearly set.

Key Facts of Diabetic Foot
- Diabetic foot is a complex disease that is generated as a result of a cluster of complications of diabetes insisting on the lower limbs.
- Diabetic foot affects one out of four diabetic patients at least once in their life and it's responsible for the vast majority of non-traumatic lower limb amputation on planet Earth.
- Peripheral neuropathy, peripheral vascular disease and infection are the concauses determining the different clinical pictures of diabetic foot and they tend to cluster so as to give rise to complex and evolutive clinical patterns.
- Skin lesions, even minor ones, have been demonstrated to be the initial determinants of lower limb amputations in diabetic patients in more than 85% of cases.
- Prevention is of paramount importance in DF since it has been demonstrated to be much more effective and less costly than treatment of acute phase of the disease.
- DF has been considered a marker of co-morbidities in diabetic patients, mainly from a cardiovascular point of view, its presence is associated to a higher mortality in diabetic patients, being probably the most risk condition related to this chronic disease.
- Although evolutive and bound to determine a very high risk, both for amputations and survivor, DF can be prevented and treated in a vast majority of cases applying a multidisciplinary approach based on International Consensus Guidelines.

Definitions

Diabetic Foot: Clinical syndrome characteristic of diabetes, the severity of which is determined by a grading ranging from alterations in the structure of the foot to ulceration or necrosis, with or without infection and/or destruction of deep tissues, associated with neurological abnormalities and various degrees of peripheral vascular disease of the lower limbs.

Diabetic peripheral neuropathy: Chronic degenerative nerve disease affecting the peripheral nerves due to diabetes mellitus.

Autonomic neuropathy: Diabetes related pathology of the autonomic nervous system affecting both the sympathetic and the parasympathetic components.

Diabetic peripheral vascular disease: Chronic evolutive obstructive disease affecting the lower limb in diabetic patients.

Skin anhydrosis: Reduction of the hydration state of the skin due to the hypotrophy of the sweat glands induced by the denervation of the autonomic nervous system.

Skin hypotrophy: Reduction in the thickness of the skin due to reduced blood supplies secondary to the peripheral vascular disease.

Summary Points

- Diabetic foot is a complex and evolutive disease which deserves aggressive preventative strategies.
- Nanotechnology-derived footsocks (Difoprev) may play a role in the management of DF.
- Difoprev is been successfully applied for the prevention of neuropathic anhydrosis.
- It has been also successfully applied in the prevention of skin hypotrophy.
- Difoprev should be considered part of the multidisciplinary strategy for the prevention and treatment of all the aspects of diabetic foot.

Acknowledgments

We acknowledge the generous contribution of LVM Technologies (Bologna, I), manufacturer of Difoprev® which supplied the authors with all the material necessary for this study.

Abbreviations

DF	:	Diabetic Foot
NLC	:	Nanostructured lipid carriers
SC	:	Stratum corneum
TEWL	:	Trans-epidermal water loss
ABPI	:	Ankle brachial pressure index
HbA1c	:	Glycated Haemoglobin

References

Banchellini, E., S. Macchiarini, V. Dini, L. Rizzo, A. Tedeschi, A. Scatena, C. Goretti, F. Campi, M. Romanelli and A. Piaggesi. 2008. Use of nanotechnology-designed foot sock in the management of pre-ulcerative conditions in the diabetic foot: results of a single, blind randomized study. Int. J. Low Extrem Wounds 7: 82–87.

Bilen, H., A. Atmaca and G. Akcay. 2007. Neuropad indicator test for the diagnosis of sudomotor function in type 2 diabetes. Adv. Ther. 24: 1020–1027.

Boulton, AJ. 2004. The diabetic foot: from art to science. The 18th Camillo Golgi Lecture. Diabetologia. 47: 1343–1353.

Boulton, A.J., L. Vileikyte, G. Ragnarson-Tenvall and J. Apelqvist. 2005. The global burden of diabetic foot disease. Review Lancet 366: 1719–1724.

Boulton, A.J., A.I. Vinik, J.C. Arezzo, V. Bril, E.L. Feldman, R. Freeman, R.A. Malik, R.E. Maser, J.M. Sosenko and D. Ziegler. 2005. American Diabetes Association. Diabetic neuropathies: a statement by the American Diabetes Association. Diabetes Care 28: 956–962.

De la Maza, A., L. Codech, O. Lopez, J.L. Parra, M. Sabes and J. Guinea. 1999. Ability of the exopolymer excreted by Pseudoalteromonas Antarctica NF3 to coat liposomes and to protect these structures against octyl glucoside. J. Biomater. Sci. Polym. Ed. 10: 557–572.

De la Maza, A., J.L. Parra, F. Congregado, N. Bozal and J. Guinea. 1998. Interaction of the glycoproteins secreted by Pseudoalteromonas Antarctica NF3 with phosphatidylcholine liposomes. Colloids Surfaces 137: 181–188.

Diamant, A.L., S.H. Babey, T.A. Hastert and E.R. Brown. 2007. Diabetes. The growing epidemic Surgeon 5: 219–230.

Hoedlftke, R.D., K.D. Bryner, G.G. Horvath, R.W. Phares, L.F. Broy and G.R. Hobbs. 2001. Redistribution of sudomotor responses is an early sign of sympathetic dysfunction in type 1 diabetes. Diabetes 50: 436–443.

Jeffcoate, W.J. and K.G. Harding. 2003. Diabetic foot ulcers. Review Lancet 361: 1545–1551.

Johnston, M.V., L. Pogach, M. Rajan, A. Mitchinson, S.L. Krein, K. Bonacker and G. Reiber. 2006. Personal and treatment factors associated with foot self-care among veterans with diabetes. J. Rehabil. Res. Dev. 43: 227–238.

Levin, J. and H. Maibach. 2005. The correlation between trans epidermal water loss and percutaneous absorption: an overview. J. Control Release 103: 291–299.

Low, P.A., L.M. Benrud-Larson, D.M. Sletten, T.L. Opfer-Gehrking, S.D. Weigand, S. O'Brien, G.A. Suarez and P.J. Dyck. 2004. Autonomic symptoms and diabetic neuropathy. A population-based study. Diabetes Care 27: 2942–2947.

Low, V.A., P. Sandroni, R.D. Fealey and P.A. Low. 2006. Detection of small-fibre neuropathy by sudomotor testing. Muscle Nerve 34: 57–61.

Muller, R.H., R.D. Petersen, A. Hommoss and J. Pardeike. 2007. Nanostructured lipid carriers (NLC) in cosmetic dermal products. Adv. Drug Deliv. Rev. 59: 522–530.

Nevot, M., V. Deroncelé, P. Messner, J. Guinea and E. Mercadé. 2006. Characterization of outer membrane vesicles released by the psychrotolerant bacterium Pseudoalteromonas Antarctica NF3 Environ. Microbiol. 8: 1523–1533.

Pecoraro, R. 1991. Chronology and determinants of tissue repair in diabetic lower extremity ulcers. Diabetes 40: 1305–1313.
Piaggesi, A., M. Romanelli, E. Schipani, F. Campi, A. Magliaro, F. Baccetti and R. Navalesi. 1999. Hardness of plantar skin in diabetic neuropathic feet. J. Diabet. Compl. 13: 129–134.
Pi-Chang, S., L. Hong-Da, E.J. Shyh-Hua, K. Yan-Chiou, C. Rai-Chi and C. Cheng-Kung. 2006. Relationship of skin temperature to sympathetic dysfunction in diabetic at-risk feet. Diab. Res. Clin. Pract. 73: 41–46.
Prompers, L., M. Huijberts, J. Apelqvist, E. Jude, A. Piaggesi, K. Bakker, M. Edmonds, P. Holstein, A. Jirkowska, D. Mauricio, G. Ragnarson Tennvall, H. Reike, M. Spraul, L. Uccioli, V. Urbancic, K. Van Acker, J. Van Baal, F. Van Merode and N. Cghaper. 2007. High prevalence of ischaemia, infection and serious comorbidity in patients with diabetic foot disease in Europe. Baseline results from the Eurodiale study. Diabetologia 50: 18–25.
Reiber, G.E. and G.J. Raugi. 2005. Preventing foot ulcers and amputations in diabetes Lancet 366: 1695–1703.
Rojanasakul, Y., L.Y. Wang, M. Bhat, D.D. Glover, C.J. Malagna and J.K.H. Ma. 1992. The transport barrier of epithelia: a comparative study on membrane permeability and charge selectivity in the rabbit. Pharm. Res. 9: 1029–1034.
Sakai, S., K. Kikuchi, J. Satoh, H. Tagami and S. Inoue. 2005. Functional properties of the stratum corneum in patients with diabetes mellitus: similarities to senile xerosis. Br. J. Dermatol. 153: 319–323.
Sung, J.K. and H.Y. Joung. 2007. Cryoprotective properties of exopolysaccharide (p-21653) produced by the Antarctic bacterium, Pseudoalteromonas Antarctica KOPRI 21653. J. Microbiol. 45: 510–514.
Urbancic-Rovan, V. 2005. Causes of diabetic foot lesions. Lancet 366: 1678–1679.
Verdier-Sévrain, S. and F. Bonté. 2007. Skin hydration: a review of its molecular mechanisms. J. Cosmet. Dermatol. 6: 75–82.
Wu, S.C. and D.G. Armstrong. 2006. The role of activity, adherence, and off-loading on the healing of diabetic foot wounds Plast Reconstr. Surg. 117: 248S-253S.
Yilmaz, E. and H. Borchert. 2006. Effect of lipid-containing, positively charged nanoemulsion on skin hydration, elasticity and erythema—an *in vivo* study. I J. Pharm. 307: 232–238.
Yosipovitch, G., E. Hodak, P. Vardi, I. Shraga, M. Karp, E. Sprecher and M. David. 1998. The prevalence of cutaneous manifestations in IDDM patients and their association with diabetes risk factors and microvascular complications. Diabetes Care 21: 4506–4509.

20

Nanosciences, Diabetes and the Patient

Martin C.R.,[1], Le L.,[2] Hunter R.,[3] Patel V.B.[4] and Preedy V.R.[5]*

Diabetes has a complex aetiology and about 90% of cases pertain to Type 2 diabetes as opposed to the less common Type 1. Other spectral conditions include gestational diabetes, insulin resistance, etc. The statistics associated with this disease are staggering. The World Health Organisation estimates that one in 20 deaths are due to diabetes world wide. Compared to 2005, the number of people with diabetes will double by 2030 which for many readers of this book will be within their life time. The patient with diabetes is exposed to numerous risk factors and various tissue systems are affected. These include the vasculature and nervous system, leading to blindness, kidney failure, loss of limbs and a variety of other conditions (Frier et al. 2008; Stirban and Tschoepe 2008; Lam 2009; Ma et al. 2009; Mattila and De 2010; Hayashi et al. 2010; Benhalima et al. 2011; Evans et al. 2011). Indeed it could be argued that diabetes can potentially affect every organ in the body. As a consequence of these wide ranging effects, diabetes is a major contributor to the global disease burden (Stovring 2009).

---

[1]School of Health, Nursing and Midwifery, University of the West of Scotland, Ayr Campus, Beech Grove, Ayr, KA8 0SR; E-mail: colin.martin@uws.ac.uk
[2]Rosemead Surgery, Maidenhead, Berkshire, SL6 8DS, UK; E-mail: organisedlan@gmail.com
[3]Cardiology Research Fellow, St Bartholomew's Hospital, London,UK;
E-mail: <ross.hunter@doctors.org.uk>
[4]Department of Biomedical Science, School of Life Sciences, University of Westminster, 115 New Cavendish Street, London, W1W 6UW; E-mail: <V.B.Patel@westminster.ac.uk>
[5]Diabetes and Nutritional Sciences, School of Medicine, Kings College London, Franklin Wilkins Buildings, 150 Stamford Street, London SE1 9NU; E-mail: victor.preedy@kcl.ac.uk
*Corresponding author

There is a great deal of research into the causes of diabetes and its complications. These focus on for example individual components and pathways (Ergul 2011), genetics (Jones et al. 2010), epidemiology (Bruno and Landi 2011) and pharmacological treatments and management (DeFronzo et al. 2011; Haller et al. 2011; Griffin et al. 2011) to name but a few. However, whilst there are technical advances in the understanding, diagnosis and treatment of diabetes, one must not lose focus of the patient.

The diabetic patient has a greater risk of impaired cognitive function (Frier 2011), depression (Wilfley et al. 2011; Lamers et al. 2011) and poor quality of life measures (Ribu 2009; Li and Ford 2009; Vijayakumar and Varghese 2009; Egede and Ellis 2010; Wilfley et al. 2011). Some specific symptoms such as nausea and vomiting or retinopathy will impact on those subjective symptoms related to quality of life measures (Mitchell and Bradley 2009; Mitchell et al. 2009; Jaffe et al. 2011; Mazhar et al. 2011). Poor quality of life may in turn impact on the physical health domains. For example diabetic patient with depression have a greater risk of being admitted to the intensive care unit (Davydow et al. 2011) and have reduced cognitive scores (Kadoi et al. 2011). The presence of diabetes as a co-morbid condition in schizophrenia and bipolar disorders also increases mortality (Vinogradova et al. 2010). The detailed interrelationships between these two-way processes between body and mind are yet to be elucidated. However, attempts are now being made to address these linkages. One contemporary example is the current Management and Impact for Long-Term Empowerment and Success (MILES) study being conducted in Australia (MILES 2011). This landmark investigation is the first nationwide, indeed continent-wide study to examine the whole spectrum of the actual experience of living with diabetes, both Type 1 and Type 2 from the patients perspective. Importantly, the data capture approach to studies such as MILES enables a comprehensive account to be generated of both the everyday lived experience of the patients but also to develop a minimum data set that allows both change over time and the effect of interventions to be evaluated. Critically, the former gives a unique and person-centred context to evaluate the complex etiological aspects of both types of diabetes. Understanding these more complex interactions ultimately facilitates, not only a greater understanding of the underlying pathological process, but also emphasises the direct role of the patients themselves in terms of the effective management of the course of their disease. Moreover, the effects of interventions can be evaluated against a comprehensive baseline dataset, the implications of which include being able to establish the relative contribution of the intervention against a whole spectrum of clinically relevant psychosocial and biological background variables. The extension of studies such as MILES to other countries using essentially the same and inclusive data collection and analysis strategy will also enable the discrete and often occluded genetic contribution to diabetes to be investigated, again within a contextually sensitive manner in which the relative contribution can be studied in detail. Borrowing from statistical techniques used in epidemiological research and psychometric evaluation of instruments such as structural equation modelling, sophisticated and causal models of etiology may be evaluated and more importantly, compared to determine, even within the context of complex explanatory models, the most parsimonious account of data.

The opportunities provided by the advances in nanomedicine offer added gravitas to the relevance and application of sophisticated and inclusive patient profiles to provide synergies between nanomedicine-derived interventions and the estimation of the contribution of nanomedicine approaches to patient experience and patient outcome. Furnished with an understanding of these more complex interactions, the true contribution of nanomedicine approaches to diabetes, its treatment and management, will then ultimately be realised. Consequently, a greater and indeed unique insight into the disease and advances into its diagnosis and treatments will contribute to reducing the impact of diabetes at all levels in every patient group, irrespective of whether the presentation is Type 1 or Type 2.

The nanosciences are a rapidly expanding area that embraces main stream sciences and medicine in a holistic manner. Chemistry and physics provide the tangible elements of this combination, whilst the disease state of diabetes provides the more subjective and less tangible components. The nanosciences enable new treatments to be developed and provides innovative platforms for the understanding of diabetes at the molecular and cellular levels. New discoveries will pave the way for the earlier diagnosis and treatment of diabetes or its numerous complications. However one must not lose sight of the mental wellbeing of the diabetic patient.

References

Benhalima, K., E. Wilmot, K. Khunti, L.J. Gray, I. Lawrence and M. Davies. 2011. Type 2 diabetes in younger adults: Clinical characteristics, diabetes-related complications and management of risk factors. Primary Care Diabetes 5: 57–62.

Bruno, G. and A. Landi. 2011. Epidemiology and costs of diabetes. Transplantation Proceedings 43: 327–329.

Davydow, D.S., J.E. Russo, E. Ludman, P. Ciechanowski, E.H. Lin, K.M. Von, M. Oliver and W.J. Katon. 2011. The association of comorbid depression with

intensive care unit admission in patients with diabetes: a prospective cohort study. Psychosomatics 52: 117–126.
DeFronzo, R.A., D. Tripathy, D.C. Schwenke, M. Banerji, G.A. Bray, T.A. Buchanan, S.C. Clement, R.R. Henry, H.N. Hodis, A.E. Kitabchi, W.J. Mack, S. Mudaliar, R.E. Ratner, K. Williams, F.B. Stentz, N. Musi and P.D. Reaven. 2011. Pioglitazone for diabetes prevention in impaired glucose tolerance. New England Journal of Medicine 364: 1104–1115.
Egede, L.E. and C. Ellis. 2010. The effects of depression on metabolic control and quality of life in indigent patients with Type 2 diabetes. Diabetes Technology and Therapeutics 12: 257–262.
Ergul, A. 2011. Endothelin-1 and diabetic complications: Focus on the vasculature. Pharmacological Research 63: 477–482.
Evans, K.K., C.E. Attinger, A. Al-Attar, C. Salgado, C.K. Chu, S. Mardini and R. Neville. 2011. The importance of limb preservation in the diabetic population. Journal of Diabetes and its Complications 25: 227–231.
Frier, B.M. 2011. Cognitive functioning in Type 1 diabetes: The Diabetes Control and Complications Trial (DCCT) revisited. Diabetologia 54: 233–236.
Griffin, S.J., K. Borch-Johnsen, M.J. Davies, K. Khunti, G.E. Rutten, A. Sandbaek, S.J. Sharp, R.K. Simmons, M. Van Den Donk, N.J. Wareham and T. Lauritzen. 2011. Effect of early intensive multifactorial therapy on 5-year cardiovascular outcomes in individuals with Type 2 diabetes detected by screening (ADDITION-Europe): A cluster-randomised trial. The Lancet 378: 156–167.
Haller, H., S. Ito, J. Izzo, J.L, A. Januszewicz, S. Katayama, J. Menne, A. Mimran, T.J. Rabelink, E. Ritz, L.M. Ruilope, L.C. Rump and G. Viberti. 2011. Olmesartan for the delay or prevention of microalbuminuria in Type 2 diabetes. New England Journal of Medicine 364: 907–917.
Hayashi, T., S. Takai and C. Yamashita. 2010. Impact of the renin-angiotensin-aldosterone-system on cardiovascular and renal complications in diabetes mellitus. Current Vascular Pharmacology 8: 189–197.
Jaffe, J.K., S. Paladugu, J.P. Gaughan and H.P. Parkman. 2011. Characteristics of nausea and its effects on quality of life in diabetic and idiopathic gastroparesis. Journal of Clinical Gastroenterology 45: 317–321.
Jones, D.A., S.L. Prior, T.S. Tang, S.C. Bain, S.J. Hurel, S.E. Humphries and J.W. Stephens. 2010 Association between the rs4880 superoxide dismutase 2 (C>T) gene variant and coronary heart disease in diabetes mellitus. Diabetes Res Clin Pract 90: 196–201.
Kadoi, Y., C. Kawauchi, M. Ide, M. Kuroda, K. Takahashi, S. Saito, N. Fujita, A. Mizutani. 2011. Preoperative depression is a risk factor for postoperative short-term and long-term cognitive dysfunction in patients with diabetes mellitus. Journal of Anesthesia 25: 10–17.
Lam, K.S.L. 2009. Circulating levels of adipocyte and epidermal fatty acid-binding proteins in relation to nephropathy staging and macrovascular complications in Type 2 diabetic patients. Diabetes Care 32: 132–134.
Lamers, F., C.C. Jonkers, H. Bosma, J.A. Knottnerus and J.T. van Eijk. 2011. Treating depression in diabetes patients: Does a nurse-administered minimal psychological intervention affect diabetes-specific quality of life and glycaemic

control? A randomized controlled trial. Journal of Advanced Nursing 67: 788–799.
Li., C. and E.S. Ford. 2009. Healthy lifestyle habits and health related quality of life in diabetes. *In:* V.R. Preedy and R.R. Watson (Eds.), Handbook of Disease Burdens and Quality of Life Measures. New York: Springer, Pages 2095–2114.
Ma, J., W. Yang, N. Fang, W. Zhu and M. Wei. 2009. The association between intensive glycemic control and vascular complications in Type 2 diabetes mellitus: A meta-analysis. Nutrition, Metabolism and Cardiovascular Diseases 19: 596–603.
Mattila, T.K. and B.A. De. 2010. Influence of intensive versus conventional glucose control on microvascular and macrovascular complications in Type 1 and 2 diabetes mellitus. Drugs 70: 2229–2245.
Mazhar, K., R. Varma, F. Choudhury, R. McKean-Cowdin, C.J. Shtir and S.P. Azen. 2011. Severity of diabetic retinopathy and health-related quality of life: The Los Angeles Latino eye study. Ophthalmology 118: 649–655.
MILES, 2010. The Diabetes MILES Study (Management and Impact for Long-term Empowerment and Success) for Australian adults with Type1 or Type 2 diabetes. http://www.diabetesmiles.org/ [Accessed: 30 August 2011].
Mitchell, J. and C. Bradley. 2009 . Measuring quality of life in macular degeneration . *In:* V.R. Preedy and R.R. Watson (Eds.), Handbook of Disease Burdens and Quality of Life Measures. New York: Springer 2633–2648.
Mitchell, J., A. Woodcock and C. Bradley. 2009. The MacDQoL individualized measure of the impact of macular disease on quality of life. *In:* V.R. Preedy and R.R. Watson (Eds.), Handbook of Disease Burdens and Quality of Life Measures. New York: Springer 247–264.
Ribu, L. 2009. Quality of life in patients with diabetic foot ulcers. In VR Preedy and RR Watson (Eds.), Handbook of Disease Burdens and Quality of Life Measures. New York: Springer Pages 2115–2134.
Stirban, A.O. and D. Tschoepe 2008. Cardiovascular complications in diabetes: targets and interventions. Diabetes Care 31 Suppl. S215–S221.
Stovring, H. 2009. The Use of pharmacoepidemiological database to assess disease burdens: application to diabetes. *In:* V.R. Preedy and R.R. Watson (Eds.), Handbook of Disease Burdens and Quality of Life Measures. New York: Springer Pages 671–684.
Vijayakumar, K. and R.T. Varghese. 2009. Quality of life among diabetic subjects: Indian perspectives . *In:* V.R. Preedy and R.R. Watson (Eds.), Handbook of Disease Burdens and Quality of Life Measures. New York: Springer, Pages 2071–2094.
Vinogradova, Y., C. Coupland, J. Hippisley-Cox, S. Whyte and C. Penny. 2010. Effects of severe mental illness on survival of people with diabetes. British Journal of Psychiatry 197: 272–277.
Wilfley, D., R. Berkowitz, A. Goebel-Fabbri, K. Hirst, C. Ievers-Landis, T.H. Lipman, M. Marcus, D. Ng, T. Pham, R. Saletsky, J. Schanuel and B.D. Van. 2011. Binge eating, mood, and quality of life in youth with Type 2 diabetes: Baseline data from the TODAY study. Diabetes Care 34: 858–860.

Index

A

$\alpha_2\beta_3$ integrin 349
$\alpha_{IIb}\beta_3$ 349
Acinar 46
Advanced glycation end-products 38, 52
Advantages 272, 279, 283
Algal polysaccharides 346
Alginate 169–171, 173, 175, 178
Amyloid deposits 43, 46
Amyloid fibrils 43, 44, 50, 51
Angiogenesis 304–306, 308, 315, 319, 320, 351, 352, 357, 359
Angiogenic inhibitor 305, 306, 308, 310, 312, 316–318, 320
Angiogenic stimulator 305, 308
Antibiotics 328, 334, 339, 340
Antidiabetic drugs 16, 29
Antidiabetic effect 242, 243
Antisense oligonucleotides 59, 60, 75
Ara-A 346, 360
Ara-C 346, 360
Au NPs 109, 110
Autoimmune 270–277, 279–281, 283, 284
Autoimmune diabetes 270, 271, 274, 276, 279, 283, 284
Autoimmune insulitis 271, 279, 280

B

Basal lamina 349
β strand 44
BB rats 273
Beta cell mass quantification 159, 161
Biocompatible 325, 327, 330, 334–336, 339, 341
Biodegradable Nanoparticles 227, 230, 231, 233, 240, 242, 245
Biomedical Application 120, 136
Blindness 380
Blood-retinal-barrier 321
Borosilicate glass capillary tip 127

C

Ca^{2+} 82–84, 86–91
Carbon nanomaterials 202–205, 208, 219
Carbon nanotubes 202–204, 211, 215, 219, 220
Carbon nanotubes 81, 82, 84, 85, 87
Cataract 41–43
CD4+ 273, 274
CD8+ 273
Cellulose 169, 170, 172, 174, 178, 183
Chemical enhancers 187, 199
Chitosan 169–171, 173, 174, 179, 183
CLIO-Tat 291–293, 300
CNT 81, 84–86, 88, 91
Cognitive function 381
Colloidal particles 17, 24
Co-morbid 381
Crystallization of insulin 49
Curcumin 326, 335–340
Cytokine gene therapies 284
Cytokine gene therapy 271, 272, 276, 281, 284

D

db/db 351
Depression 381
Derivatives 169–172, 175
Designing 231, 233
Diabetes 3, 5–9, 14–16, 20, 21, 24, 26, 28–30, 81, 82, 84, 89, 90, 121, 122, 125, 126, 134, 136–138, 270–277, 279–281, 283, 284
Diabetic FOOT 365, 370, 374–377
Diagnosis 381, 382
Diatoms 346
Disulfide bonds 168, 170, 177, 178, 181
Dithizone 292, 293
Drawbacks 279
Drug delivery 3, 8, 10, 227, 228, 230, 231, 244
Drug delivery systems 14, 15, 31

E

Efficacy of insulin 49
EGF 357, 360
Electrochemical 99–106, 108, 110–114, 116
Electrochemical sensors 203, 216
Electromechanical coupling 50, 52
Electrospinning 325, 326, 329–334, 339, 340
Encapsulation 48, 50
Epidemiology 381
Ets1 352, 353
Extracellular matrix (ECM) 326, 327

F

Fibrillin 41, 42
Fibrinogen 349
Fibroblasts 328, 332, 341
FM-AFM 37, 44, 45
Fucoidan 346
Fullerene 91
Fullerenes 202–204, 208, 211–214
Functional/multi component 99, 103
Functionalization 137
Functionalization of Nanoparticles 176
Functionalized SWNT 87

G

GAD 272, 274–276, 284
Gene therapy 313–315, 320
Glipizide 227–229, 233, 238, 241–245
Glucagon 46, 51
Glucose 79, 81–84, 86, 88–91
Glucose monitoring 3, 4, 9
Glucose Sensing 48
Glucose Sensors 98–104, 108
Glucose transporters 89
GLUT4 83, 84, 90
GOx 99, 101–116
Graphitic nanofibers 202, 203, 210, 211, 219
Growth factors 327, 328, 334, 340

H

H-2Kd 294, 295, 300
Heat shock proteins 276
High-speed AFM 37, 50
Human Adipocytes and frog Oocytes 123, 134
Hyaluronic acid 349
Hyperglycemia 273, 280

I

IFN-γ 273, 274, 276
IKK 353
IL-1 273, 274, 353, 354, 356, 357, 360
IL-2 273, 274, 276
IL-4 274, 276, 277, 279, 280, 283
IL-5 274
IL-10 270, 271, 274, 276–284
Immobilized metal ion affinity chromatography 203, 210, 220
Immune response 249, 252, 256, 265
In vitro cell labeling 145
In vivo cell imaging 145, 160
In vivo cell labeling 145, 146
Insulin 14, 16, 20, 21, 24, 26–30, 270–275, 277, 283, 274
Insulin delivery 3, 5–7, 9, 10
Insulin Dependent Diabetes Mellitus (IDDM) 277
Insulin enrichment 209, 215
Insulin resistance 81, 82, 89, 380
Insulin-loaded carrier 197, 199
Insulitis 287–289, 292, 296, 297
Insulitis diagnosis 161
Interfering RNA (siRNA) 82, 91
Interleukin-10 (IL-10) 270
Intracellular Glucose measurement 129, 131
Intrinsic viscosity 346
Iontophoresis 187, 189, 190, 192–194, 196, 197, 199
Islet amyloid polypeptide 43
Islet cells 271, 272, 274
Islet cells transplantation 271

K

Kidney failure 380

L

Layer-by-layer 258
Layer-by-layer assembly 48
Lectins 168, 170, 177
Lipid drug conjugates 15, 21
Lipid matrices 14, 16, 30
Lipid-anchoring 263
Liposomes 15, 16, 24–27, 30, 309, 318
Long circulating nanoparticles 242
LPS 356, 360

M

MALDI-TOF-MS 205, 207, 208, 210, 220
MAPK 352, 359, 360

Matrices 325, 330–336, 339–341
Matrix 203, 205, 208, 213–215, 217, 219, 220
Mechanical properties of collagen 39
MELDI 208, 209, 214, 215, 220
Microalga 346
MILES 381, 382
MMP3 352
MMP9 352
Mouse models 351
mRDH bandage 350
Mucins 168, 169, 173, 175–178, 181, 182
Mucoadhesion 167–169, 171, 174, 176–179, 181, 182
Mucoadhesive polymer 168, 169
Mucoadhesive properties 166, 167, 170, 172, 173, 175, 176, 178–181
Mucus 166–172, 175–178, 180–183
Mucus diffusing nanoparticles 171, 175, 176, 181
Mucus glycoproteins 168, 170
Mucus-polymer interactions 169
Multiple low-dose streptozotocin (MLD-STZ) 273, 280
Muscle fibers 84, 86–87, 90

N

NAG 351, 358, 360
Nanocarriers 59, 60, 67–69, 71–75
Nanocomponents 99, 102
Nanodrugs 61, 63–65, 67, 73
Nano-encapsulation 248, 249, 251, 257, 258, 261, 264, 265
Nanofibers 325, 326, 328–341
Nanomaterials 99, 100, 102, 103, 111, 114
Nanomedicine 3, 29, 59–61, 63, 121, 136, 165–169, 181, 182
Nanonetworks 104
Nanoparticle 151–156, 158, 162, 165–167, 171–183, 270–272, 279–284, 304–306, 309, 311–316, 319–321
Nanostructured lipid carriers 30
Nanotechnology 3, 7, 9, 10, 59–61, 63, 64, 73, 75, 121, 122
Neovascularization 304, 306, 307, 317–321
Neuropathy 366, 368, 370, 372, 374–376
NFκB 353, 356, 360
Nitric-oxide 48
NOD mice 273, 276, 277
NOD mouse 273
Non-specific signal 146, 148
Non-viral vectors 272, 277, 279

O

ob/ob 351
Oral administration of insulin 165–167, 173, 179

P

Pancreatic islet transplantation 275, 279
Pancreatitis 46
Paracellular pathway 178, 181
Particle size 232, 233, 237, 238, 240, 241
Pegylated PLGA-NP 241
PGlcNAc nanofibers 346, 349, 350
Pharmacokinetics 49
Phospholipid micelles 15, 16, 22, 23
PI3K/Akt 353
Pigment epithelium-derived factor (PEDE) 308, 320, 321
Platelets 349, 350
PLGA nanoparticles 235, 236, 239, 240, 242
PLGA-mPEG NP 241
Podocytes 46, 47
Poly(acrylic acid) 171, 183
Poly(fumaric-co-sebacic) 179
Polyanhydrides 179, 180
Polymer precipitation 234
Pressure ulcers 325, 327, 330, 331
Prevention 366, 374–377
Protein expression 84, 86, 88
Protein profiling 209
Protein-membrane Interactions 46
Psychometric evaluation 382

R

Rac 353
Radiopharmaceuticals 146, 149, 160, 161
Red blood cells 41, 42, 51
Release kinetics 234, 240
Retina 306–310, 312, 314–321
Retinopathy 304–307, 314, 319–321, 381

S

Second Generation Sulfonylurea 227, 229, 243, 244
Self-emulsifying drug delivery systems 15, 31
Sensitivity of imaging 148
Sialic acid 168, 173, 177
Single-walled carbon nanotubes (SWNT) 81, 85–88, 90, 91

siRNA 81, 82, 84–88, 90, 91
Skeletal muscle 81, 82, 84–88, 90
sNAG 351, 352, 354–356, 358, 360
Solid lipid nanoparticles 15, 17, 18, 30
Solid phase extraction 203, 212, 220
Solvent extraction 228, 234, 235
Sonophoresis 187, 189, 190, 192, 195, 198, 199
Spill over 144, 146, 147, 156, 160
SPIO 290, 299, 300
Staphylococcus aureus 354
Stealth nanoparticles 236, 242
Structural changes 39, 41
STZ-induced diabetes 279
Surface Erosion 171, 179
Synthetic polymers 329
Syvek Patch 350

T

T cells 271–274, 276, 277, 283
Techniques 227, 228, 232, 234, 237–239, 242
TGFβ 357, 360
T_H1 273, 274, 276, 277
T_H2 270, 273, 274, 276, 277, 283
Therapeutic strategies 228
Thiol groups 168, 177–179
Thromboxane 349
Thyroid disease 38
Tight junctions 170, 173, 174, 178
Time-lapse AFM 43, 44, 46, 47, 49
Tissue regeneration 328, 329, 331, 339
TNF-α 273, 274, 277
Toll-like receptors 354, 356
Transdermal drug delivery 187, 189, 195, 199, 200
Transfection 81, 84, 86–88
Transient receptor potential channel 84
Transplanted islets 249

Treatment of type 1A diabetes 271, 274, 277
TRP 91
TRPC 84, 91
Type 1A diabetes 270–277, 280, 283, 284
Type 2 diabetes 272
Type II diabetes mellitus 227, 244
Tyrosine phosphatase 272, 274

U

Ulceration 366, 368, 369, 373–376

V

Vascular endothelial growth factor (VEGF) 305, 308, 320, 352, 353, 357, 359, 360
Viral or nonviral vector 271, 272, 277, 279, 283
Von Willebrand factor 349

W

Wheat germ agglutinin 177, 183
Wound 325–328, 300, 331, 333–336, 338, 340, 341
Wound dressings 325, 327, 330, 331, 333, 341
Wound healing 3, 8–10

X

X-ray scattering 348

Y

Young's moduli 42

Z

ZnO nanostructure 123, 128

About The Editors

Lan-Anh Le BSc, MBBS, MRCGP qualified from Guys, King's & St Thomas' Medical School, London in 2001. This is part of King's College London, currently ranked 21st in the QS World University Rankings. Her undergraduate degree was a BSc in Medical Anthropology from University College, London in 1998, with her thesis on the psychological and cultural aspects of anorexia. University College London is currently ranked 4th in the QS World University Rankings. She has a background in diabetes research. She worked in Australia in both urban and rural contexts and has won awards for her contributions to the study of diabetes care in the Aboriginal population. Prior to becoming a principal General Practitioner, she undertook a period of specialist training in diabetes. As a practising general practitioner her areas of special interest include diabetes, cardiology and dermatology. Her current practice is the diabetes lead in the primary care locality, advising practice and local policy for the region. She is dedicated to teaching having taught medical students, lectured regularly at Charing Cross Hospital, London and is a General Practitioner trainer, training the General Practioners of the future. In addition to her academic, teaching and advisory roles she remains a practicing clinician, caring for patients with diabetes on a daily basis.

Ross J. Hunter AKC, BSc, MBBS, MRCP, PhD trained in medical sciences at King's College London (Times University ranking 11th in UK). He spent a further year at Imperial College London (Times University ranking 3rd in UK) and was awarded his BSc in Cardiovascular medicine in 1998. Since returning to his medical training at King's College School of Medicine, he has remained an honorary research fellow at The Department of Nutritional Sciences, researching the effect of different nutritional states and alcoholism on the cardiovascular system. He was awarded his bachelor of medicine & surgery (MBBS) with distinction in 2001. He trained in general medicine in London and Brighton and was made a member of the Royal College of Physicians (UK) in 2005. He trained as a Registrar in cardiology and general internal medicine from 2005–2008 in the London Deanery. Since 2008 he has been a research fellow at the Department of Cardiology & Electrophysiology at St Bartholomew's Hospital London, conducting clinical research and clinical trials in cardiology and electrophysiology, and was awarded his PhD in 2011. He was a young investigator of the year finalist at both the Heart Rhythm Society (USA) and Heart Rhythm UK in 2011. He has published over 60 scientific articles of various kinds and is Editor of 4 books.

Victor R. Preedy BSc, PhD, DSc, FIBiol, FRCPath, FRSPH is Professor of Nutritional Biochemistry, King's College London, Professor of Clinical Biochemistry, King's College Hospital and Director of the Genomics Centre, King's College London. Presently he is a member of the King's College London School of Medicine. Professor Preedy graduated in 1974 with an Honours Degree in Biology and Physiology with Pharmacology. He gained his University of London PhD in 1981 when he was based at the Hospital for Tropical Disease and The London School of Hygiene and Tropical Medicine. In 1992, he received his Membership of the Royal College of Pathologists and in 1993 he gained his second doctoral degree, i.e., DSc, for his outstanding contribution to protein metabolism in health and disease. Professor Preedy was elected as a Fellow to the Institute of Biology in 1995 and to the Royal College of Pathologists in 2000. Since then he has been elected as a Fellow to the Royal Society for the Promotion of Health (2004) and The Royal Institute of Public Health (2004). In 2009, Professor Preedy became a Fellow of the Royal Society for Public Health. In his career Professor Preedy has carried out research at the National Heart Hospital (part of Imperial College London) and the MRC Centre at Northwick Park Hospital. He has collaborated with research groups in Finland, Japan, Australia, USA and Germany. He is a leading expert on the mechanisms of disease and has lectured nationally and internationally. He has published over 570 articles, which includes over 165 peer-reviewed manuscripts based on original research, 90 reviews and 20 books.

Color Plate Section

Chapter 2

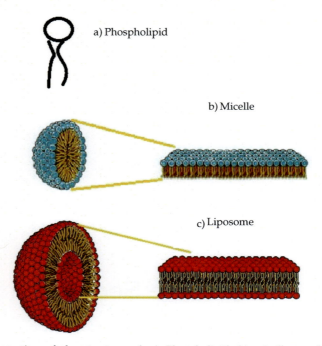

Fig. 5. Illustration of the structure of: a) Phospholipid; b) micelles. and c)liposome (Unpublished). Phospholipid, usually represented by the head with lipophilic character and tails with hydrophilic character. In micelles (b) there is only one phospholipid layer and in lipossomes (c) there is at least one phospholipid bilayer that resembles the cell membranes Both systems are association colloids formed by amphiphilic compounds. They consist of two clearly distinct regions with opposite affinities toward a given solvent. Hydrophobic fragments of amphiphilic molecules form the core of a micelle, which can solubilize poorly soluble drugs, while hydrophilic fragments form micelle's corona. In aqueous systems, non-polar molecules are solubilized within the micelle core, polar molecules will be adsorbed on the micelle surface and substances with intermediate polarity will be distributed along surfactant molecules in intermediate positions.

Chapter 4

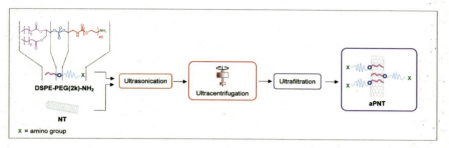

Fig. 5. Fabrication of functionalized PEGylated carbon nanotubes. Pristine (non-functionalized) carbon nanotubes (NTs) were ultrasonicated with phospholipids terminated with amino-functionalized 2-kDa mw PEG chains {1,2-distearoyl-*sn*-glycero-3-phosphoethanolamine-*N*-[amino(polyethylene glycol)$_{2000}$]} [DSPE-PEG(2k)-NH$_2$], fractionated by stepwise ultracentrifugation to isolate short hydrophilic amino-terminated PEGylated NTs and purified by ultrafiltration to remove free phospholipids (aPNT).

Chapter 5

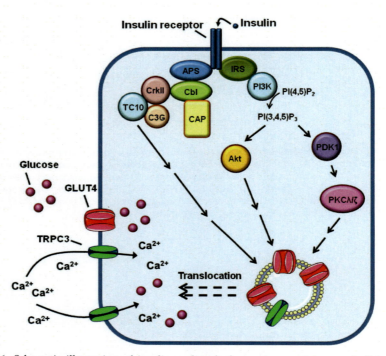

Fig. 1. Schematic illustration of insulin-mediated glucose transport in muscle/fat cells. Following insulin binding to its receptor, a series of signaling reactions initiate translocation of GLUT4-containing vesicles to the cell surface. At the same time, extracellular Ca^{2+} enters the cell via TRPC3, resulting in localized increases that facilitate docking/fusion/insertion of GLUT4 in the membrane, where it accelerates glucose transport. Single arrow reflects a direct effect. Multiple arrows in a pathway reflect two or more steps. For the sake of simplicity several established steps have been omitted. (Permission to publish; adapted from Lanner et al. Curr Opin Pharmacol 2008).

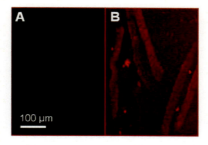

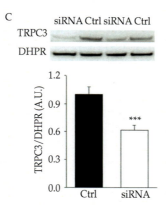

Fig. 3. Carbon nanotubes successfully transfected adult skeletal muscle fibers. (A) Non-transfected FDB fibers exposed to only SWNT demonstrating little autofluorescence. (B) Fibers transfected with SWNT-siRNA constructs labelled with Cy5; all fibers in the field of view show fluorescence indicating successful entry of the labelled SWNT-siRNA. Images were taken 48 hours post transfection. (C) Representative western blots of TRPC3 expression obtained after 48h incubation without (Ctrl) or with TRPC3-siRNA; the voltage-gated Ca^{2+} channel (dihydropyridin receptor, DHPR) was used as loading control. (Permission from the publisher to reproduce image; Lanner et al. FASEB Journal 2009).

Chapter 6

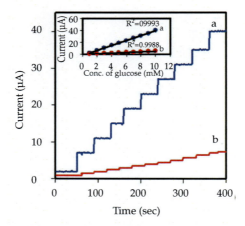

Fig. 2. (A) Chronoamperometric current responses of electrodes: (a) GOx/Au/CS–IL–MWNT(SH)/ITO and (b) GOx/MWNT(SH)/ITO to successive addition of 1 mM glucose. Inset: Calibration plot of concentration of glucose (1–10 mM) vs. current; (a) GOx/Au/CS–IL–MWNT(SH)/ITO and (b) GOx/MWNT(SH)/ITO.

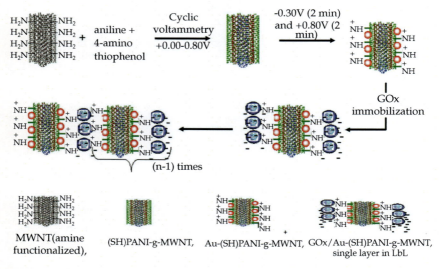

Scheme 1. Fabrication of LbL based biosensor comprising of multi-components, MWNT, PANI(SH), Au particles and GOx.

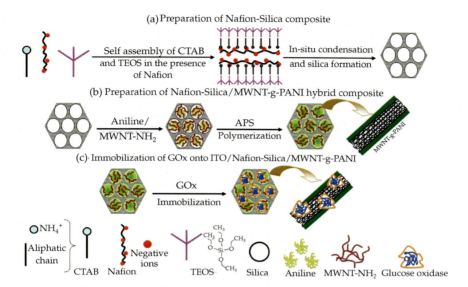

Scheme 2. Fabrication of Nafion-silica/MWNT-g-PANI/GOx biosensor.

Chapter 7

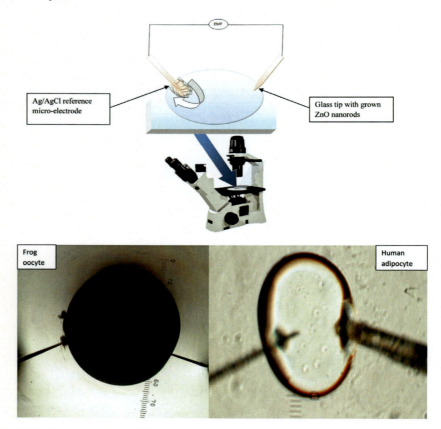

Fig. 2. Experimental setup of intracellular measurements. (a) Schematic diagram illustrating the selective intracellular glucose measurement setup. (b) Microscope images of a single frog (*Xenopus laevis*) oocyte and a single human fat cell (adipocyte) during measurements with a functionalised ZnO-nanorod coated probe as a working electrode and with an Ag/AgCl reference microelectrode.

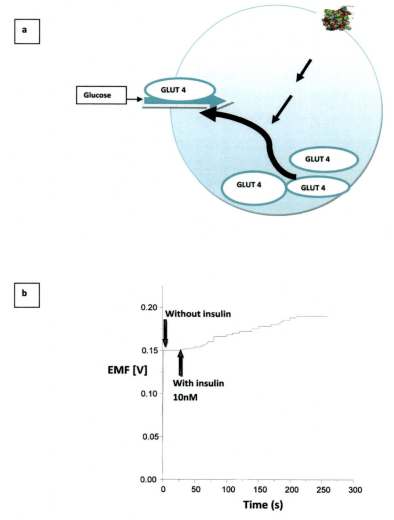

Fig. 4. activation of glucose uptake and output response with insulin. (a) Intracellular mechanism for insulin-induced activation of glucose uptake. (b) Output response with respect to time for intracellularly positioned electrodes when insulin is applied to the extracellular solution.

Chapter 10

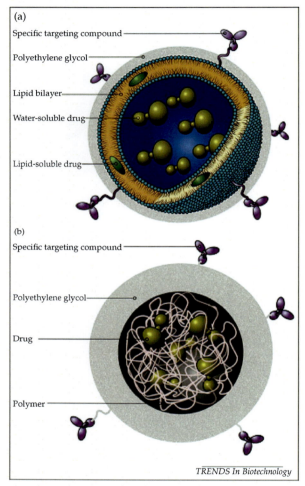

Fig. 1. Nanoparticles for non-parenteral drug delivery: (a) Liposome nanoparticle. (b) Polymer nanoparticle. This figure shows the basic structure of nanoparticles used for drug delivery. Drugs are encapsulated in the core of the nanoparticles or mixed with the polymer. The whole particles were guided by the specific targeting compounds to the targets and thus release the drug specifically. (Antosova et al. 2009).

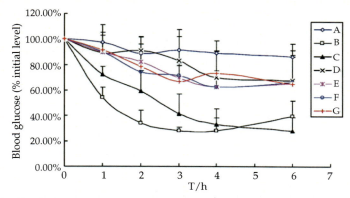

Fig. 5. The blood glucose levels of various treatment groups (n=5). This figure implies the hypoglycemic effects of six different treatment groups.(A) is used as the Negative control. While under the treatment of (B) Free insulin administrated by subcutaneous injection (1.0 IU/kg), the blood glucose decreased largely, which may lead to glucopenia. so is the result of group (C) ILV4 combined with iontophoresis and microneedles. The groups of (D) Free Insulin solution combined with iontophoresis and microneedles, (E) ILV4 combined only with microneedles, (F) ILV4 combined only with Iontophoresis, and (G) ILV4 based onpassive penetration all show similar effects in decreasing the blood glucose. The Glucose levels of streptozotocin-induced diabetic rats were normalized against the initial (0 h) value (Chen et al. 2009).

Chapter 14

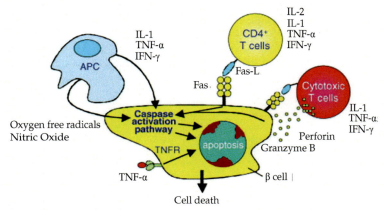

Fig. 1. Mechanisms of β cell destruction in type 1 diabetes. (With permission from Elsevier, Kawasaki et al. 2004).

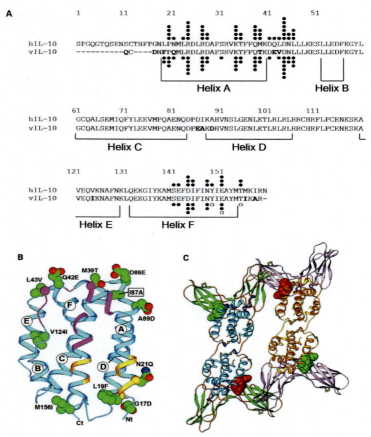

Fig. 2. Sequence and Structure of vIL-10. (A) Structure-based sequence alignment based on vIL-10/sIL-10R1 and hIL-10/sIL-10R1 complexes. vIL-10 amino acids that differ from the hIL-10 sequence are shown in bold. The approximate amount of surface area buried by residues in viral (vIL-10 and vIL-10$_{A87I}$) and human IL-10 is shown with black circles according to the following code; (5Å^2 < one circle ≤ 20Å^2), (20Å^2 < two circles ≤ 40Å^2), (40Å^2 < three circles ≤ 60Å^2), (60Å^2 < four circles ≤ 80Å^2), (80Å^2 < five circles ≤ 100Å^2), (>100Å^2 = six circles). Residues that bury surface area only in vIL-10 or only in vIL-10$_{A87I}$ are shown by black squares and open circles, respectively. If two markers are shown, the greatest position of each marker reflects the amount of surface area buried in the respective complexes. (B) Ribbon diagram of one vIL-10 domain. The side chains of vIL-10 residues that differ from hIL-10 are shown. vIL-10 residues that bury surface area in the site Ia and Ib interfaces are shown in purple and gold, respectively. (C) 2:4 vIL-10/sIL-10R1 complex. The site II interface is located between the cyan/green and gold/purple 1:2 vIL-10/sIL-10R1 complexes. Residues that form the putative IL-10R2 binding site are shown on the vIL-10s in space filling representation. IL-10R2 binding sites are colored green if they are accessible to the IL-10R2 chain or red if the site is occluded in the site II interface. (With Permission from Elsevier, Yoon et al. 2005).

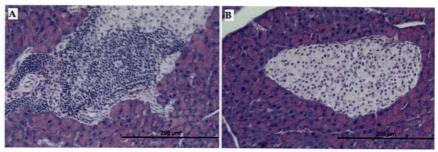

Fig. 5. Protective ability of IL-10 gene therapy in pancreas of Streptozotocin-treated mice. Light micrographs of histological section of pancreas from (A) streptozotocin-treated animal indicating infiltration by the immune cells and (B) streptozotocin-treated animal, also treated with IL-10 plasmid loaded on PLGA/50%E100 nanoparticles at a dose of 50μg of DNA exhibiting protection of pancreatic islets from immune destruction (Reproduced from Basarkar 2007).

Chapter 15

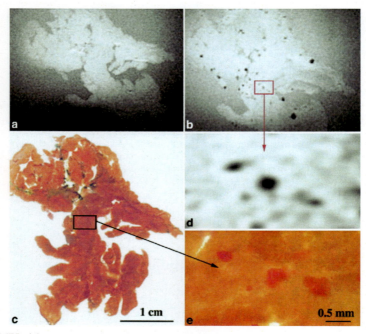

Fig. 1. MRI of the pancreas after adoptive transfer of diabetic splenocytes labeled with CLIO-Tat into NOD.scid mice. T_1-weighted image showing physiological details of the pancreas (a), T_2-weighted image, showing infiltration of labeled cells into pancreatic islets (b). Dithizone stained pancreas (c). Correlation between MR images (d) and staining for DTZ (e) shows the presence of infiltrated cells labeled with CLIO-Tat (dark dots on d) in the pancreatic islets (purple DTZ staining on e). Magnification bars: a,b,c, 1 cm; d,e, 0.5 mm. Reproduced by permission of John Wiley and Sons from Magnetic Resonance in Medicine, 2002, 47(4): 751–758.

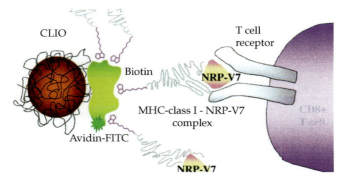

Fig. 3. Schematic representation of CLIO-NRP-V7 probe. Biotinylated NRP-V7/H-2Kd complex was coupled to CLIO particles modified with avidin. CLIO-NRP-V7 probe is recognized by the TCR on NRP-V7–reactive CD8+ T-cells. Copyright 2004 American Diabetes Association. From Diabetes, Vol. 53, 2004; 1459–1466. Reproduced by permission of The American Diabetes Association.

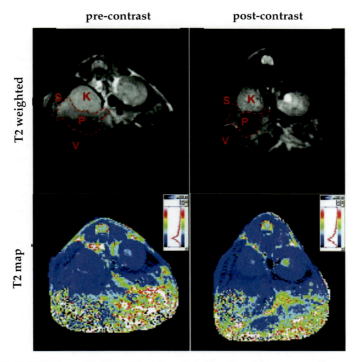

Fig. 4. MRI of NRP-V7-specific CD8+ T lymphocyte infiltration of the pancreas. Representative T_2-weighted image (top) and its corresponding multiecho T_2 map (bottom) before (left) and 24 h after (right) intravenous injection of MN-NRP-V7. The pancreas is outlined in red. V = stomach, S = spleen, K = kidney, P = pancreas. Reproduced by permission of John Wiley and Sons from Magnetic Resonance in Medicine, 2008, 59(4): 712–720.

Chapter 19

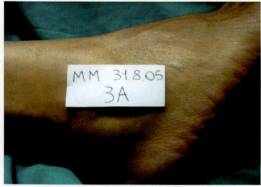

Fig. 2. A case report of a patient treated with the Difoprev® system. a) the patient's skin was dehydrated and inelastic, and a score of 1 was attributed. The same patient after 6 weeks of treatment; b) the skin is normally hydrated and the score attributed was 3 (compares the text in the *Patients and Methods* section).
Unpublished material of the author.